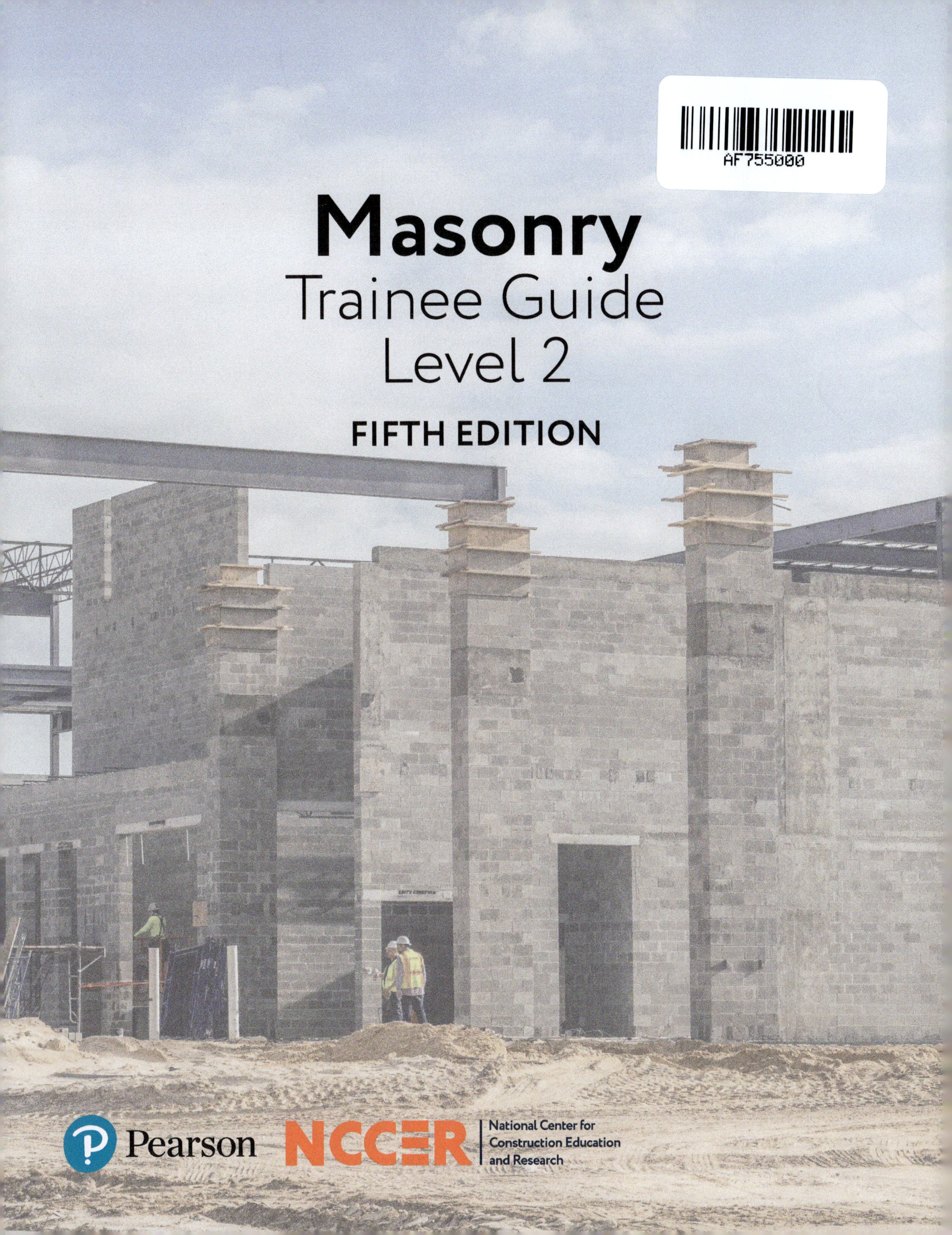

Masonry

Trainee Guide
Level 2

FIFTH EDITION

Pearson
NCCER National Center for Construction Education and Research

NCCER
President and Chief Executive Officer: Boyd Worsham
Vice President of Innovation and Advancement: Jennifer Wilkerson
Chief Learning Officer: Lisa Strite
Senior Curriculum Leader: Chris Wilson
Senior Manager of Production: Graham Hack
Masonry Project Manager: Patrick Bruce
Technical Writing Manager: Gary Ferguson
Technical Writers: Mitch Ryan, Jo Ann Bartusik
Art Manager: Bree St Germain
Technical Illustrators: Judd Ivines, Liza Wailes
Production Artist: Chris Kersten
Permissions Specialist: Sherry Davis
Project Coordinator: Emmah Walker

Pearson
Product Manager—Associations: Andy Dunaway
Senior Digital Content Producer: Shannon Stanton
Associate Project Manager: Monica Perez-Kim
Cover Designer: Mary Siener
Rights and Permissions: Kaitlyn Coddington
Composition: Happenstance Type-O-Rama

Content Technologies: Gnostyx
Printer/Binder: Lakeside Book Company
Cover Printer: Lakeside Book Company
Text Typefaces: Palatino LT Pro, Helvetica Neue

Cover Image

This Scherer Construction project for Tower Hill Insurance Group in Alachua, FL, features CMU pilaster columns with formwork set for a cast in place pour. Visible in the photo are partially completed masonry walls, temporary formwork, and scaffolding used to support the ongoing work. Crew members are wearing required personal protective equipment (PPE), including hard hats and high-visibility vests, in compliance with safety protocols. The image highlights standard jobsite conditions during masonry installation and the coordination of materials, labor, and safety measures in a typical construction environment.

10 9 8 7 6 5 4 3 2 1

1 2025

Pearson

NCCER | National Center for Construction Education and Research

Paperback
ISBN-13: 978-0-13-546628-5

Hardcover
ISBN-13: 978-0-13-546629-2

PREFACE

To the Trainee

Masons are recognized as premier craftworkers on any construction site. Although masonry is one of the world's oldest crafts, masons also use 21st-century technology on the job. Using brick, block, or stone, and binding them with mortar, masons build durable structures with optimized energy performance. With the support of the Mason Contractor Association of America (MCAA) and the North Carolina Masonry Contractors Association, NCCER's program has been designed and revised by subject matter experts from across the nation and industry to update the curriculum with modern techniques.

Our three levels present an apprentice and K–12 classroom approach to the masonry field and help keep you knowledgeable, safe, and effective on the job and in the classroom. We wish you the best as you begin an exciting and promising career. This newly revised masonry curriculum will help you enter the workforce with the knowledge and skills needed to perform productively in either the residential or commercial market

New with *Masonry Level 2*

NCCER is proud to present *Masonry Level 2*. This new edition includes updates to the curriculum that will help engage you and provide the best training possible. There is also updated art throughout. Learners will receive extensive information on reinforced masonry as well as various climate conditions that impact the work of masons. These updates will assist the learner with training for real-world conditions and the success of completing a project to satisfy their end user and customer.

We wish you success as you progress through this training program. If you have any comments on how NCCER might improve this textbook, please complete the User Update form using the QR code on this page. NCCER appreciates and welcomes its customers' feedback. You may submit yours by emailing **support@nccer.org**. When doing so, please identify feedback on this title by listing *#MasonryL2* in the subject line.

Our website, **www.nccer.org**, has information on the latest product releases and training.

SCAN ME

NCCER Standardized Curricula

NCCER is a not-for-profit 501(c)(3) education foundation established in 1996 by the world's largest and most progressive construction companies and national construction associations. It was founded to address the severe workforce shortage facing the industry and to develop a standardized training process and curricula. Today, NCCER is supported by hundreds of leading construction and maintenance companies, manufacturers, and national associations. The NCCER Standardized Curricula was developed by NCCER in partnership with Pearson, the world's largest educational publisher.

Some features of the NCCER Standardized Curricula are as follows:

- An industry-proven record of success
- Curricula developed by the industry, for the industry
- National standardization providing portability of learned job skills and educational credits
- Compliance with the Office of Apprenticeship requirements for related classroom training (CFR 29:29)
- Well-illustrated, up-to-date, and practical information

NCCER maintains a secure online database that provides certificates, digital badges, transcripts, and wallet cards to individuals who successfully complete programs under an NCCER-accredited organization or through one of NCCER's self-paced online programs. This system also allows individuals and employers to track and verify industry-recognized credentials and certifications in real time.

For information on NCCER's credentials, contact NCCER Customer Service at 888.622.3720 or visit **www.nccer.org**.

Digital Credentials

Show off your industry-recognized credentials online with NCCER's digital credentials! Transform your knowledge, skills, and achievements into digital credentials that you can share across social media platforms, send to your network, and add to your resume. For more information, visit **www.nccer.org**.

Cover Image Provider

Scherer Construction is a full-service design-build and general contracting firm with a strong reputation for excellence, integrity, and innovation. Since 1984, Scherer has been delivering high-quality commercial, industrial, and institutional projects across Florida and the SE United States. Our team is committed to building lasting relationships through transparent communication, customized solutions, and an unwavering dedication to safety and client satisfaction. From concept to completion, Scherer Construction builds with purpose and pride—creating spaces that serve, inspire, and endure.

DESIGN FEATURES

Content is organized and presented in a functional structure that allows you to access the information where you need it.

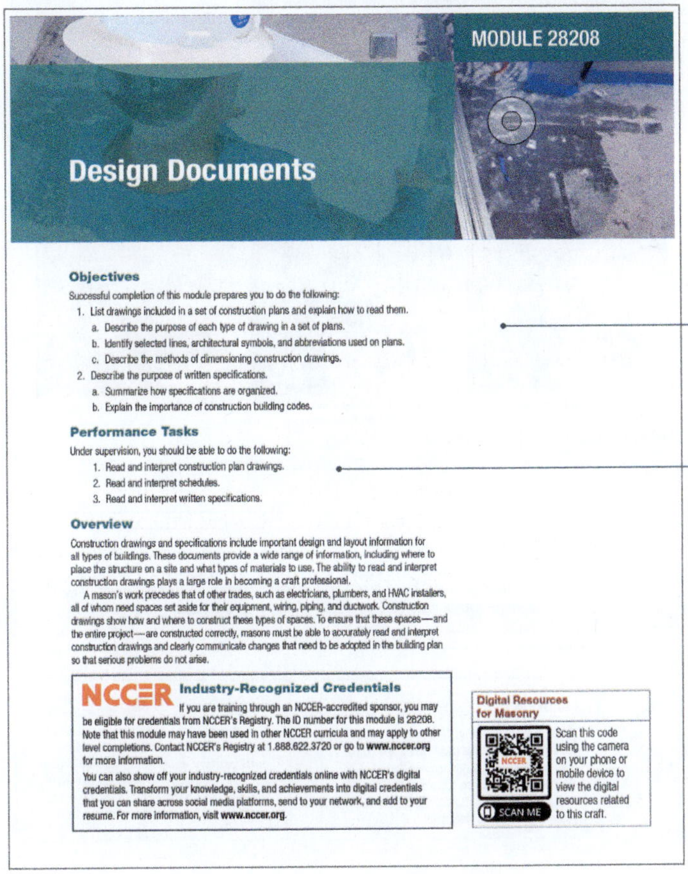
MODULE 28208

Design Documents

Objectives

Successful completion of this module prepares you to do the following:

1. List drawings included in a set of construction plans and explain how to read them.
   a. Describe the purpose of each type of drawing in a set of plans.
   b. Identify selected lines, architectural symbols, and abbreviations used on plans.
   c. Describe the methods of dimensioning construction drawings.
2. Describe the purpose of written specifications.
   a. Summarize how specifications are organized.
   b. Explain the importance of construction building codes.

Performance Tasks

Under supervision, you should be able to do the following:

1. Read and interpret construction plan drawings.
2. Read and interpret schedules.
3. Read and interpret written specifications.

Overview

Construction drawings and specifications include important design and layout information for all types of buildings. These documents provide a wide range of information, including where to place the structure on a site and what types of materials to use. The ability to read and interpret construction drawings plays a large role in becoming a craft professional.

A mason's work precedes that of other trades, such as electricians, plumbers, and HVAC installers, all of whom need spaces set aside for their equipment, wiring, piping, and ductwork. Construction drawings show how and where to construct these types of spaces. To ensure that these spaces—and the entire project—are constructed correctly, masons must be able to accurately read and interpret construction drawings and clearly communicate changes that need to be adopted in the building plan so that serious problems do not arise.

NCCER **Industry-Recognized Credentials**

If you are training through an NCCER-accredited sponsor, you may be eligible for credentials from NCCER's Registry. The ID number for this module is 28208. Note that this module may have been used in other NCCER curricula and may apply to other level completions. Contact NCCER's Registry at 1.888.622.3720 or go to **www.nccer.org** for more information.

You can also show off your industry-recognized credentials online with NCCER's digital credentials. Transform your knowledge, skills, and achievements into digital credentials that you can share across social media platforms, send to your network, and add to your resume. For more information, visit **www.nccer.org**.

Digital Resources for Masonry

SCAN ME

Scan this code using the camera on your phone or mobile device to view the digital resources related to this craft.

You can navigate Masonry using color-coded tabs on the upper righthand corner.

The Objectives list the skills and knowledge you need in order to complete the module successfully.

The Performance Tasks give you an opportunity to apply your knowledge to real-world tasks.

Section Openers provide a visual organizational structure for the information. Objectives and Performance Tasks are broken out for each section.

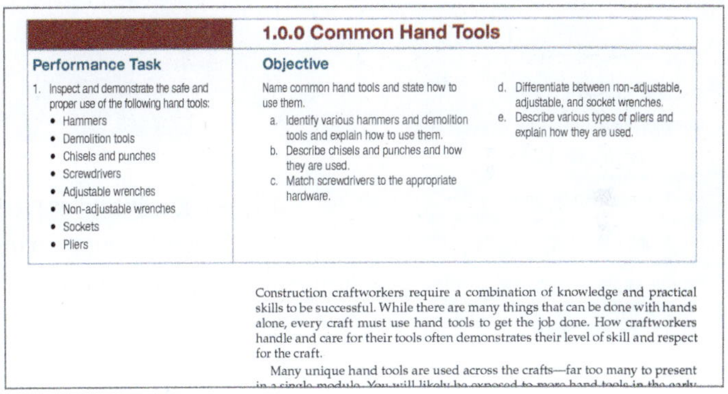
1.0.0 Common Hand Tools

Performance Task

1. Inspect and demonstrate the safe and proper use of the following hand tools:
   - Hammers
   - Demolition tools
   - Chisels and punches
   - Screwdrivers
   - Adjustable wrenches
   - Non-adjustable wrenches
   - Sockets
   - Pliers

Objective

Name common hand tools and state how to use them.

a. Identify various hammers and demolition tools and explain how to use them.
b. Describe chisels and punches and how they are used.
c. Match screwdrivers to the appropriate hardware.
d. Differentiate between non-adjustable, adjustable, and socket wrenches.
e. Describe various types of pliers and explain how they are used.

Construction craftworkers require a combination of knowledge and practical skills to be successful. While there are many things that can be done with hands alone, every craft must use hand tools to get the job done. How craftworkers handle and care for their tools often demonstrates their level of skill and respect for the craft.

Many unique hand tools are used across the crafts—far too many to present

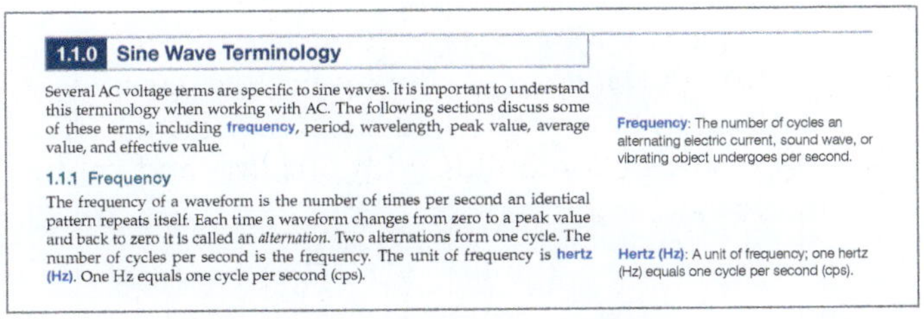
1.1.0 Sine Wave Terminology

Several AC voltage terms are specific to sine waves. It is important to understand this terminology when working with AC. The following sections discuss some of these terms, including **frequency**, period, wavelength, peak value, average value, and effective value.

1.1.1 Frequency

The frequency of a waveform is the number of times per second an identical pattern repeats itself. Each time a waveform changes from zero to a peak value and back to zero it is called an *alternation*. Two alternations form one cycle. The number of cycles per second is the frequency. The unit of frequency is **hertz (Hz)**. One Hz equals one cycle per second (cps).

Frequency: The number of cycles an alternating electric current, sound wave, or vibrating object undergoes per second.

Hertz (Hz): A unit of frequency; one hertz (Hz) equals one cycle per second (cps).

Trade Terms appear on the page adjacent to the text where they are first presented.

Step-by-step math equations help make the concepts clear and easy to grasp.

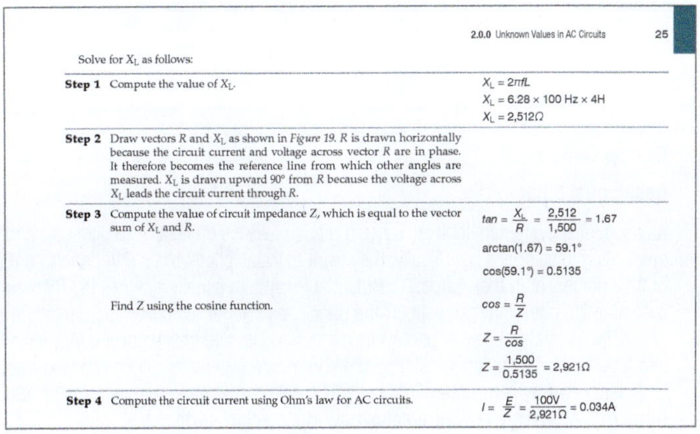
2.0.0 Unknown Values in AC Circuits 25

Solve for X_L as follows:

Step 1 Compute the value of X_L.

$X_L = 2\pi fL$

$X_L = 6.28 \times 100\ \text{Hz} \times 4\text{H}$

$X_L = 2{,}512\Omega$

Step 2 Draw vectors R and X_L as shown in *Figure 19*. R is drawn horizontally because the circuit current and voltage across vector R are in phase. It therefore becomes the reference line from which other angles are measured. X_L is drawn upward 90° from R because the voltage across X_L leads the circuit current through R.

Step 3 Compute the value of circuit impedance Z, which is equal to the vector sum of X_L and R.

$tan = \frac{X_L}{R} = \frac{2{,}512}{1{,}500} = 1.67$

$\arctan(1.67) = 59.1°$

$\cos(59.1°) = 0.5135$

Find Z using the cosine function.

$cos = \frac{R}{Z}$

$Z = \frac{R}{cos}$

$Z = \frac{1{,}500}{0.5135} = 2{,}921\Omega$

Step 4 Compute the circuit current using Ohm's law for AC circuits.

$I = \frac{E}{Z} = \frac{100\text{V}}{2{,}921\Omega} = 0.034\text{A}$

QR codes link directly to videos that highlight current content.

SUCCESS STORIES: *Credentials Matter: Start Building Your Career Today*

Scan this code using the camera on your phone or mobile device to view this video.

Important information is highlighted, illustrated, and presented to facilitate learning.

Placement of images near the text description and details such as callouts and labels help you absorb information.

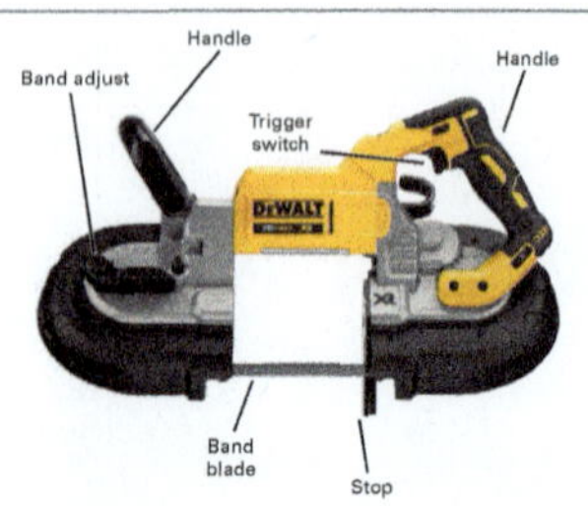

Figure 27 Portable handheld band saw.
Source: Image property of Stanley Black & Decker. Used with permission

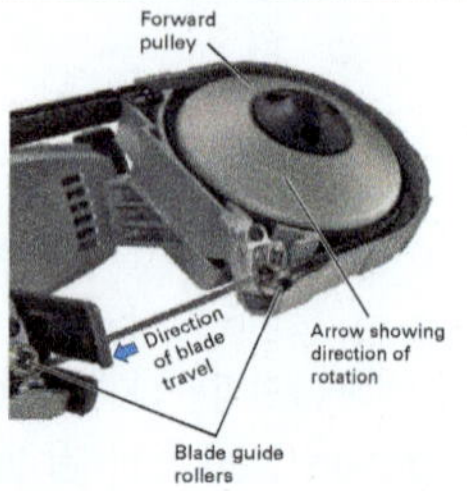

Figure 28 Band saw pulley and blade guide rollers.
Source: Courtesy of Milwaukee Electric Tool Corporation

blade length is determined by referencing the manufacturer's documentation. Like most blades, its coarseness is rated in teeth per centimeter or teeth per inch (TPI). As a general rule, higher TPIs produce smoother cuts. Thicker materials require coarser blades. However, if the blade is too coarse for the material, the individual teeth may begin to break off. *Figure 28* **shows how the blade is routed** around the pulleys and through the blade guides.

While some band saws have multiple speeds, most do not. The portable band saw generally cuts best at a low speed since higher speeds cause the blade's teeth to rub rather than cut. This can create heat through friction, which reduces the life of the blade.

Preparing Drills with Keyless Chucks

Most cordless drills use a keyless chuck. While the steps for preparing a cordless drill are similar, there are some small differences. Follow the steps below when preparing to use drills with keyless chucks:

Step 1 Disconnect the drill from its power source by removing the battery pack before loading a bit.
Step 2 As shown in (*Figure 7A*), open the chuck by turning it counterclockwise until the jaws are wide enough to insert the bit shank.
Step 3 Insert the bit shank into the chuck opening (*Figure 7B*). Keeping the bit centered in the opening, turn the chuck by hand until the jaws grip the bit shank.
Step 4 Tighten the chuck securely with your hand so that the bit does not move (*Figure 7C*). You are now ready to use the cordless drill.

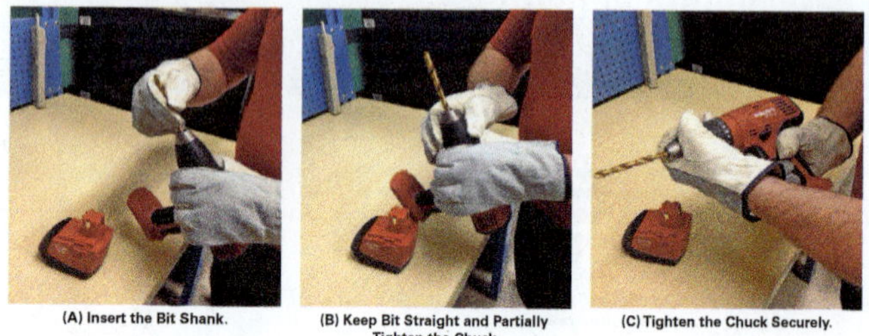

(A) Insert the Bit Shank. **(B) Keep Bit Straight and Partially Tighten the Chuck.** **(C) Tighten the Chuck Securely.**

Figure 7 Loading the bit on a keyless chuck.
Source: Cianbro Corporation

Warning boxes stress potentially dangerous situations, while Caution boxes alert to dangers that may cause damage to equipment. Note boxes provide additional information on a topic.

WARNING!

A portable band saw always cuts in the direction of the user. For that reason, workers must be especially careful to avoid injury when using this type of saw. Always wear appropriate PPE and stay focused on the work.

CAUTION

Never assume anything. It never hurts to ask questions, but disaster can result if you don't ask. For example, do not assume that an electrical power source is turned off. First ask whether the power is turned off, then check it yourself to be completely safe.

NOTE

This training alone does not provide any level of certification in the use of fall arrest or fall restraint equipment. Trainees should not assume that the knowledge gained in this module is sufficient to certify them to use fall arrest equipment in the field.

Did You Know?

Lightning Rods

An interesting fact about grounding is that a lightning rod (air terminal) isn't meant to bring a bolt of lightning to ground. To do this, its conductors would have to be several feet (1 m or more) in diameter. The purpose of the rod is to dissipate the negative static charge that would cause the positive lightning charge to strike the house.

These boxed features provide additional information that enhances the text.

Orthographic Drawings

Orthographic drawings are used for elevation drawings. They show straight-on views of the different sides of an object with dimensions that are proportional to the actual physical dimensions. In orthographic drawings, the designer draws lines that are scaled-down representations of real dimensions. Every 12 inches, for example, may be represented by ¼ inch on the drawing. Similarly, in an example using metric measurements with a ratio of 1:2, every 30 millimeters may be represented by 15 millimeters on the drawing.

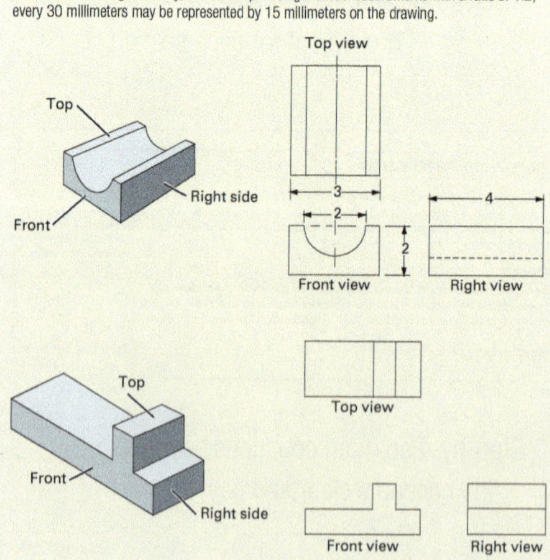

Around the World

Predicting Weather

Barometers measure atmospheric air pressure. Generally, an increased pressure indicates that the weather is pleasant. A reduced pressure indicates that the weather is cloudy and/or rainy.

Changes in barometric pressure are used to predict changes in weather conditions. For example, if it is raining outside, but the barometer is rising, it indicates that better weather is on the way. In the midst of a hurricane, the barometric pressure is an indicator of intensity. Extremely low pressures exist in powerful hurricanes. While traditional barometers work very well, sophisticated weather stations capture far more data and can transmit the information wirelessly.

Going Green

Residential Solar

As residential customers look for ways to reduce reliance on fossil fuels like coal and oil, they are turning to solar energy. Usually, they want to install photovoltaic (PV) panels on the roofs of their homes, or in their yards. A residential PV system usually includes the PV panels, a mounting structure that can support the panels, an inverter to convert the direct current (DC) electricity generated by solar photovoltaic modules into alternating current (AC) electricity, and a battery (or batteries) for storing the energy. Most states require a licensed electrician for solar energy system installation, but you should check the local requirements. (Source: **https://www.energy.gov/eere/solar/how-does-solar-work**)

Going Green looks at ways to preserve the environment, save energy, and make good choices regarding the health of the planet.

Cornerstone of Craftsmanship

How did you choose a career in the industry?

I was working at a gas station and wanted to make more money. So, I applied for a welder helper position, and then when they asked about a millwright helper, I asked, "What's that?" True story! I had no idea what it was. Alan Landry, the plant manager at the local chemical plant, got me into the millwright craft.

What types of training have you been through?

I've been through ABC Millwright training, Total Quality Management (TQM) programs, Rand McNally mechanical seals training, Lubritech lubrication system training, and laser alignment classes. There are many others that I've attended over the years to both broaden and sharpen my skills.

How important is education and training in construction?

It's very important! Nobody is born knowing how to do all this stuff. Knowledge can't be taken away from someone, so always take any training offered to you. The only other way to learn is hands-on, but classrooms and curriculums set the standard, and that's how you know you're learning the right way. Hands-on training and practice then become an extension of the classroom, instead of the only learning opportunity.

How important are NCCER credentials to your career?

My early training came before NCCER curricula was available, but I am an NCCER-certified instructor. Right now, many plants in my state are hiring rotating equipment people and many require the NCCER training to be eligible to be hired. I think that's a good indicator of how important the credentials have become.

How has training/construction impacted your life?

Without training, I certainly wouldn't be as successful as I have been. It is essential for career growth.

What kinds of work have you done in your career?

I've done just about everything from being a simple oiler to a project manager and served in many possible positions in between those two.

Tell us about your current job.

I'm currently a mechanical planner, and my primary task is to develop the scope for specific jobs. These are detailed, step-by-step plans of what we are going to do with a piece of equipment—take it down and/or put it back together—and I actually happen to have a scope on my desk right now. They outline what is required, what parts need to be ordered and acquired in advance, and the labor hours required.

Kim T. Panepinto
Mechanical Planner
Industrial Parts Specialties, LLC

Cornerstone of Craftsmanship sections feature career stories from people working in related fields.

Review questions at the end of each section and module allow you to measure your progress.

2.0.0 Section Review

1. A straight pull contains two raceways. One of the raceways has a trade size of 3" and one has a trade size of 2". The length of the box *must* be _____.
   a. 16"
   b. 24"
   c. 26"
   d. 32"

2. Where angle or U pulls are made, to determine the distance between raceway entries enclosing the same conductor, you need to multiply the trade size of the largest raceway by _____.
   a. two
   b. four
   c. six
   d. eight

3. If possible, pull boxes should be installed _____.
   a. high on the wall for security
   b. at a height/location that makes it easy to pull conductors
   c. behind wall coverings for a neater look
   d. with extra knockouts left open for air flow

Section Review questions can be found at the end of each section to test your knowledge of the content. Review Questions at the end of each module are provided to reinforce the knowledge you have gained.

Module 26207-23 Review Questions

1. In which *NEC*® article is cable tray installation primarily addressed?
   a. *NEC Article 300*
   b. *NEC Article 392*
   c. *NEC Article 517*
   d. *NEC Article 550*

2. Which type of tray is *best* suited for corrosive areas and atmospheres?
   a. Stainless steel cable tray
   b. Coated aluminum cable tray
   c. Nonmetallic cable tray
   d. Basket tray

3. When a cable tray has a solid bottom, it is referred to as _____.
   a. basket tray
   b. ladder tray
   c. trough
   d. raceway

4. Cable tray is generally manufactured in _____.
   a. 2' and 4' lengths
   b. 6' and 8' lengths
   c. 6' and 12' lengths
   d. 12' and 24' lengths

Additional information can be found in Appendixes at the back of the book.

APPENDIX 29202A

Example of a Three-View Detail Drawing

Figure A01 is an example of an object that may be encountered in welding. *Figure A02* is an example of a three-view detail drawing representing that same object.

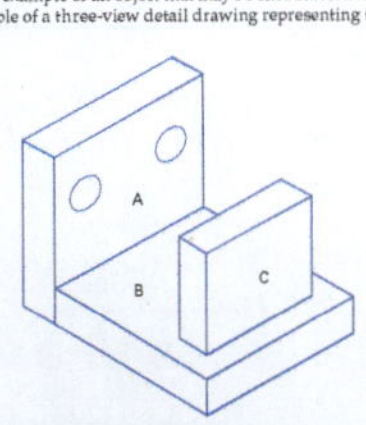

Details			
Part	Thickness	Length	Width
A	¾"	3½" (Tall)	3½"
B	¾"	4" (Long)	3½"
C	¾"	2½" (Tall)	2½"

Hole Dimensions and Location
- Hole diameter is ¾"
- Hole centers are 1¼" from the top edge of block "A"
- Hole centers are ¾" from the outside edges of block "A"

Figure A01

Some modules have corresponding Appendixes that provide supplementary information or activities to enhance your understanding of the material.

NCCERCONNECT

This interactive online course is a unique web-based supplement that provides a range of visual, auditory, and interactive elements to enhance training. Also included is a full eText.

Visit **www.nccerconnect.com** for more information!

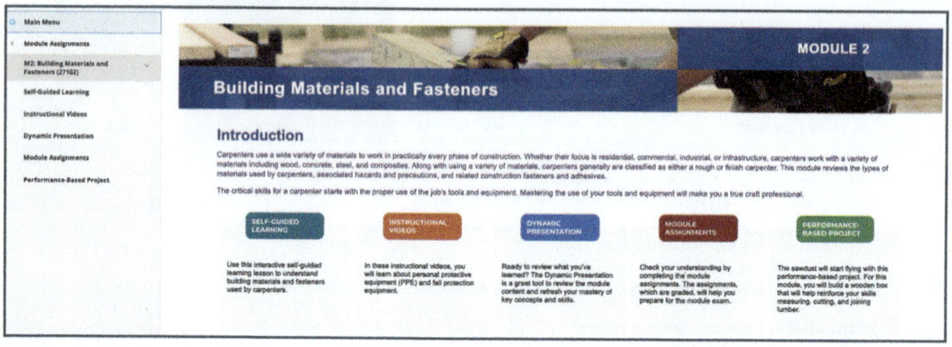

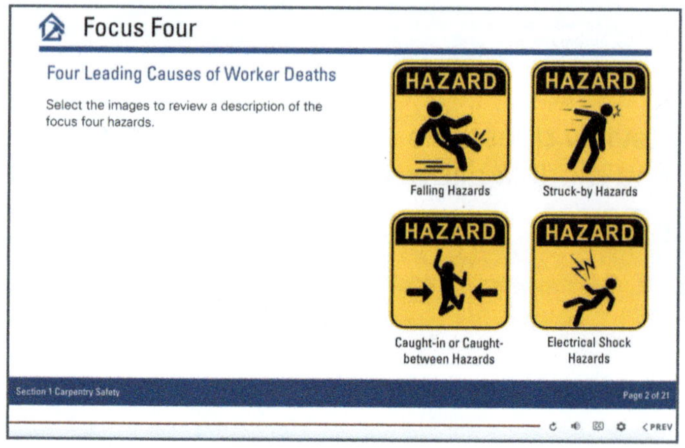

Use the interactive self-guided learning lesson to understand key concepts and terms needed for a career in construction.

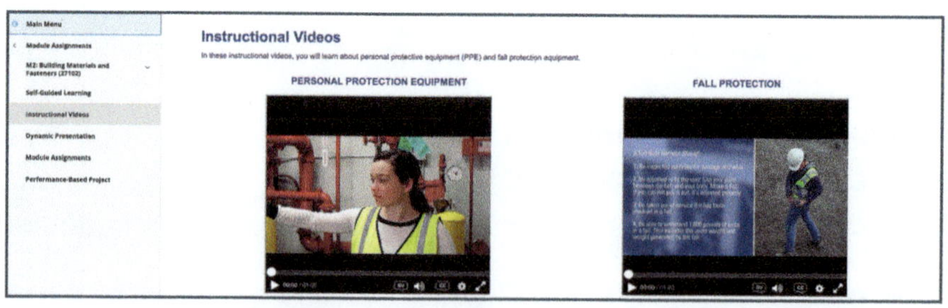

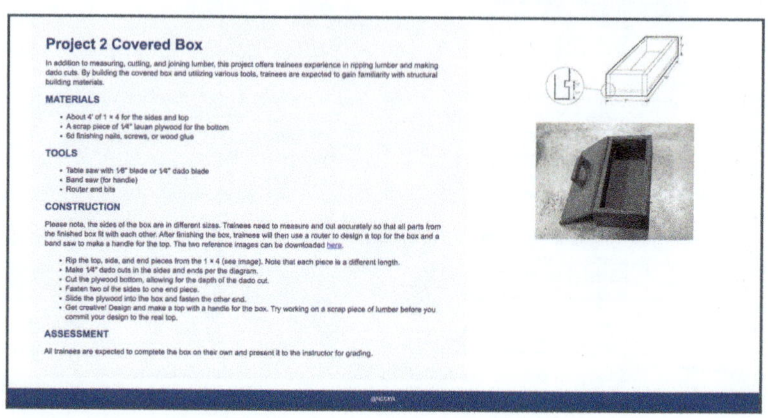

ACKNOWLEDGMENTS

This curriculum was revised as a result of the farsightedness and leadership of the following sponsors:

DRP Masonry
GG Masonry
IMS Masonry
Lang Masonry
Masonry Contractors Association of America
McGee Brick
North Carolina Masonry Contractors Association
Telligent Masonry
WBF Construction

This curriculum would not exist were it not for the dedication and unselfish energy of those volunteers who served on the Authoring Team. A sincere thanks is extended to the following:

Jeff Buczkiewicz
Jameel Ervin
Kale Hallman
Heath Holdaway
Josh Hoskinson
Melonie Leslie
Jason Roberts
Ryan Shaver
Donnie Williams

A final note: This book is the result of a collaborative effort involving the production, editorial, and development staff at Pearson Education, Inc., and NCCER. Thanks to all of the dedicated people involved in the many stages of this project.

NCCER PARTNERS

To see a full list of NCCER Partners, please visit:

www.nccer.org/about-us/partners.

SCAN ME

You can also scan this code using the camera on your phone or mobile device to view these partnering organizations.

CONTENTS

MODULE 28208 Design Documents

MODULE 28203 Reinforced Masonry

MODULE 28204 Masonry Openings and Metalwork

MODULE 28205 Advanced Laying Techniques

MODULE 28206 Effects of Climate on Masonry

Masonry Appendixes

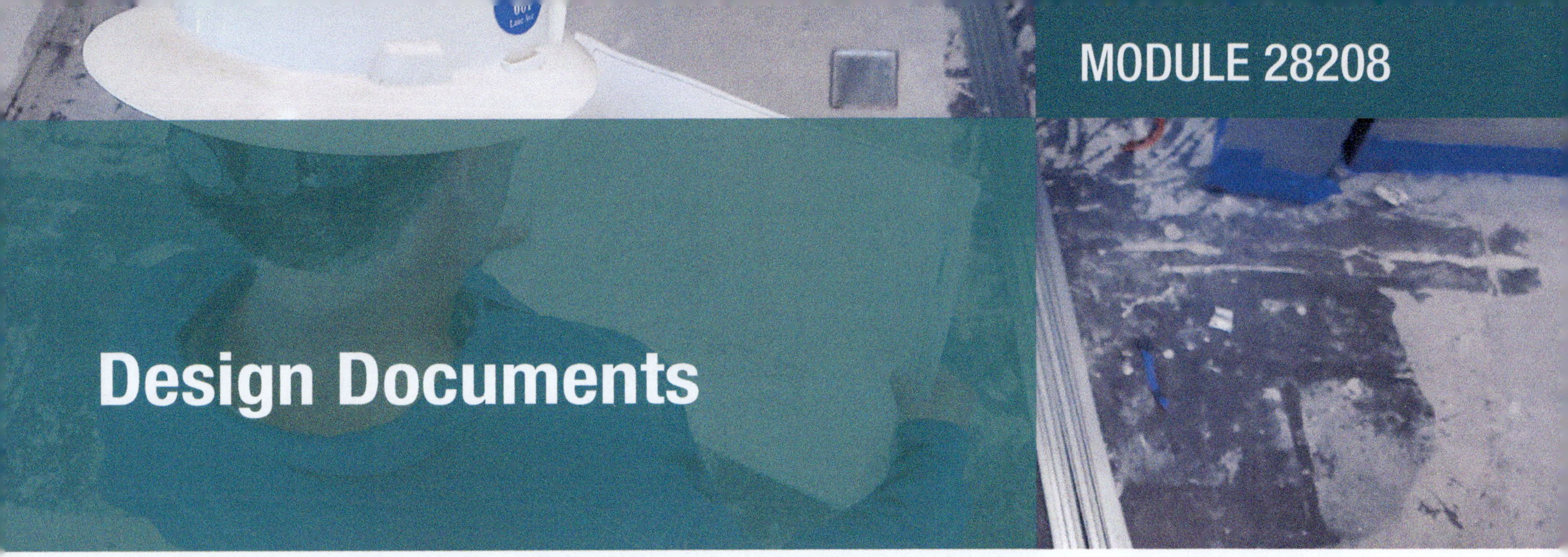

MODULE 28208

Design Documents

Objectives

Successful completion of this module prepares you to do the following:

1. List drawings included in a set of construction plans and explain how to read them.
   a. Describe the purpose of each type of drawing in a set of plans.
   b. Identify selected lines, architectural symbols, and abbreviations used on plans.
   c. Describe the methods of dimensioning construction drawings.
2. Describe the purpose of written specifications.
   a. Summarize how specifications are organized.
   b. Explain the importance of construction building codes.

Performance Tasks

Under supervision, you should be able to do the following:

1. Read and interpret construction plan drawings.
2. Read and interpret schedules.
3. Read and interpret written specifications.

Overview

Construction drawings and specifications include important design and layout information for all types of buildings. These documents provide a wide range of information, including where to place the structure on a site and what types of materials to use. The ability to read and interpret construction drawings plays a large role in becoming a craft professional.

A mason's work precedes that of other trades, such as electricians, plumbers, and HVAC installers, all of whom need spaces set aside for their equipment, wiring, piping, and ductwork. Construction drawings show how and where to construct these types of spaces. To ensure that these spaces—and the entire project—are constructed correctly, masons must be able to accurately read and interpret construction drawings and clearly communicate changes that need to be adopted in the building plan so that serious problems do not arise.

NCCER **Industry-Recognized Credentials**

If you are training through an NCCER-accredited sponsor, you may be eligible for credentials from NCCER's Registry. The ID number for this module is 28208. Note that this module may have been used in other NCCER curricula and may apply to other level completions. Contact NCCER's Registry at 1.888.622.3720 or go to **www.nccer.org** for more information.

You can also show off your industry-recognized credentials online with NCCER's digital credentials. Transform your knowledge, skills, and achievements into digital credentials that you can share across social media platforms, send to your network, and add to your resume. For more information, visit **www.nccer.org**.

Digital Resources for Masonry

SCAN ME

Scan this code using the camera on your phone or mobile device to view the digital resources related to this craft.

1.0.0 Construction Drawings

Performance Tasks

1. Read and interpret construction plan drawings.
2. Read and interpret schedules.

Objective

List drawings included in a set of construction plans and explain how to read them.

a. Describe the purpose of each type of drawing in a set of plans.
b. Identify selected lines, architectural symbols, and abbreviations used on plans.
c. Describe the methods of dimensioning construction drawings.

Symbols: Drawings that represent a material or component on a plan.

Construction plans: Detailed documents, both written and visual, that outline how to complete a construction project.

A special language of lines, **symbols**, abbreviations, and notes is used on construction drawings to deliver the architect's message as clearly as possible. This special language is used across all the different types of drawings that make up a typical set of **construction plans**.

These plans can include drawings from architects and structural engineers and may include information about HVAC (heating, ventilating, and air conditioning), electrical, plumbing, and other trades. This module describes the different types of drawings, what they show, and how to read them by interpreting the lines, dimensioning, symbols, and abbreviations.

1.1.0 Construction Plans

Request for information (RFI): A document used during the construction process to clarify and resolve questions and information gaps in plans, drawings, specifications, and agreements.

Scale: The ratio between the size of a drawing of an object and the size of the actual object.

Computer-aided design (CAD): The making of a set of construction drawings with the aid of a computer.

Construction plans or drawings are detailed written and visual documents that outline how to complete a construction project. They convey to skilled craftworkers the information needed to construct a specific building or structure. The construction drawings, together with the specifications, fully describe the building or structure about to be built.

Failure to follow construction plans can result in mistakes that, depending on their severity, can expose you and your employer to legal liability. If there are any questions or if information is missing from the construction plans, a **request for information (RFI)** should be issued.

An RFI is a document used during the construction process to clarify and resolve questions and information gaps in plans, drawings, specifications, and agreements. It is a formal document in which parties, such as the contractor and designer, clarify information in the construction documents as questions and clarification issues arise.

The set of construction drawings, drawn to **scale** by an architect and/or engineer, show all the information and dimensions necessary to build or remodel a structure. While in the past these drawings were created by hand, today they are created using various **computer-aided design (CAD)** software applications.

The drawings are then printed on large-format printers called *plotters.* Copies of the original drawings are made for contractors and other vendors. These copies are commonly called *prints.* Most prints are produced on large engineering copiers (similar to photocopiers). Most prints show the building and other construction information using black lines on white paper.

A great deal of work goes into creating a set of construction drawings before they can be used by the various contractors involved in the construction process. The path to completing a set of construction documents is referred to as the *architectural design process* (*Figure 1*) and typically includes the following phases:

- *Predesign*—This is the information gathering phase in which goals and objectives are reviewed and the full project scope is determined.

- *Schematic design*—Project requirements are reviewed and rough study sketches and drawings are produced to illustrate the basic design concepts and options.
- *Design development*—The schematic design is developed further, with the addition of materials and general structural details and the creation of more detailed site plans, floor plans, elevations, and sections.
- *Construction documents*—Once design development drawings are approved by the client, the final construction documents are produced. Final drawings and specifications will be submitted to the appropriate authority to receive approval for a building permit. Construction documents are considered legal documents.
- *Bidding*—Construction drawing sets and specifications are released to various trade contractors to bid on the project. The contracts are awarded to the appropriate contractors.
- *Construction administration*—The architect and/or a general contractor provide construction oversight to ensure that the project is being constructed based on the drawings and specifications.

Figure 1 Architectural design process.

While paper prints are still used on many construction projects, paperless drawings, often referred to as *digital plans*, are becoming increasingly common. Many construction collaboration software applications are available that allow users to access the project's construction drawings using any computer, tablet, or smartphone. These applications allow users to share, view, notate, and mark up drawings as well as collaborate with others involved in the project.

Blueprints

The term *blueprint* is derived from a print reproduction method that was used in the past. The original drawing was placed over a light-sensitive sheet of paper, and the drawing and paper were exposed to ultraviolet light. The light-sensitive paper was then placed in an ammonia vapor bath to develop the prints. A true blueprint shows the details of the structure as white lines on a blue background or as dark blue lines over a light blue background. Even though the true blueprinting process is rarely (if ever) used today, the term *blueprint* is still used when referring to copies made from original drawings. However, with the advent of CAD and, most recently, building information modeling (BIM), drawings are created on a computer and reproduced using large-format plotters, printers, and copiers. Today, construction drawings are simply called *drawings* or *prints*.

Construction drawings, also referred to as **blueprints**, consist of several different kinds of drawings assembled into a set (*Figure 2*). To convey complete information about a structure to the reader, various drawings in the set illustrate the structure using a variety of different views. A set of drawings also includes sheets that contain relevant written information, such as window or door schedules.

Blueprints: The traditional name and method of reproducing technical drawings.

Drawing sets can differ based on whether they are for a residential project such as a single-family home or a townhouse, or they are for a commercial project such as an office building or a large-scale multifamily housing complex. Commercial projects are typically much larger and more complex than residential projects.

There are differences in the building materials used, the codes that must be followed, the equipment required, and the construction methods themselves. Consequently, plans and drawings for a commercial project are also more complicated and require many more drawings to convey the information needed to complete the project.

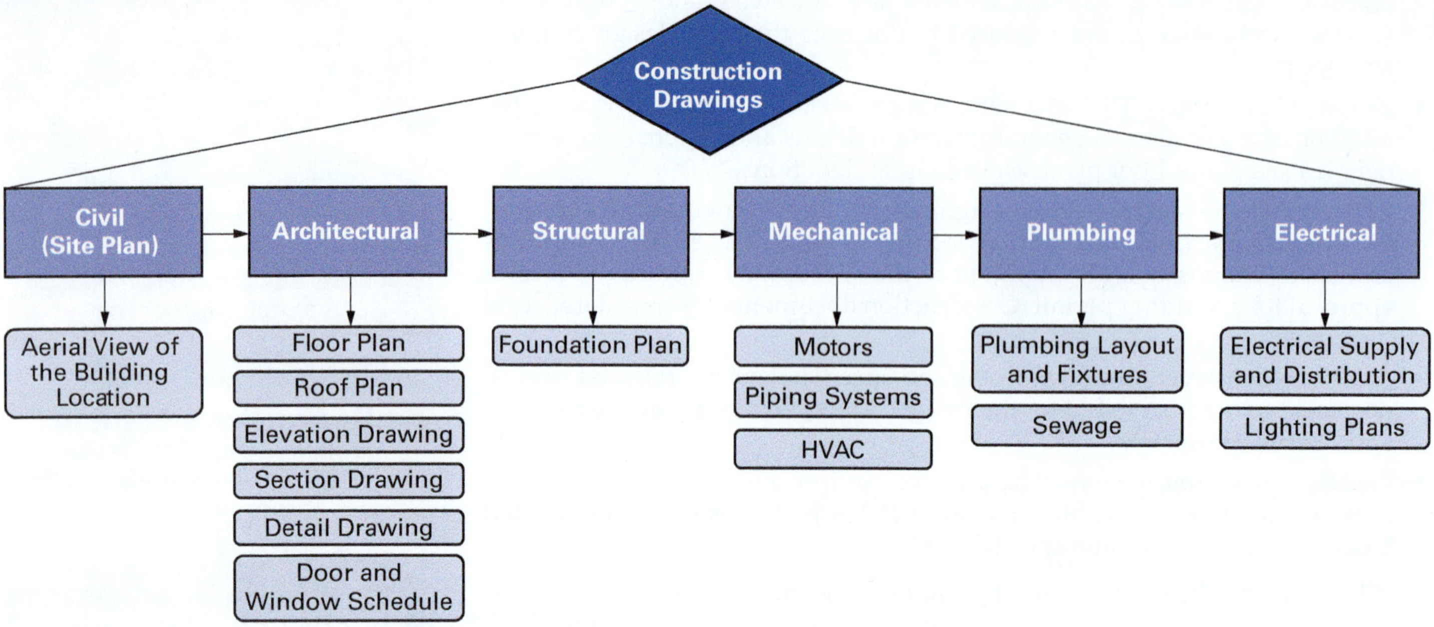

Figure 2 Typical organization of a construction drawing set.

The types of written information and views contained in residential and commercial drawing sets typically include the following:

Residential drawing set:

- Site plan
- Elevations
- Floor plans
- Foundation plan
- Roof framing plan
- Section views
- Framing and utility plans
- Details and notes
- Schedules
- Electrical plans

Commercial drawing set:

- Civil plan (site plan)
- Architectural plans
- Structural plans
- Mechanical, electrical, and plumbing plans

1.1.1 Title Sheets, Title Blocks, and Revision Blocks

A title sheet is typically placed at the beginning of a set of drawings or at the beginning of a major section of drawings. The title sheet provides an index to the other drawings, a list of abbreviations used on the drawings and their meanings, a list of symbols used on the drawings and their meanings, and various other project data such as the project location, size of the land parcel, and building size. The title sheet(s) should be carefully reviewed to understand the specific symbols and abbreviations used throughout the set of drawings. These symbols and abbreviations may vary from plan to plan.

Building Information Modeling

Architectural plans have been used over the centuries to pictorially describe buildings and structures before they are built. In the past, draftspersons would draw these plans by hand. Today, most construction drawings are generated using CAD and architectural software. The CAD designer, working with architects and engineers, creates the drawings on the computer. The drawings can then be printed or plotted onto paper.

Some CAD software is capable of building information modeling (BIM). BIM allows designers to prepare digital models of an entire structure and all its components. BIM allows people to virtually walk or even fly through a building before it has been built to see how it will look when it has been completed. This allows designers to identify problems and conflicts with various components in time to correct them before construction has begun or is completed. BIM also allows designers to estimate the costs of constructing the building and even the cost of facility maintenance over time.

Source: Shutterstock

A **title block** (*Figure 3*) is commonly placed on each sheet in a set of drawings. It is usually located in the lower-right corner or along the right edge of the sheet.

Title block: A part of a drawing sheet that includes some general information about the project.

The title block communicates several types of information. It typically contains the name of the firm that prepared the drawings, the owner's name and address, and the name of the project. It also provides locator information, such as the title of the sheet, the drawing or sheet number, the date the sheet was prepared, the scale, and the initials or names of the people who prepared and checked the drawings.

The design and construction phases of a project are very fluid. After construction has started for certain portions of a project, design changes to other portions of the project are not uncommon, even if the drawing package has already been submitted. In those instances, the affected drawings need to be revised to show the design changes. These revisions need to be tracked to ensure that the appropriate drawings are being used in the field.

Figure 3 Title block.

Presentation Drawings

Presentation drawings are three-dimensional drawings that show a building from a desirable vantage point to display its most interesting features. They do not provide detailed information for construction purposes. They are used mainly by an architect or a contractor to sell the design of a house or building to a prospective customer. The drawing shown here was included as a presentation drawing for NCCER's headquarters in Alachua, Florida.

A revision block (*Figure 4*) is typically shown on each sheet in a set of drawings, usually in the upper- or lower-right corner of the sheet near or within the title block. A revision block is used to record any changes (revisions) to the drawing.

An entry in the revision block usually contains the revision number or letter (keyed to a location on the plan), a brief description of the change, the date, and the initials of the person making the revision(s). When using drawings, it is essential to note the revision designation on each drawing and use only the latest issue; otherwise, costly mistakes will result. If you doubt the revision status of a drawing, check with your supervisor to ensure that the most recent version of the drawing is being used.

PROJ	NO	REVISION	RVSD	CHKD	APPD	DATE
3483	01	Released for construction		APD	NWS	July 2024
3483	02	Deleted part of line 12037		APD	NWS	Aug 2024
3483	03	⚠ Added welding symbol		APD	NWS	Aug 2024

Figure 4 Revision block.

1.1.2 Site Plans

Site plans are typically considered **civil plans**. They show the aerial view of the building location on the construction site.

Man-made and topographical (natural) features and other relevant project information are shown on a site plan, including the information needed to properly locate the structure on the site. Man-made features include roads, sidewalks, utilities, and buildings. Topographical features include trees, streams, springs, and existing ground contours. Project information includes the building outline, general utility information, proposed sidewalks, parking areas, roads, landscape information, proposed contours, and any other information that conveys what is to be constructed or changed on the site. A prominently displayed north arrow is included for orientation purposes on site plans. Sometimes a site plan contains a large-scale map of the overall area that indicates where the project is located on the site. For most residential projects and smaller commercial projects, site plans are often only included within the architectural drawings. Architectural site plans are like a map of the construction site, showing how the building and any other structures will be located on the site. For larger, more complex projects, site plans are also included within the civil drawings. Civil site plans show information on grading, landscaping, storm sewer piping, parking, and other site details. *Figure 5* shows an example of a typical architectural site plan, and *Figure 6* shows an example of a typical civil site plan.

Civil plans: Drawings that show the location of the building on the site from an overhead view, including contours, trees, construction features, and dimensions.

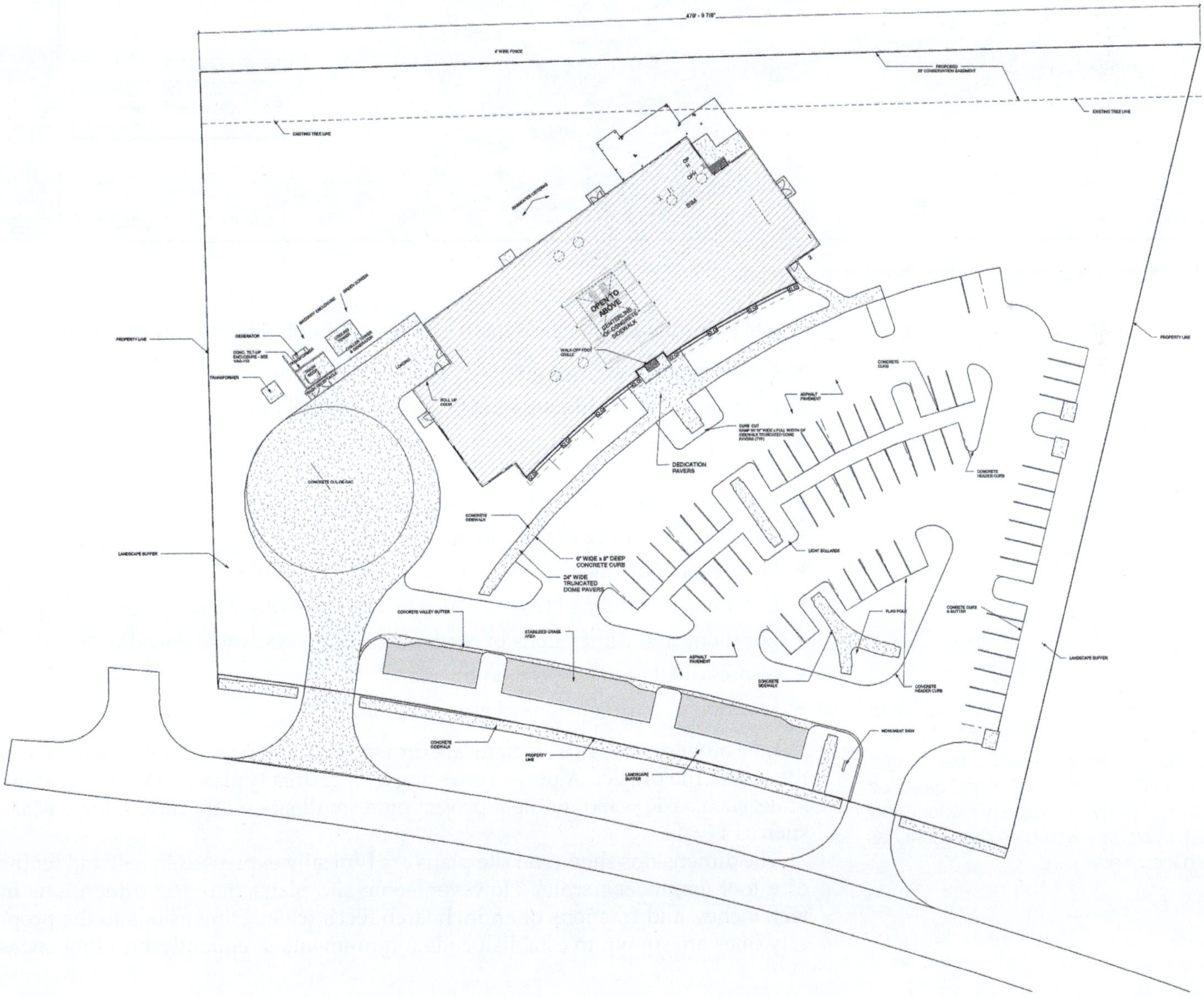

Figure 5 Typical architectural site plan.

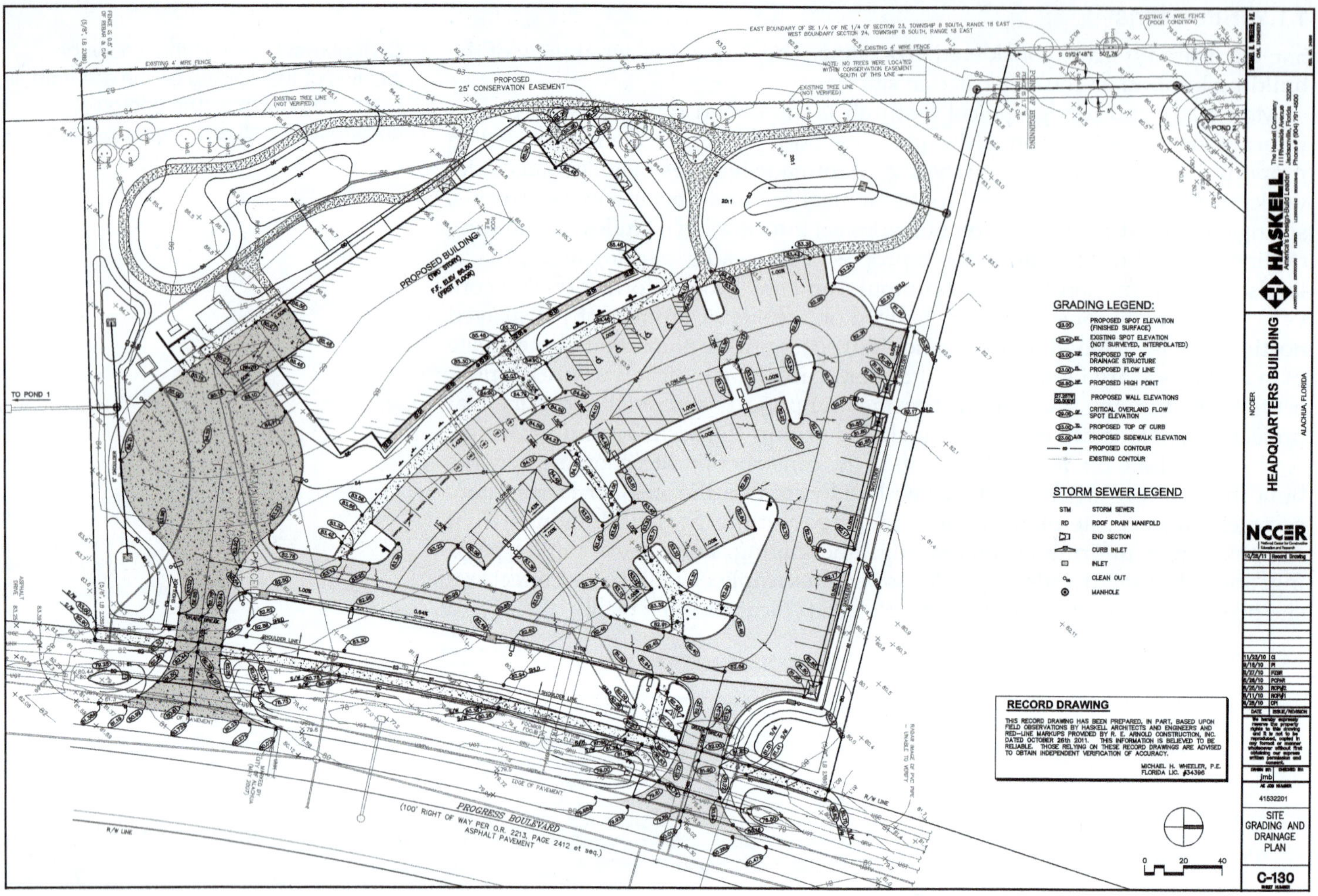

Figure 6 Typical civil site plan showing topographical features.

Property lines: The recorded legal boundaries of a piece of property.

Monuments: Physical structures that mark the location of every survey point.

Easement: A legal right-of-way provision on another person's property (e.g., the right of a neighbor to build a road or of a public utility to install water and gas lines on the property). A property owner cannot build on an area where an easement has been identified.

Site plans usually include the following types of detailed information:

- Coordinates of control points or property corners
- Direction and length of **property lines** or control lines
- Descriptions, or references to descriptions, of all control and property **monuments**
- Location, dimensions, and elevations of the structure on-site
- Finish and existing grade contours
- Locations of utilities
- Locations of existing elements such as trees and other structures
- Locations and dimensions of roadways, driveways, and sidewalks
- Names of all roads shown on the plan
- Location and dimensions of any **easement**

Like other drawings, site plans are drawn to scale. The scale used depends on the size of the project. A project covering a large area typically will have a small scale, such as 1" = 100', while a project on a small site might have a large scale, such as 1" = 10'.

The dimensions shown on site plans are typically expressed in feet and tenths of a foot (engineer's scale). However, some site plans state the dimensions in feet, inches, and fractions of an inch (architect's scale). Dimensions to the property lines are shown to establish code requirements. Frequently, building codes require that nothing be built on certain portions of the land. For example, a local building code may have a **front setback** requirement that dictates the minimum distance that must be maintained between the street and the front of a structure.

Front setback: The distance from the property line to the front of the building.

Typically, side yards have an easement to allow for access to rear yards and to reduce the possibility of fire spreading to adjacent buildings. A property owner cannot build on an area where an easement has been identified. Examples of typical easements are the right of a neighbor to build a road; the right of a public utility to install water, gas, or electric lines on the property; or an area set aside for groundwater drainage.

Site plans show finish grades (also called *elevations*) for the site based on data provided by a surveyor or engineer. It is necessary to know these elevations for grading the lot and for construction of the structure. Finish grades are typically shown for all four corners of the lot and other points within the lot. Finish grades or elevations are also shown for the corners of the structure and relevant points within the building.

All the finish-grade references shown are keyed to a reference point, called a **benchmark** or *job datum*. A benchmark is a reference point established by the surveyor on or close to the property, usually at one corner of the lot. At the site, this point may be marked by a plugged pipe driven into the ground, a brass marker, or a wooden surveyor stake.

Benchmark: A point established by the surveyor on or close to the building site. It is used as a reference for determining elevations during the construction of a building.

Surveyor stakes are markers that surveyors use in surveying projects to prepare jobsites and mark out property boundaries, building locations, and benchmarks. The location of the benchmark is shown on the site plan with a grade figure next to it. This grade figure may indicate the actual elevation relative to sea level, or it may be an arbitrary elevation of, for example, 100.00' or 500.00'. All other grade points shown on the site plan, therefore, are relative to the benchmark.

A site plan usually shows the **finish floor elevation (FFE)** of the building. This is the level of the first floor of the building relative to the jobsite benchmark. For example, if the benchmark is labeled 100.00' and the FFE indicated on the plan is marked 105.00', the FFE is 5' above the benchmark. During construction, many important measurements are taken from the FFE point.

Finish floor elevation (FFE): The uppermost surface of a floor once construction is complete and prior to any finishes or coverings being applied. It is used as a reference plane for other vertical dimensions.

1.1.3 Foundation Plans

The **foundation plan** for a residential structure is usually contained within the architectural plans; for commercial and industrial projects, it is usually contained within the structural plans. Foundation plans provide information about the locations and dimensions of footings, grade beams, foundation walls, stem walls, piers, equipment footings, and windows and doors. The specific information shown on the plan is determined by the type of construction involved: a full-basement foundation, a crawl space, or a concrete slab-on-grade foundation.

Foundation plan: A drawing that shows the layout and elevation of the building foundation.

The following types of information are commonly shown on foundation plans for full-basement and crawl space foundations:

- Locations of the inside and outside of the foundation walls
- Locations of the footings for foundation walls, columns, posts, chimneys, and fireplaces
- Wall openings for windows, doors, crawl space access, and vents
- Walls for entrance platforms (stoops)
- Floor joist direction, size, and spacing
- Locations of stairways
- Notations for the strength of concrete used for various parts of the foundation and floor
- Notations for the composition, thickness, and underlying material of the basement floor or crawl space surface
- Locations of furnaces and other equipment

The following types of information are commonly shown on foundation plans for slab-on-grade foundations:

- Size and shape of the slab
- Exterior and interior footing locations
- Locations of fireplaces, floor drains, and HVAC ductwork
- Notations for slab thickness
- Notations for wire mesh reinforcing, fill, and vapor barrier materials

Not to Scale (NTS): Describes drawings that show relative positions and sizes only, without scale.

Drawings may be printed or copied to show details more clearly. Always label drawings as **Not to Scale (NTS)** whenever they are printed or copied. Doing so lets the reader know that dimensions cannot be accurately obtained from these versions.

As with other types of plans, commercial foundation plans (*Figure 7*) typically include much more detail than their residential counterparts. Commercial foundation plans include much of the same information commonly shown on residential foundation plans (*Figure 8*), but due to their complex nature and the heavy loads involved, they also include additional information detailing structural features such as additional reinforced concrete, base plates, columns, beams, and lintels.

Going Green

Insulating Concrete Forms

Insulating concrete forms (ICFs) are a type of concrete forming system in which the concrete is sandwiched between two insulating foam panels. The foam panels remain in place after the concrete has set. ICFs offer many advantages, including low amounts of air infiltration and lower heating and cooling loads, thus making the building very energy efficient.

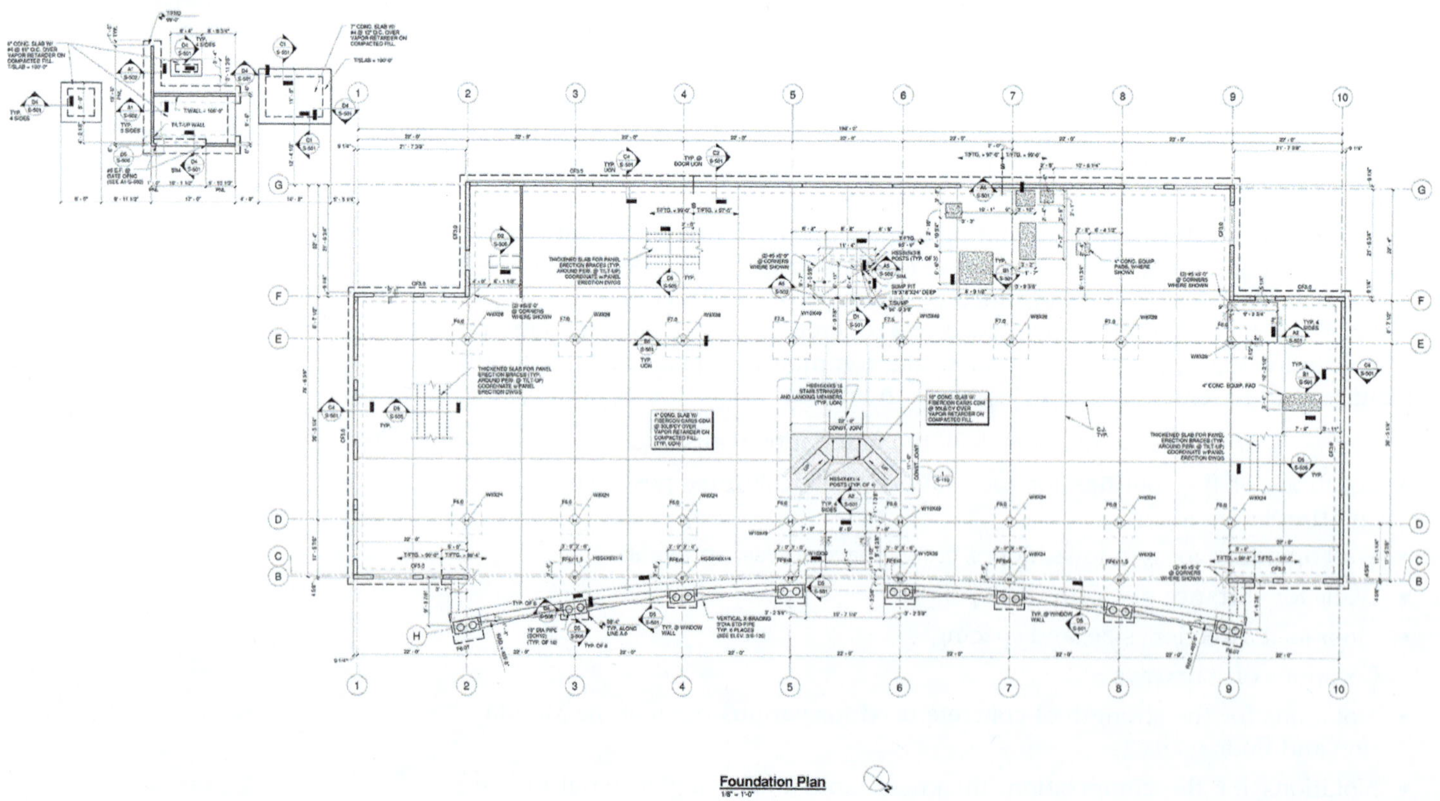

Figure 7 Commercial foundation plan (NTS).

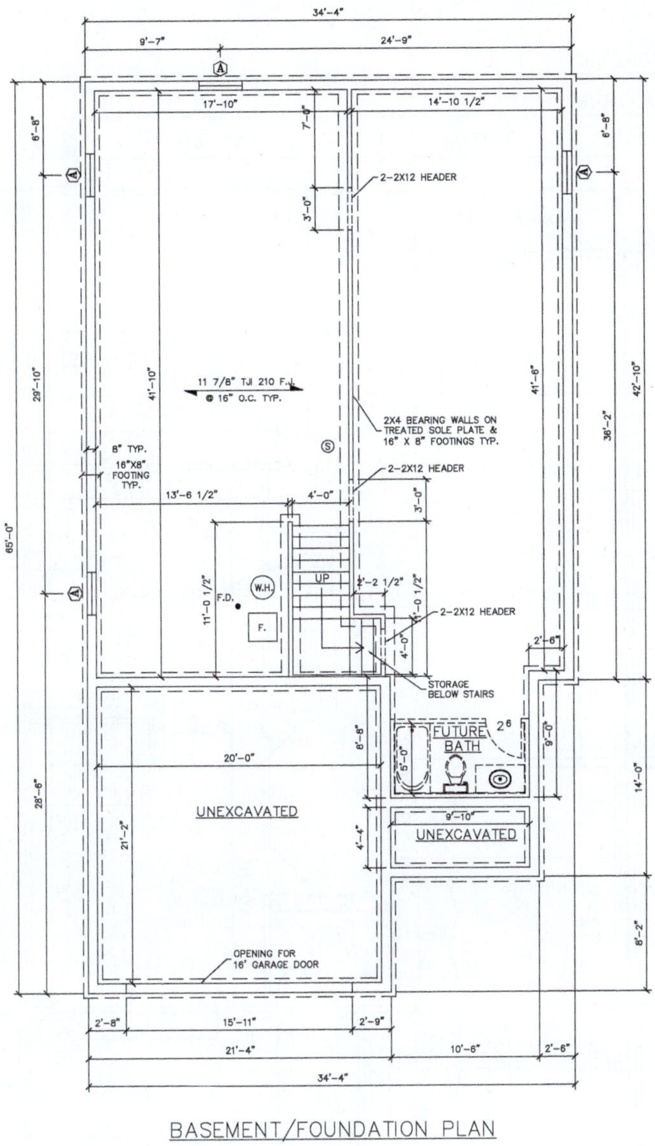

Figure 8 Residential foundation plan (NTS).

1.1.4 Floor Plans

Floor plan: A drawing that provides an overhead view of the layout of each room.

Plan view: A drawing that represents a view looking down on an object.

The **floor plan** is found in the architectural plans and is the primary drawing of the entire set. For a floor **plan view**, an imaginary line is cut horizontally across the structure at varying heights so that all the important features such as windows, doors, and plumbing fixtures can be shown.

For multistory buildings, separate floor plans are typically provided for each floor. However, if several floors have the same layout (as in a hotel), one drawing may be used to show all the floors that are similar. *Figure 9* shows an example of a basic floor plan. The following types of information are commonly shown on floor plans:

- Outside walls, including the locations and dimensions of all exterior openings
- Types of construction materials
- Locations of interior walls and partitions
- Locations and swing of doors
- Stairways
- Locations of windows
- Locations of cabinets, electrical and mechanical equipment, and fixtures
- Locations of cutting plane lines

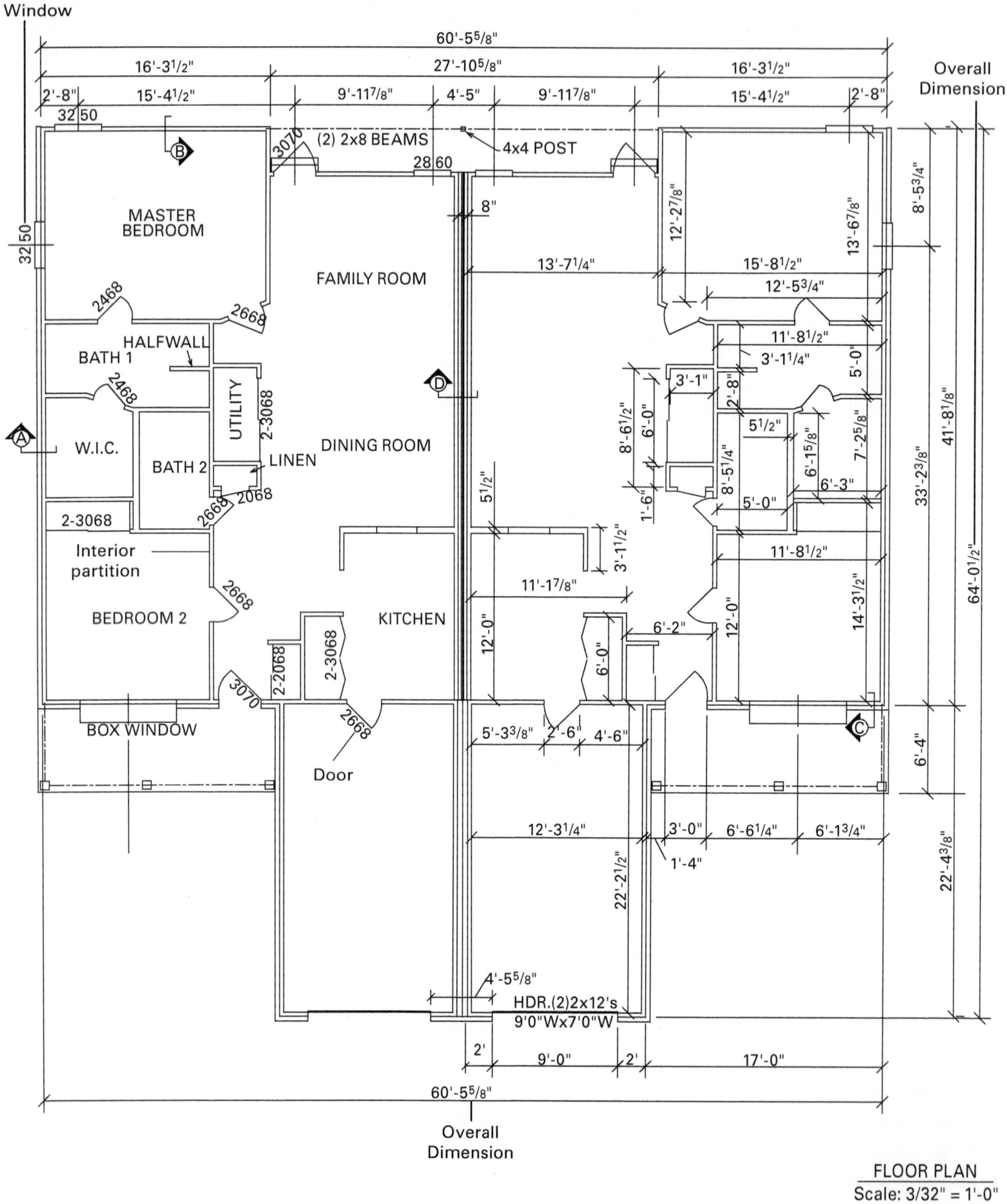

Figure 9 Basic residential floor plan (NTS).
Source: John Hoerlein

Each door or window shown on a floor plan for a commercial building is typically accompanied by a number, letter, or both. This number/letter is an identifier that refers to a door or window schedule that describes details such as the corresponding size, type of material, and model number for the specific door or window. Door and window schedules are discussed in more detail later in this section. Residential floor plans often show door sizes directly on the floor plan.

1.1.5 Roof Plans

Roof plans can be considered architectural or structural drawings, depending on the size and complexity of the project. They provide information about the roof slope, roof drain placement, and other pertinent information regarding items such as ornamental sheet metalwork, gutters, and downspouts. Where applicable, the roof plan may also show information on the locations of air conditioning units, exhaust fans, and other ventilation equipment. Not all sets of drawings include roof plans.

Roof plan: A drawing of the view of the roof from above the building.

Some drawing sets may include a ceiling plan that shows the locations of supply diffusers, exhaust grilles, access panels, structural components, and other mechanicals. Some drawing sets have reflected ceiling plans, which show the details of the ceiling as though it were reflected by a mirror on the floor. Ceiling plans show features of the ceiling while keeping those features in proper relation to the floor plan. For example, if a vertical pipe runs from floor to ceiling in a room and is drawn in the upper-left corner of the floor plan, it is also shown in the upper-left corner of the reflected ceiling plan of that same room.

Floor Plans

As shown here, a floor plan is actually a horizontal section of the building with the cutting plane through the windows and doors.

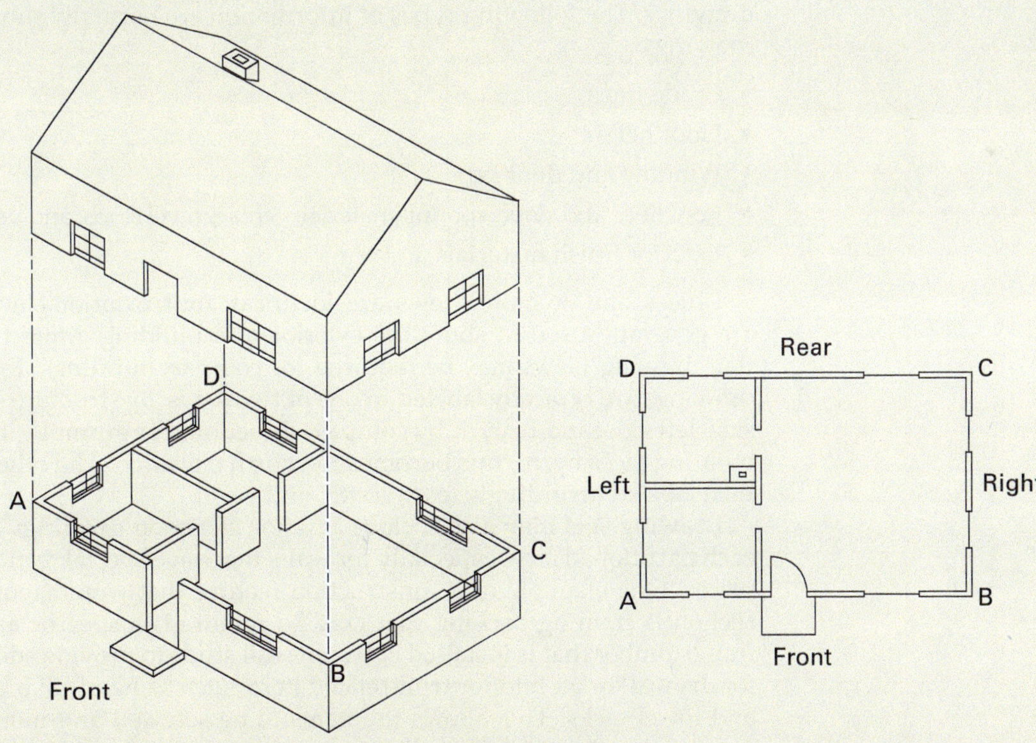

Going Green

Green Roofs

A green roof is a layer of vegetation grown on a rooftop. Green roofs provide shade and remove heat from the air through evapotranspiration, reducing temperatures of the roof surface and the surrounding air. On hot summer days, the surface temperature of a green roof can be cooler than the air temperature, whereas the surface of a conventional rooftop is considerably hotter.

Green roofs can be installed on a wide range of buildings, from industrial facilities to private residences. They can be as simple as a 2" covering of hardy groundcover or as complex as a fully accessible park complete with trees. Green roofs are becoming popular in the United States. Advantages of green roofs include the following:

- Reduced energy costs
- Reduced air pollution
- Better stormwater management

1.1.6 Elevation Views

Elevation view: A drawing providing a view from the front or side of a structure.

An **elevation view** provides a view of the side of a structure or component. The structure's features are projected onto a vertical plane. Typically, elevation views are found within the architectural plans and are used to show the exterior features of a structure so that the general size and shape of the structure can be determined.

Elevation views clarify much of the information on the floor plan. For example, a floor plan shows where the doors and windows are located on the outside walls; an elevation view of the same wall shows actual representations of these doors and windows. *Figure 10* shows an example of basic elevation view drawings. The following types of information are normally shown on elevation drawings:

- Grade lines
- Floor height
- Window and door types
- Roof lines and slope, roofing material, vents, gravel stops, and projection of eaves
- Exterior finish materials and trim

Unless one or more views are identical, four exterior elevation drawings are generally used to show the exterior of a building. More than four elevation drawing views may be required for complex buildings. Exterior elevation drawings are typically labeled in one of two ways: by structure side (front, right side, left side, and rear) or by compass direction. For example, if the front of the building faces north, this becomes the north elevation. The other elevations are then labeled accordingly (east, south, and west).

Drawing sets may also include interior elevation drawings for the walls in each partitioned area, especially for walls that have special features such as fireplaces. One example of a construction feature that would require an interior section is framing backing that may be required in specific areas. Backing is frame lumber that is installed between wall studs to provide additional support for drywall or an interior trim-related item, such as handrail brackets, cabinets, and towel racks. This allows the items to be screwed and mounted into solid wood rather than the drywall.

1.1.7 Section Drawings

Section drawing: A cross-sectional view of a specific location, showing the inside of an object or building.

Structural plans: A set of engineered drawings used to support the architectural design.

A **section drawing** or section view (*Figure 11*) shows the interior of a wall or other structural feature. Section drawings are used in both architectural and **structural plans**. They are drawn as if a cut has been made through a feature at a certain location. The location of the cut and the direction from which the section is to be viewed are shown on the related plan view.

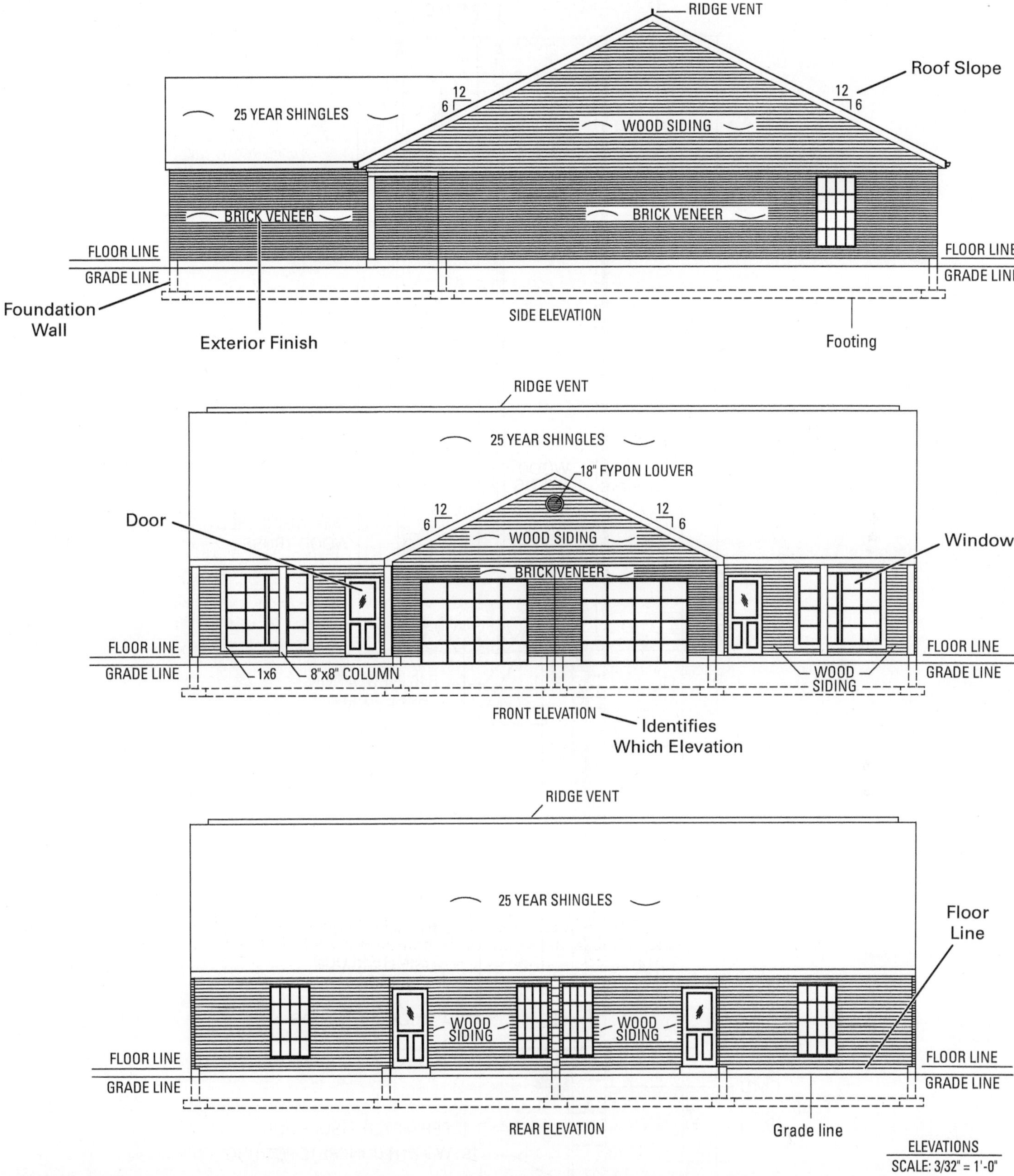

Figure 10 Elevation views (NTS).
Source: John Hoerlein

Section drawings that show a view made by cutting through the length of a structure are referred to as *longitudinal sections,* while those showing the view of a cut through the width or narrow portion of the structure are referred to as *transverse sections.*

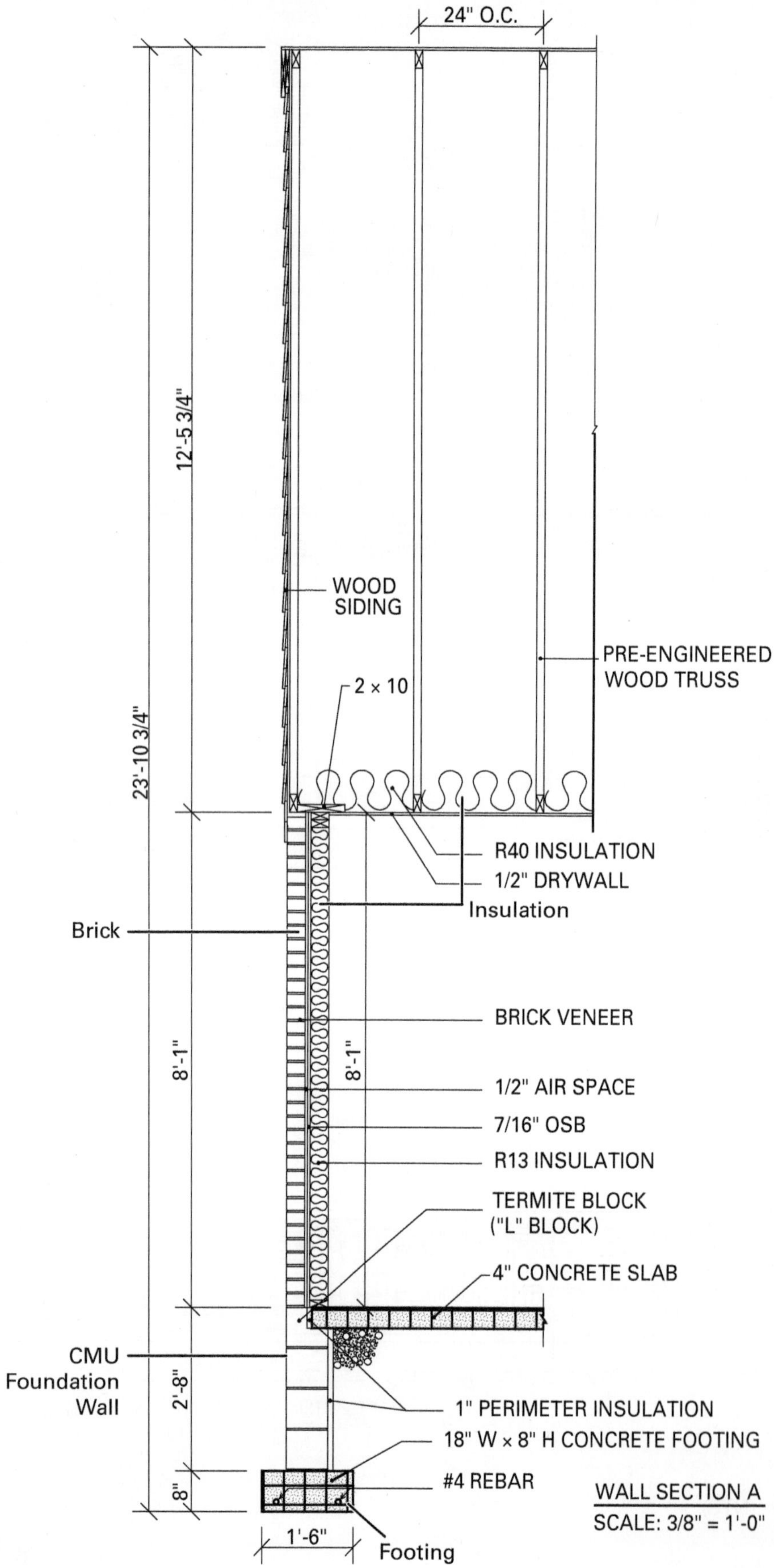

Figure 11a Section drawing (NTS, 1 of 3).
Source: John Hoerlein

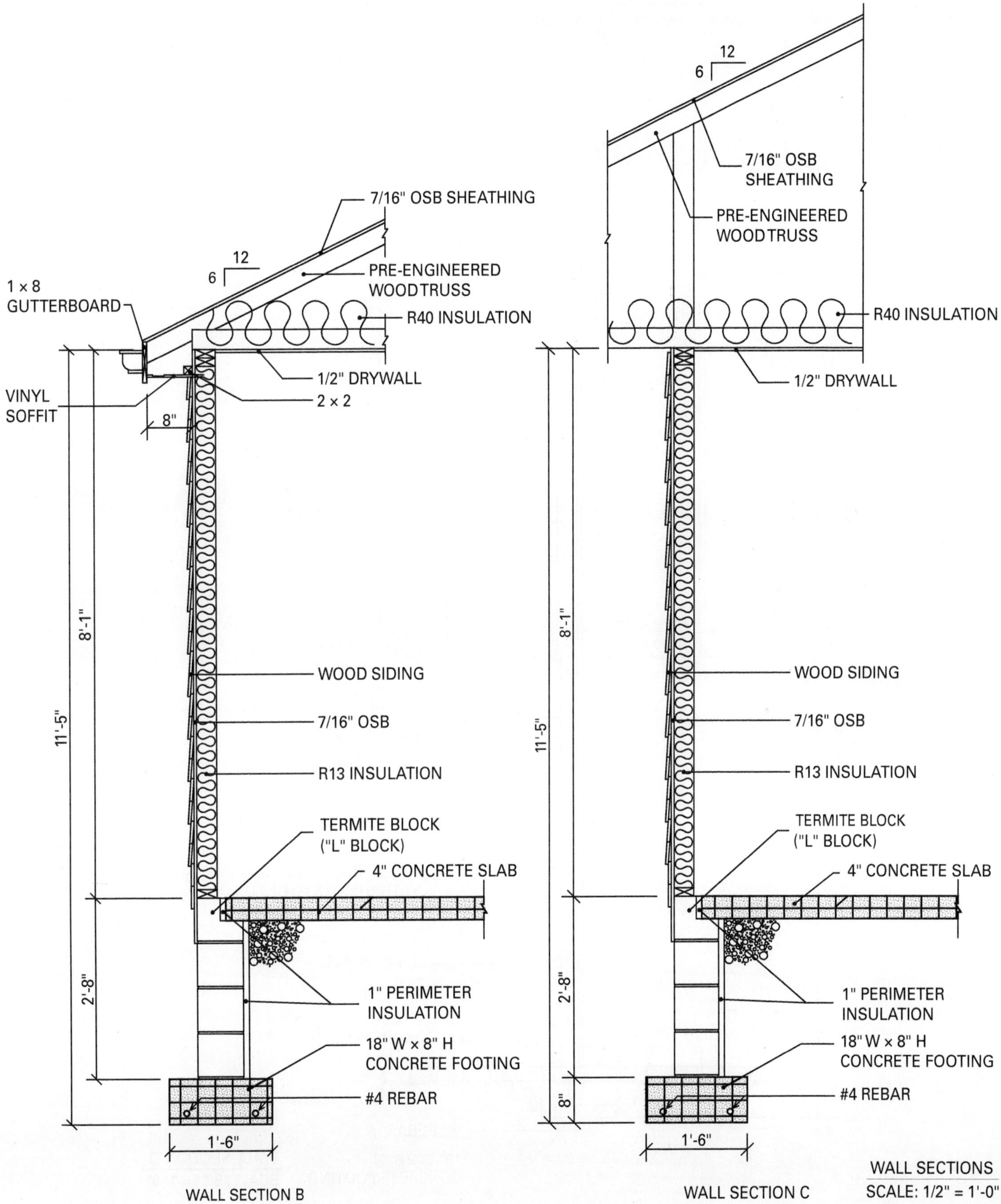

Figure 11b Section drawing (NTS, 2 of 3).
Source: John Hoerlein

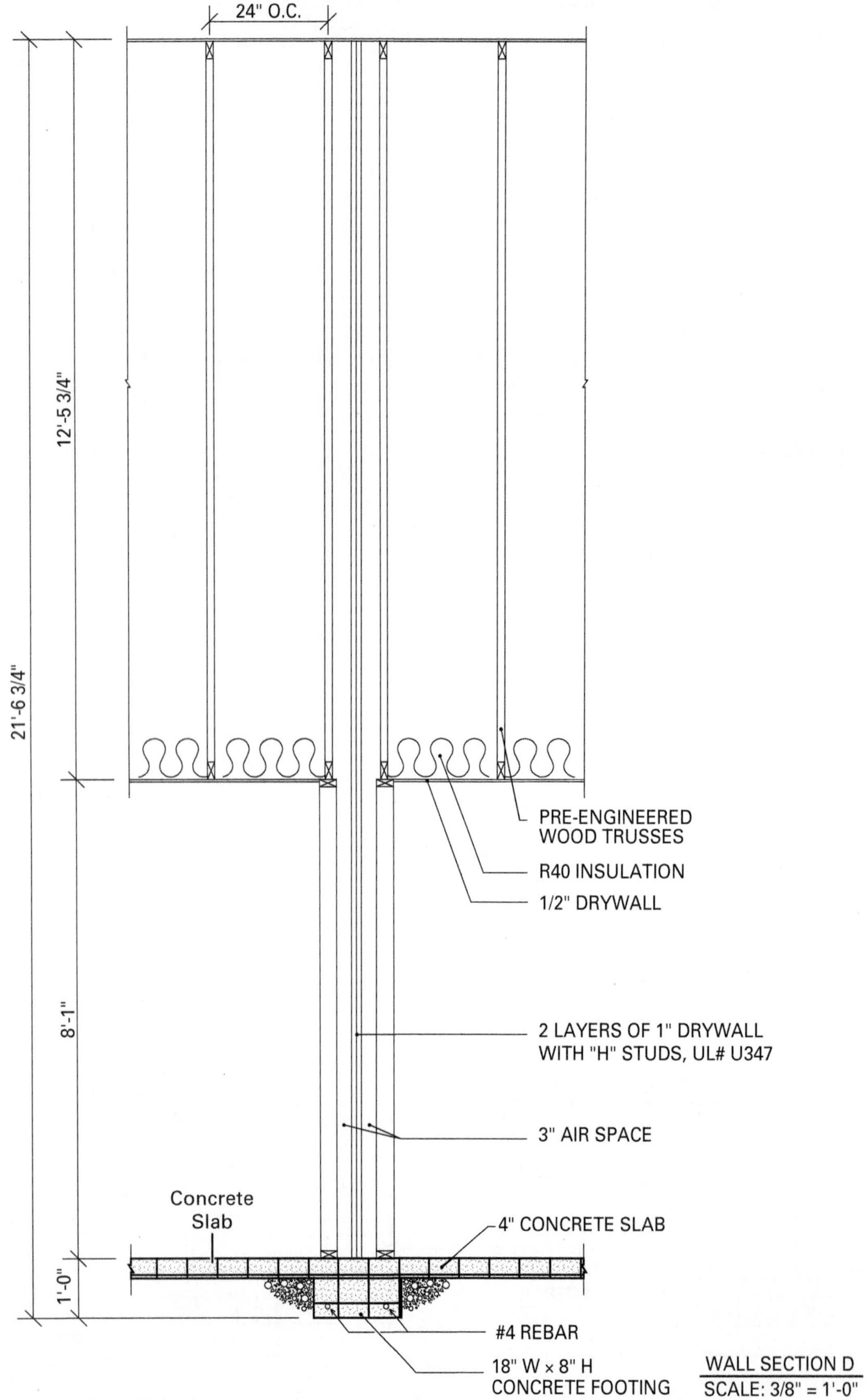

Figure 11c Section drawing (NTS, 3 of 3).

Source: John Hoerlein

To show greater detail, section drawings are normally drawn to a larger scale than plan views. The types of information commonly shown on a section drawing include the following:

- Details of construction and information about stairs, walls, chimneys, or other parts of construction that may not show clearly on a plan view
- Floor levels in relation to grade
- Wall thickness at various locations
- Anchors and reinforcing steel

1.1.8 Detail Drawings

Detail drawings are enlargements of special features of a building or of equipment installed in a building. They are typically drawn to a larger scale to make the details clear. Additionally, they are often placed on the same sheet where the feature appears in the plan. *Figure 12* shows a series of detail drawings.

Typically, detail drawings are used for the following objects or situations:

- Footings and foundations, including anchor bolts, reinforcing joints, and control joints
- Beams, floor joists, bridging, and other support members
- Sills, floor framing, exterior walls, and vapor barriers
- Floor height, thickness, expansion, and reinforcing
- Interior walls
- Windows, exterior and interior doors, and doorframes
- Roofs, cornices, soffits, and ceilings
- Gravel stops, fascia, and flashing
- Fireplaces and chimneys
- Stairways and stair assemblies
- Millwork, trim, ornamental iron, and specialty items

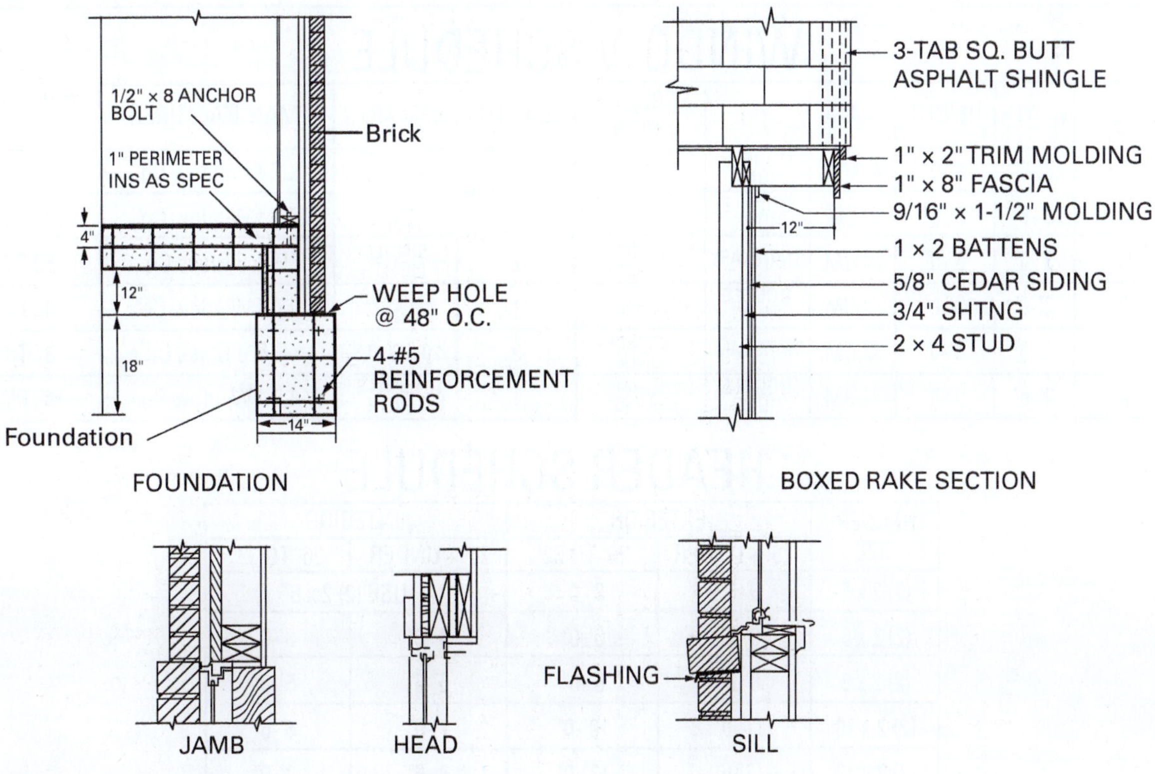

Figure 12 Detail drawings.

1.1.9 Schedules

Wall type: Wall types are defined by their different construction materials and methods.

Schedules are also commonly used with different types of drawings. Schedules are tables that describe and specify the various types and sizes of construction materials used in a building (*Figure 13*). Door schedules, window schedules, **wall type** schedules, and finish schedules are the most common schedules included on drawings. For commercial projects, additional schedules are provided for mechanical equipment and controls, plumbing fixtures, lighting fixtures, and any other equipment that needs to be listed separately.

Door, window, header, and finish schedules are of particular importance to a mason. Door and window types are identified on the various plan and elevation drawings by numbers and/or letters. The door and window schedules list

DOOR SCHEDULE

DOOR	WIDTH	HEIGHT	THICK-NESS	MAT'L	TYPE	STORM DOOR	QTY.	THRES-HOLD	REMARKS	MANUFACTURER
2068	2'-0"	6'-8"	1 ⅜"	Wood-Ash	Hollow-core	NO	5	None	Oil Stain	LBJ Door Co.
2468	2'-4"	6'-8"	1 ⅜"	Wood-Ash	Hollow-core	NO	1	None	Oil Stain	LBJ Door Co.
2668	2'-6"	6'-8"	1"	Wood-Ash	Cafe	NO	1 pr.	None	Oil Stain	LBJ Door Co.
2668	2'-6"	6'-8"	1 ⅜"	Wood-Ash	Sliding Pocket	NO	1	None	Oil Stain	LBJ Door Co.
2668	2'-6"	6'-8"	1 ⅜"	Wood-Ash	Hollow-core	1 Screen	6	Alum.	Screen door in garage	LBJ Door Co.
2868	2'-8"	6'-8"	1 ¾"	Metal Clad	Fireproof	YES	1	Alum.	Paint	LBJ Door Co.
3068	3'-0"	6'-8"	1 ¾"	Wood-Ash	Solid-core	NO	1	None	Oil Stain	LBJ Door Co.
3668	3'-6"	6'-8"	1 ¾"	Wood-Ash	Solid-core	YES	1	Alum.	Marine Varnish	LBJ Door Co.
6066	6'-0"	6'-6"	½"	Glass/Metal	Sliding	YES	1 pr.	Alum.	Sliding Screen	LBJ Door Co.
6068	6'-0"	6'-8"	1 ¼"	Wood-Ash	Bi-Fold	NO	2 sets	None	Oil Stain	LBJ Door Co.
1956	1'-9"	5'-6"	½"	Glass/Metal	Sliding Shower door	NO	2 sets	None	Frosted Glass	LBJ Door Co.

WINDOW SCHEDULE

SYMBOL	WIDTH	HEIGHT	MAT'L	TYPE	SCREEN & DOOR	QUANTITY	REMARKS	MANUFACTURER	CATALOG NUMBER
A	3'-8"	3'-0"	ALUM.	DOUBLE HUNG	YES	2	4 LIGHTS, 4 HIGH	LBJ Window Co.	141 PW
B	3'-8"	5'-0"	ALUM.	DOUBLE HUNG	YES	1	4 LIGHTS, 4 HIGH	LBJ Window Co.	145 PW
C	3'-0"	5'-0"	ALUM.	STATIONARY	STORM ONLY	2	SINGLE LIGHTS	H & J Glass Co.	59 PY
D	2'-0"	3'-0"	ALUM.	DOUBLE HUNG	YES	1	4 LIGHTS, 4 HIGH	LBJ Window Co.	142 PW
E	2'-0"	6'-0"	ALUM.	STATIONARY	STORM ONLY	2	20 LIGHTS	H & J Glass Co.	37 TS
F	3'-6"	5'-0"	ALUM.	DOUBLE HUNG	YES	1	16 LIGHTS, 4 HIGH	LBJ Window Co.	143 PW

HEADER SCHEDULE

HEADER SIZE	EXTERIOR		INTERIOR	
	26' + UNDER	26' TO 32'	26' + UNDER	26' TO 32'
(2) 2 x 4	3'-6"	3'-0"	USE (2) 2 x 6	
(2) 2 x 6	6'-6"	6'-0"	4'-0"	3'-0"
(2) 2 x 8	8'-6"	8'-0"	5'-6"	5'-0"
(2) 2 x 10	11'-0"	10'-0"	7'-0"	6'-6"
(2) 2 x 12	13'-6"	12'-0"	8'-6"	8'-0"

Figure 13 Examples of door, window, and header schedules.

these identifier numbers or letters and describe the corresponding size, type of material, and model number for each different type of door or window used in the structure.

Wall type schedules are used to call out different wall types such as interior, exterior, load-bearing, and priority. Wall types are defined by their different construction materials and methods. For example, a **priority wall** is a full-height wall—usually a fire-rated wall—where drywall cannot be put in place if ductwork or other specialty trade work has already been installed. This means the drywall contractor must work on a priority wall before other trades can do their work.

In a finish schedule (*Figure 14*), each room is identified by name or number. The material and finish for each part of the room (walls, floor, ceiling, base, and trim) are designated, along with any clarifying remarks.

Priority wall: A full-height wall (usually a fire-rated wall) where drywall cannot be put in place if ductwork or other specialty trade work has already been installed. Drywall contractors must work on a priority wall before other trades can do their work.

1.1.10 Life Safety Plan

A **life safety plan**, sometimes referred to as a *fire protection plan*, is a floor plan of a building that identifies each fire and life safety system. It also details how building occupants will be alerted to an emergency situation and evacuated from a building.

Life safety plan: A floor plan of a building that identifies each fire and life safety system.

Items commonly shown on a life safety plan include the following:

- Fire-rated walls (one-hour, two-hour, etc.)
- Fire and smoke doors

ROOM FINISH SCHEDULE

ROOMS	FLOOR				CEILING				WALL				BASE				TRIM			REMARKS
	CARPET	CERAMIC TILE	RUBBER TILE	CONCRETE	ACOUSTIC TILE	DRYWALL	PAINT	CERAMIC TILE	DRYWALL	PAINT	WALLPAPER	CERAMIC TILE	WOOD	RUBBER	CERAMIC TILE	STAIN	WOOD	STAIN	PAINT	
ENTRY		✓			✓				✓	✓	✓		✓			✓	✓	✓		See owner for all painting
HALL	✓				✓				✓	✓			✓			✓	✓	✓		
BEDROOM 1	✓				✓				✓	✓	✓		✓			✓	✓	✓		See owner for grade of carpet
BEDROOM 2	✓				✓				✓	✓			✓			✓	✓	✓		See owner for grade of carpet
BEDROOM 3	✓				✓				✓	✓			✓			✓	✓	✓		See owner for grade of carpet
BATH 1	✓	✓			✓			✓	✓	✓	✓	✓	✓			✓	✓	✓		Wallpaper 3 walls around vanity
BATH 2		✓			✓			✓	✓	✓	✓	✓			✓		✓	✓		Water-seal tile Wallpaper w/wall
UTIL + CLOSETS	✓		✓			✓	✓		✓	✓			✓			✓	✓	✓	✓	Use off-white flat latex
KITCHEN			✓		✓				✓	✓				✓			✓	✓		
DINING	✓				✓				✓	✓	✓		✓			✓	✓	✓		
LIVING	✓				✓				✓	✓			✓			✓	✓	✓		See owner for grade of carpet
GARAGE				✓		✓	✓					✓		✓			✓	✓		

Figure 14 Example of a finish schedule.

- Fire and smoke damper locations
- Smoke barriers
- Fire alarm pull station locations
- Fire extinguisher locations
- Building exit locations
- Hazardous-area markings

1.1.11 Structural Drawings

Structural drawings are created by a structural engineer and accompany the architect's plans. These types of drawings are usually drawn for large commercial structures such as office buildings or manufacturing facilities. Structural drawings show critical requirements for structural elements of the building, including columns, floor and roof systems, stairs, and load-bearing walls. It is important for masons to carefully read and implement the instructions shown on structural drawings, because they contain details on the following:

- Heights of finished floors and walls
- Heights and bearings of girders, beams, and trusses
- Locations of bearing materials
- Connection details (e.g., beam to column, beam to beam, joist to beam)
- Column or pier locations
- Locations, sizes, and spacing of anchor bolts
- Stairways

1.1.12 Mechanical, Electrical, and Plumbing Plans

For many construction jobs, information for the mechanical, electrical, and plumbing (MEP) craftworkers is included on the basic floor plans. However, for complex commercial projects, MEP information is generally shown on separate plan drawings, as shown in *Figure 15.*

Plumbing plans show the layout of water supply and sewage disposal systems and fixtures. Plumbing system plans generally have a separate plumbing riser diagram that shows the layout and identification of the piping and fixtures. A plumbing legend is usually located on the plan.

Mechanical plans show details for the installation of the HVAC and other equipment. Such plans usually include a refrigerant piping schematic that shows the types and sizes of the piping and identification of the fittings. The plan may also include a legend defining the HVAC symbols used on the drawing.

Electrical plans show the details for installing the electrical system and equipment. They indicate the locations of the meters, distribution panels, fixtures, switches, and special items, along with specifications for load capacities and wire sizes. The plan may also include a legend defining the symbols used in the drawing.

Using Mechanical Drawings

While masons usually work with architectural and structural drawings, useful notes and views are included on drawings in other sections, especially the mechanical section. For example, items not typically shown on architectural plans that may be included on mechanical drawings include exposed HVAC ductwork, heating convectors, and fire sprinkler piping.

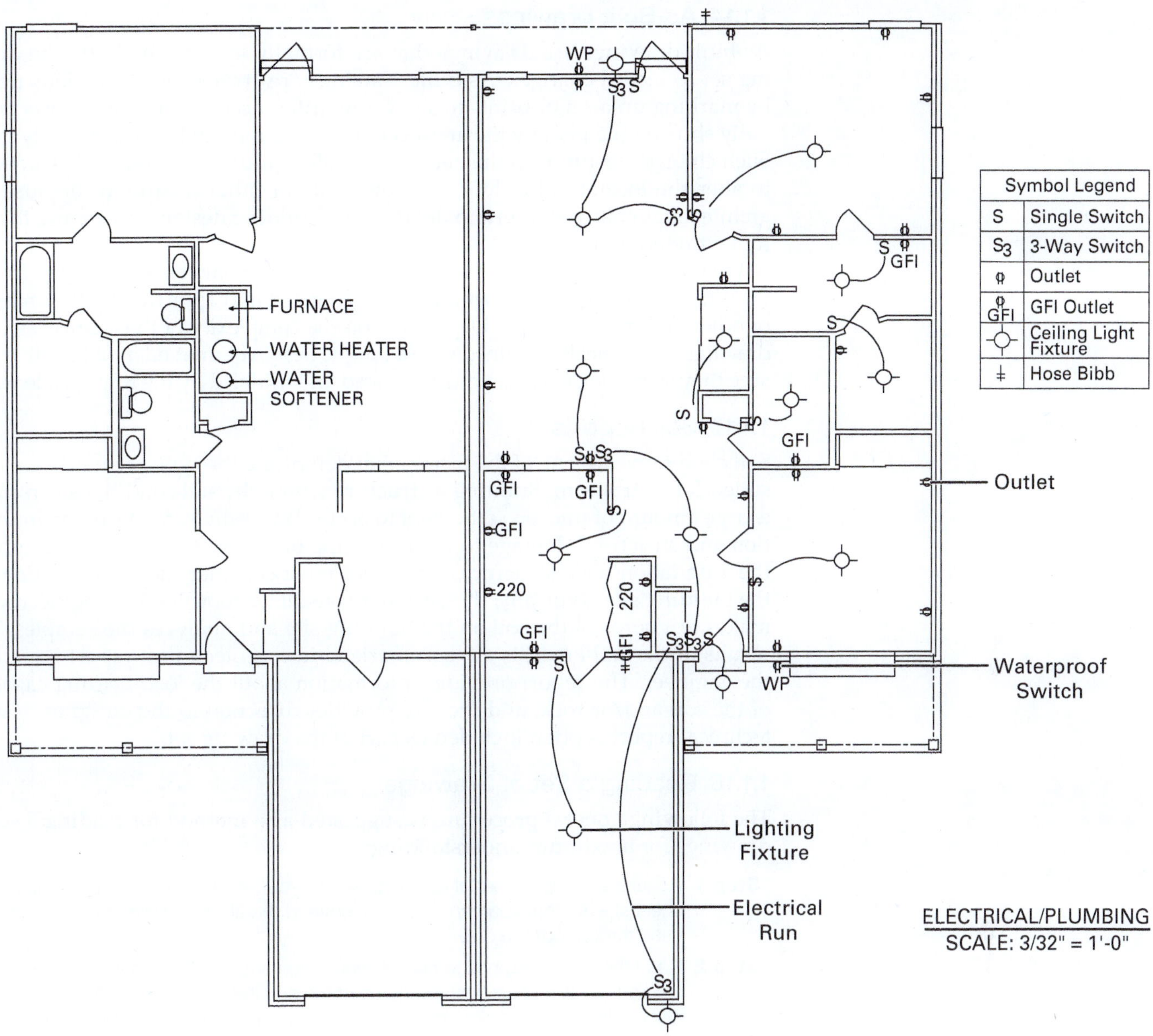

Figure 15 Example of an electrical/plumbing plan (NTS).
Source: John Hoerlein

1.1.13 Shop Drawings

Shop drawings are specialized drawings that show how to fabricate and install components of a construction project. One type of shop drawing that may be created after an engineer designs the structure is a detail drawing that shows the locations of all holes and openings and provides notes specifying how the part is to be made. Assembly instructions may be included. This type of drawing is used principally for structural steel members.

Another type of shop drawing, commonly referred to as a *submittal drawing*, pertains to special items or equipment purchased for installation in a building. Submittal drawings are usually prepared by equipment manufacturers. They show overall sizes, details of construction, methods of securing the equipment to the structure, and all pertinent data the architect and contractor need to know for the final placement and installation of the equipment.

Shop drawings produced by a contractor or fabricator are usually submitted to the owner or architect for approval and for revisions or corrections. The design drawing is often put on the same sheet as the shop drawing.

1.1.14 As-Built Drawings

As-built drawings are drawings that are formally incorporated into the drawing set to record changes. As-built drawings are created by the various trades by marking up a set of prints to show any differences between what was originally shown on a plan by the architect or engineer and what was actually built. Such changes result from the need to relocate equipment to avoid obstructions; to alter the location of a door, window, wall, or other feature; or because the architect has changed a certain detail in the building design in response to customer preferences.

On many jobs, any such changes to the design can only be made after a formal document called a *change order* has been generated and approved by the architect or other designated person. Depending on the complexity of the change, as-built drawings are typically outlined with a unique design or marked in red ink to make sure they stand out. Changes must be dated and initialed by the responsible party.

1.1.15 Soil Reports

Soil conditions are one of the factors that determine the type of foundation best suited for a structure. Building a structure where the soil conditions can cause a large amount of uneven settlement to occur can result in cracks in the foundation and structural damage to the rest of the building. Therefore, in designing the foundation for a structure, a designer must consider the soil conditions of the building site. Typically, the designer consults a geotechnical engineer, who makes test bores of the soil on the building site and analyzes the samples. The results of the soil analysis are summarized in a geotechnical report issued by the engineer. The report provides information about the load-bearing capacity of the soil and/or rock, and thereby provides direction to the designer. A geotechnical report is often included as part of the drawing set.

1.1.16 Reading a Set of Drawings

The following general procedure is suggested as a method for reading a set of drawings for maximum understanding:

Step 1 Acquire a complete set of drawings and specifications — including the title sheet(s) — so that you can better understand the abbreviations and symbols used throughout the drawings.

Step 2 Read the title block, noting critical information such as the scale, date of the last revision, drawing number, and name of the architect or engineer. After you use a sheet from a set of drawings, be sure to refold the sheet with the title block facing up.

Step 3 Locate the north arrow. Always orient yourself to the structure. Knowing where north is enables you to more accurately describe the locations of walls and other parts of the building.

Step 4 Be aware that the drawings in the set work together as a group. An architect or engineer draws multiple plans, elevations, and sections because more than one type of view is required to communicate the design concept. Learn how to use more than one drawing, when necessary, to locate needed information.

Step 5 Check the list of drawings in the set, and note the sequence of the various plans. Some drawing sets have an index on the front cover. A set can consist of the following categories of prints:

- Civil
- Architectural
- Structural
- Mechanical
- Electrical
- Plumbing

Step 6 Study the site plan to determine the location of the building to be constructed and the various utilities, the roadways, and any easements.

Step 7 Check the floor plan to determine the orientation of the building. Observe the locations and details of entries, corridors, offsets, and any special features.

Step 8 Check the foundation plan for the sizes and types of footings, reinforcing steel, and load-bearing substructures.

Step 9 Check the floor and wall construction and other details relating to exterior and interior walls.

Step 10 Study the features that extend for more than one floor, such as plumbing and vents, stairways, elevator shafts, and heating and cooling ductwork.

Step 11 Study the plumbing and/or mechanical plans for details of any masonry work pertaining to the installation of the heating, cooling, and plumbing equipment. You must coordinate this type of work with members of the HVAC and plumbing trades.

Step 12 Study the electrical plans for details of any masonry work necessary to install the electrical system, fixtures, or special equipment. You must coordinate this type of work with members of the electrical trade.

Step 13 Check the notes on various pages, and compare the specifications against the construction details.

Step 14 Thumb through the sheets of drawings until you are familiar with all the plans and structural details.

Step 15 Recognize applicable symbols and their relative locations in the plans. Note any special construction details or variations that will affect the masonry work.

1.2.0 Lines, Symbols, and Abbreviations

Construction drawings use a system of lines, symbols, and abbreviations to show how a structure should be built. This section describes how to use this system to ensure that your project adheres to the specifications shown in the construction plans.

1.2.1 Lines on Drawings

Lines are drawn wide or narrow, dark or light, and broken or unbroken, with each type of line conveying a specific meaning. *Figure 16* shows the most common lines used on construction drawings.

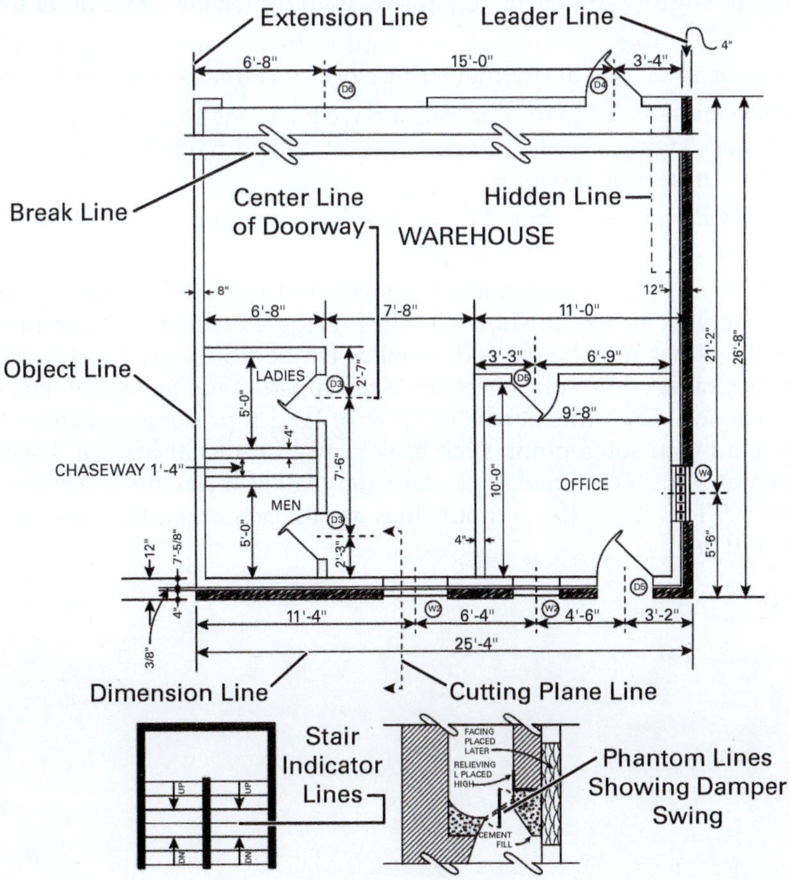

Figure 16 Lines used on construction drawings (NTS).

Following are descriptions of each type of line:

- *Object line*—A heavy line used to show the main outline of a structure, including exterior walls, interior partitions, porches, patios, sidewalks, parking lots, and driveways.
- *Dimension and extension lines*—Lightweight lines used to provide the dimensions of an object. An extension line is extended from an object at both ends of the object to be measured. To avoid confusion, the extension lines should not touch the object line. A dimension line is drawn at right angles between the extension lines and a number placed above, below, or to the side of it to indicate the length of the object being measured. Sometimes a gap is provided in the dimension line and the dimension is shown in the gap.
- *Center line*—Used to designate the center of an area or object and to provide a reference point for dimensioning. Center lines are typically used to indicate the centers of objects such as columns, posts, footings, and door openings.
- *Cutting plane line*—Used to indicate an area that is being cut away and shown in a section view so that the interior features can be seen. The arrows at the ends of the cutting plane line indicate the direction in which the section is viewed. Letters identify the section view of that specific part of the structure. More elaborate methods of labeling cutting plane lines are used in larger sets of plans (*Figure 17*), where many sections are being used. The section view may be on the same page as the cutting plane line or on another page.
- *Break line*—Used to indicate that an object or area is not being shown in its entirety.
- *Leader line*—Used to connect a note or dimension to a related part of the drawing. Leader lines are usually curved or at an angle from the feature being distinguished to avoid confusion with dimensions and other lines.
- *Hidden line*—Used to indicate an outline that is invisible to an observer because it is covered by another surface or object that is closer to the observer.
- *Phantom line*—Used to indicate alternative positions of moving parts, such as a damper's swing, or adjacent positions of related parts. It may also be used to represent repeated details.
- *Stair indicator line*—A short line with an arrowhead that shows the ascent or descent of a stairway on a floor plan.
- *Contour lines*—Drawn on a site plan (covered in the *Site Plans* section) and **topographical survey** to show changes in the elevation and contour of the land. The lines may be dashed or solid. Dashed lines are used to show the natural grade, while solid lines show the finish grade to be achieved during construction. Each line across the plot of land represents a different elevation relative to some point, such as sea level or a local feature. Each line is drawn at a uniform change in elevation (called a *contour interval*), such as every 10'. The closer the contour lines are to each other, the steeper the terrain will be.

Topographical survey: An accurate and detailed drawing of a place or region that depicts all the natural and man-made physical features, showing their relative positions and elevations.

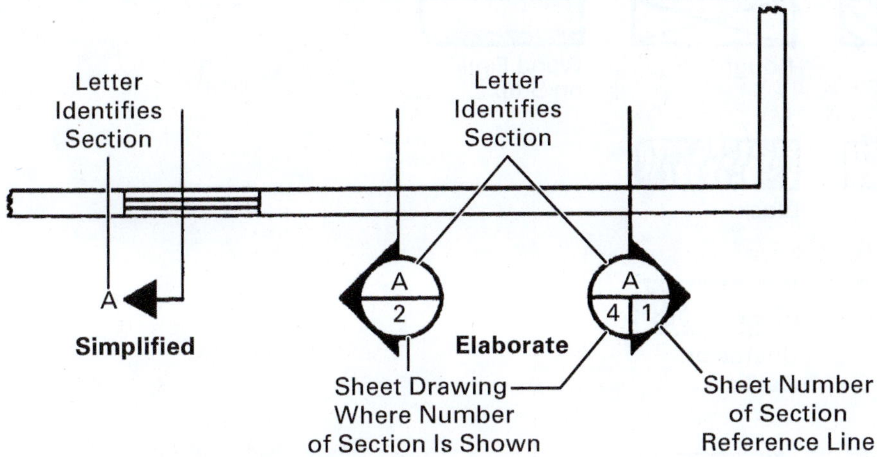

Figure 17 Methods of labeling cutting plane lines.

1.2.2 Architectural Symbols

Construction drawings use symbols to indicate different kinds of materials, fixtures, and structural members. While some symbols are standard, the meanings of others can vary from location to location. A set of drawings generally includes a sheet that identifies the specific symbols used in the set and their meanings.

Most engineers and architects use symbols adopted by the American Institute of Architects (AIA) and the American National Standards Institute (ANSI). However, many firms use their own modified symbols to suit their specific needs. For this reason, drawings typically include a symbol list or legend for identification purposes.

Some symbols are not intended to show part of an object but are provided to guide you through the drawings themselves. Examples of this type of symbol include door and window designators that reference door and window schedules where additional information is provided. Other symbols are used to show the orientation of an object, such as the north arrow, indicating the orientation of a structure on the building site. Two of the most common types of architectural symbols are material symbols and window and door symbols.

NOTE

When using any set of drawings, always refer first to the sheet containing the symbol legend for reading and interpreting the drawings.

Material Symbols

Typically, materials are shown on drawings in two ways (*Figure 18*). One way that materials are shown is as a plan view, in which the material is presented as you would see it when looking down on it. The second way materials are shown is as an elevation view or section view, in which the symbol shows the material roughly as it would look when you are facing it.

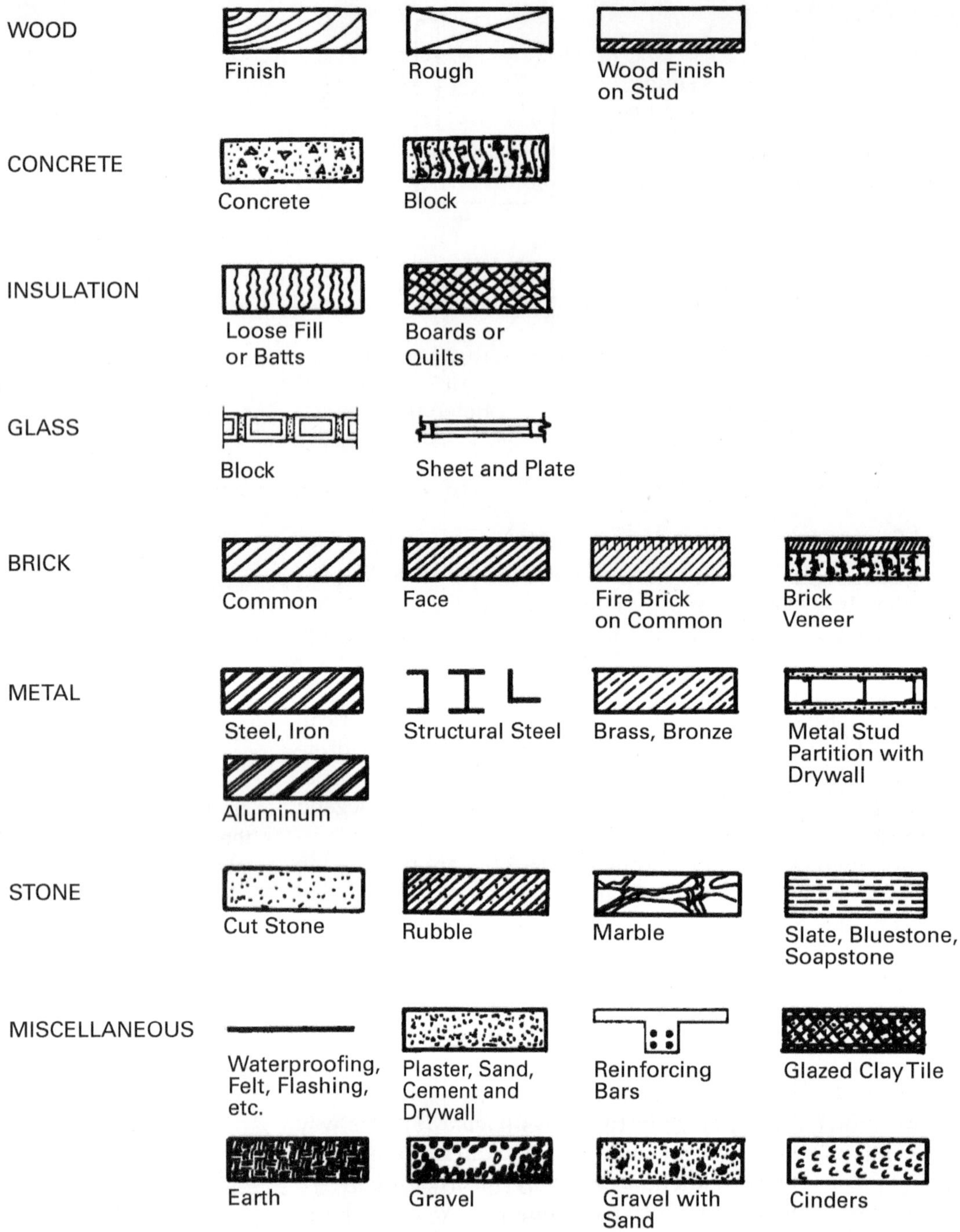

(A) General Plan Symbols

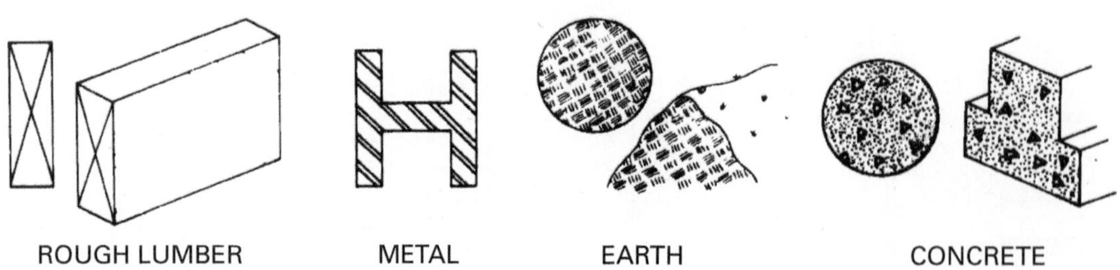

(B) Section View Symbols

Figure 18 Typical material symbols.

Window and Door Symbols

Floor plans show a great deal of information in a small drawing area; therefore, symbols are used to convey needed information to a mason. Window and door symbols (*Figure 19*) are usually shown with only a centerline measurement to locate them from some reference point. However, more information about windows and doors is typically shown in separate window and door schedules contained in the drawing set. Window and door symbols are shown on drawings as either elevation views or plan views.

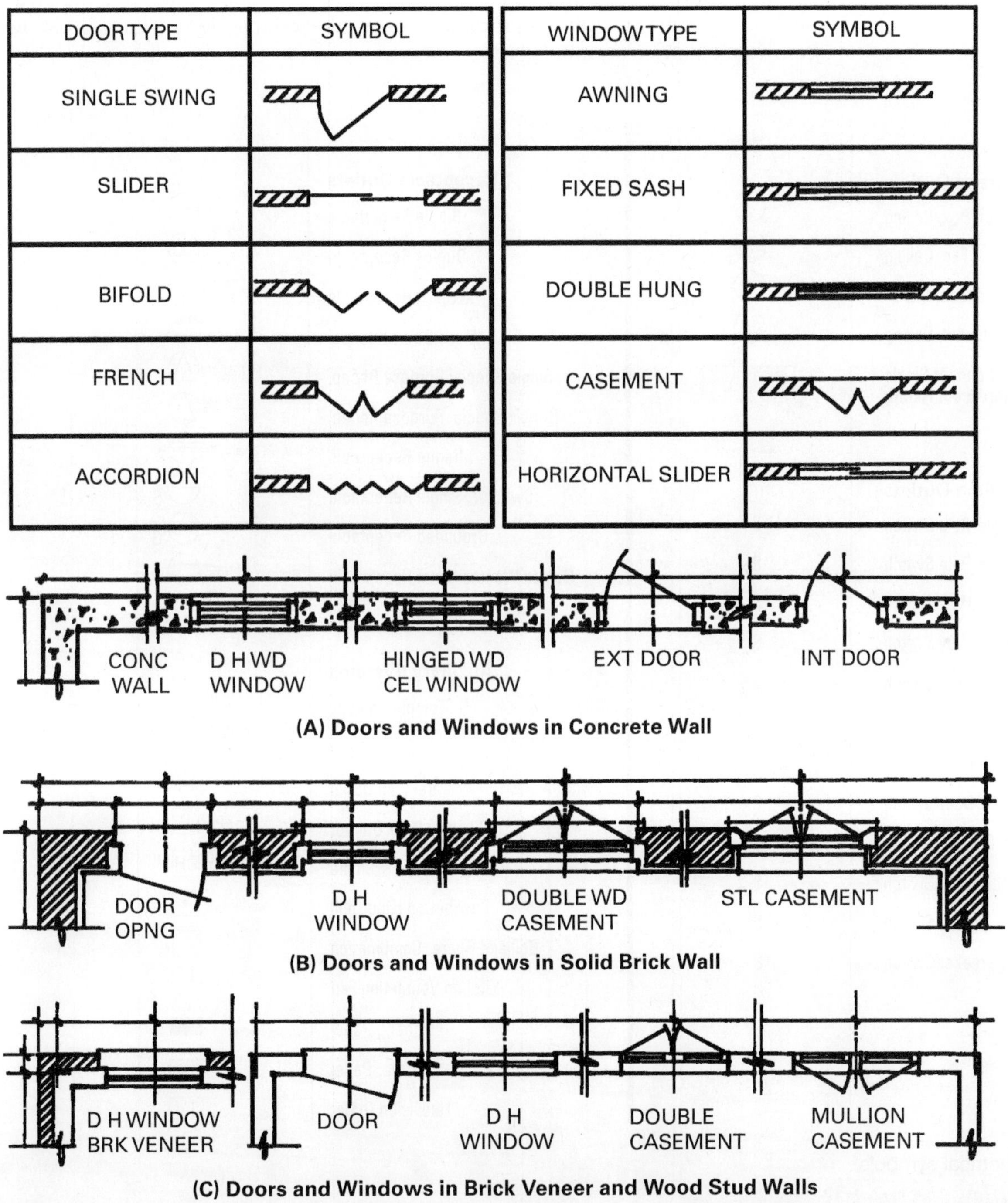

Figure 19 Window and door symbols.

1.2.3 Trade Symbols

Each trade has its own symbols and abbreviations. Electrical, plumbing, and HVAC symbols show the types of equipment to be installed, such as switches, lavatories, and warm air supply ducts. Masons must understand these symbols so that they are aware of site conditions, structural elements, mechanical systems, and other important features that affect the construction of the project.

Electrical Symbols

Electrical symbols indicate the locations of electrical panels, subpanels, outlets, switches, and wire runs for an electrician. They also show auxiliary hardware such as buzzers, telephone jacks, and computer connections. *Figure 20* shows common electrical symbols.

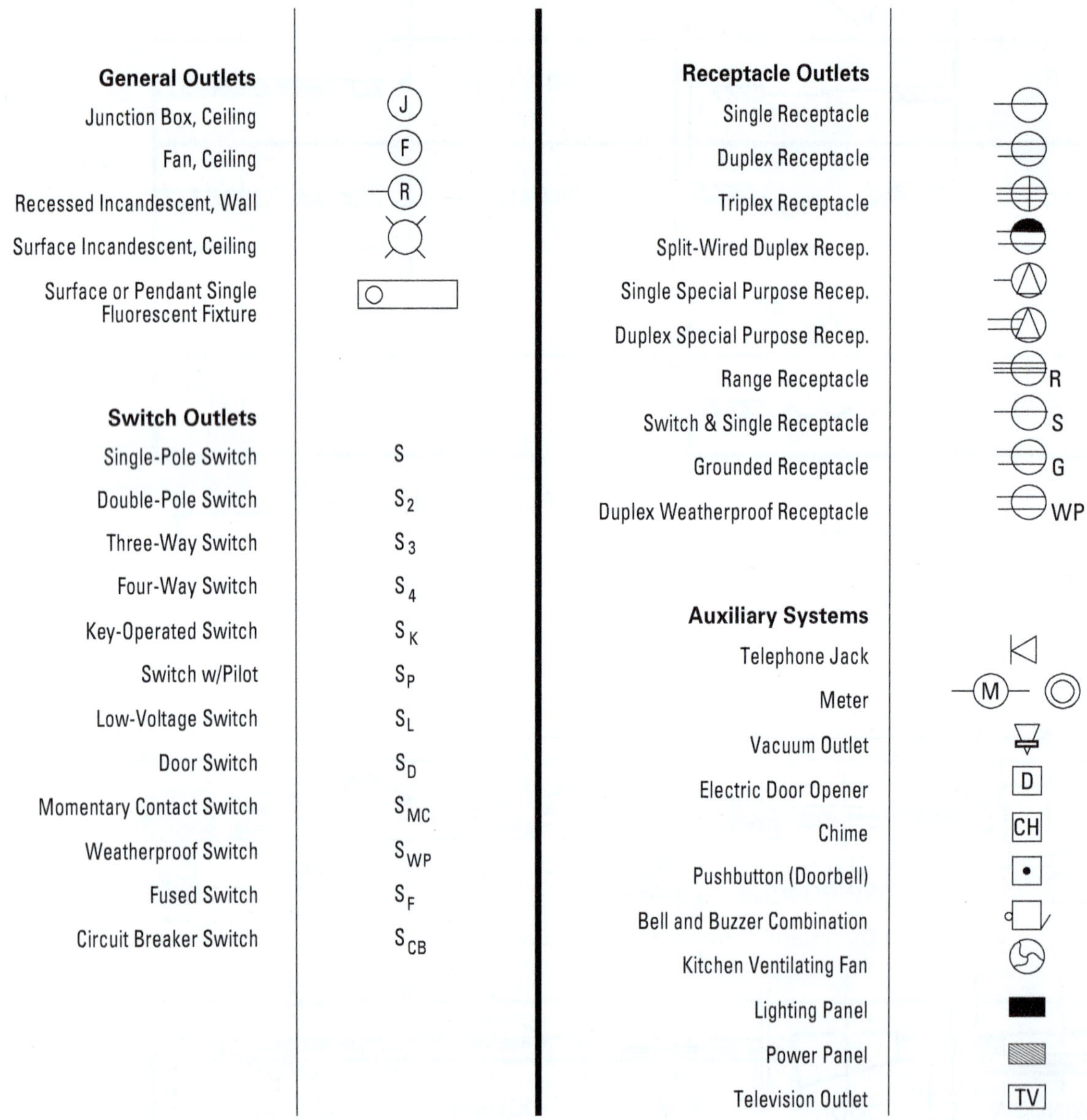

Figure 20 Common electrical symbols.

Plumbing Symbols

Plumbing symbols indicate the hardware and fixtures required for the building. Floor plans usually indicate only where the different fixtures should be located and plumbed.

Figure 21 shows some common fixture symbols as they would appear on a floor plan. It also shows some common piping symbols that would be used on special plumbing diagrams.

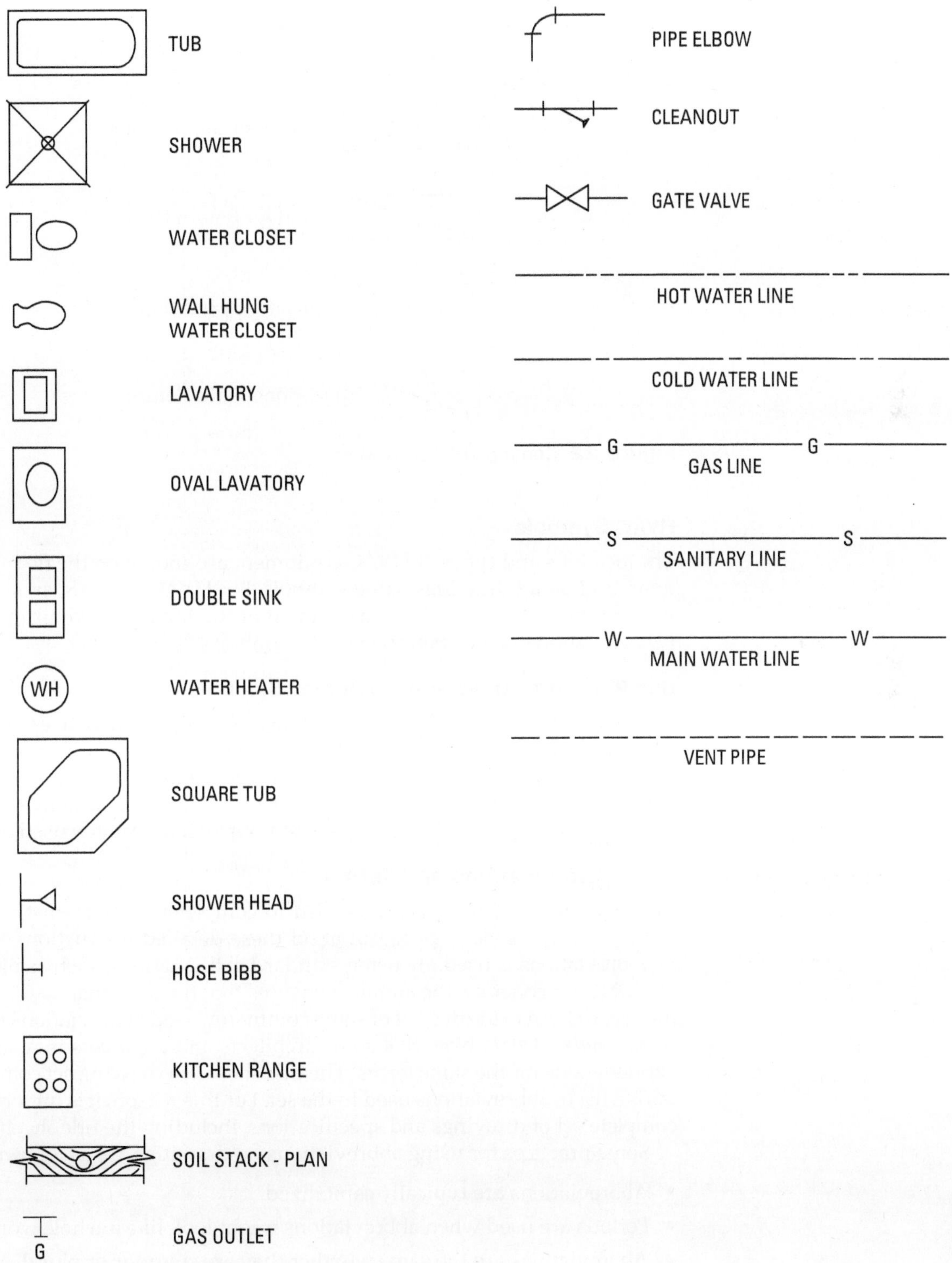

Figure 21 Common plumbing symbols.

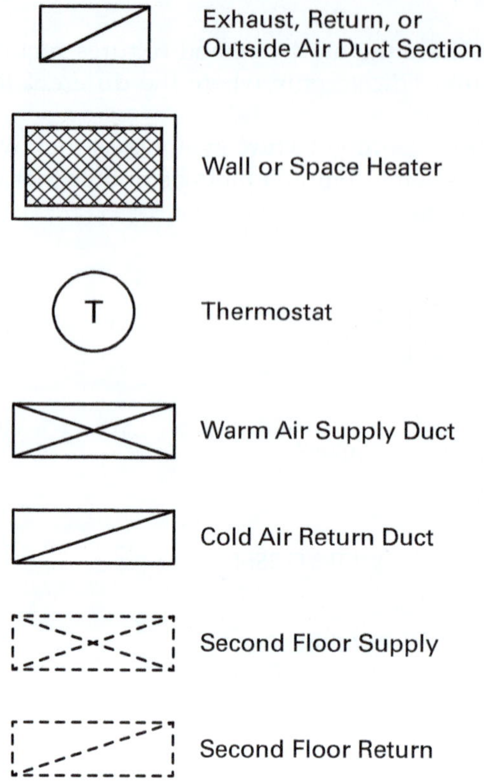

Figure 22 Common HVAC symbols.

HVAC Symbols

The locations and types of HVAC equipment are shown on the floor plan(s), sections, and detail drawings using symbols. For HVAC systems, symbols are also used to show the direction of movement of the hot and cold air in the system. *Figure* 22 shows some common HVAC symbols.

Site Plan and Topographical Survey Symbols

Site plans, also known as *plot plans,* show the positions and sizes of all relevant structures on the site and the features of the terrain. It is difficult to show the amount of information required on these drawings if symbols are not used. A good site plan includes a legend of the symbols used, especially those that are nonstandard. *Figure* 23 shows some of the symbols commonly used on site plans.

1.2.4 Abbreviations and Terms

Many written instructions are needed to complete a set of construction drawings. It is impossible to print out all of these detailed instructions, so a system of abbreviations is used. By using standard abbreviations, such as BRK for brick or CONC for concrete, the architect ensures that the drawings will be accurately interpreted. An extensive list of some commonly used abbreviations is contained in *Appendix 28208B.* Note that some architects and engineers may use different abbreviations for the same terms. The title sheet in a drawing set commonly contains a list of abbreviations used in the set. For this reason, it is important to get a complete set of drawings and specifications, including the title sheet(s).

Some practices for using abbreviations on drawings are as follows:

- Abbreviations are typically capitalized.
- Periods are used when abbreviations might look like a whole word.
- Abbreviations are the same whether they are singular or plural.
- Several terms have the same abbreviation and can only be identified from the context in which they are found.
- Many abbreviations are similar.

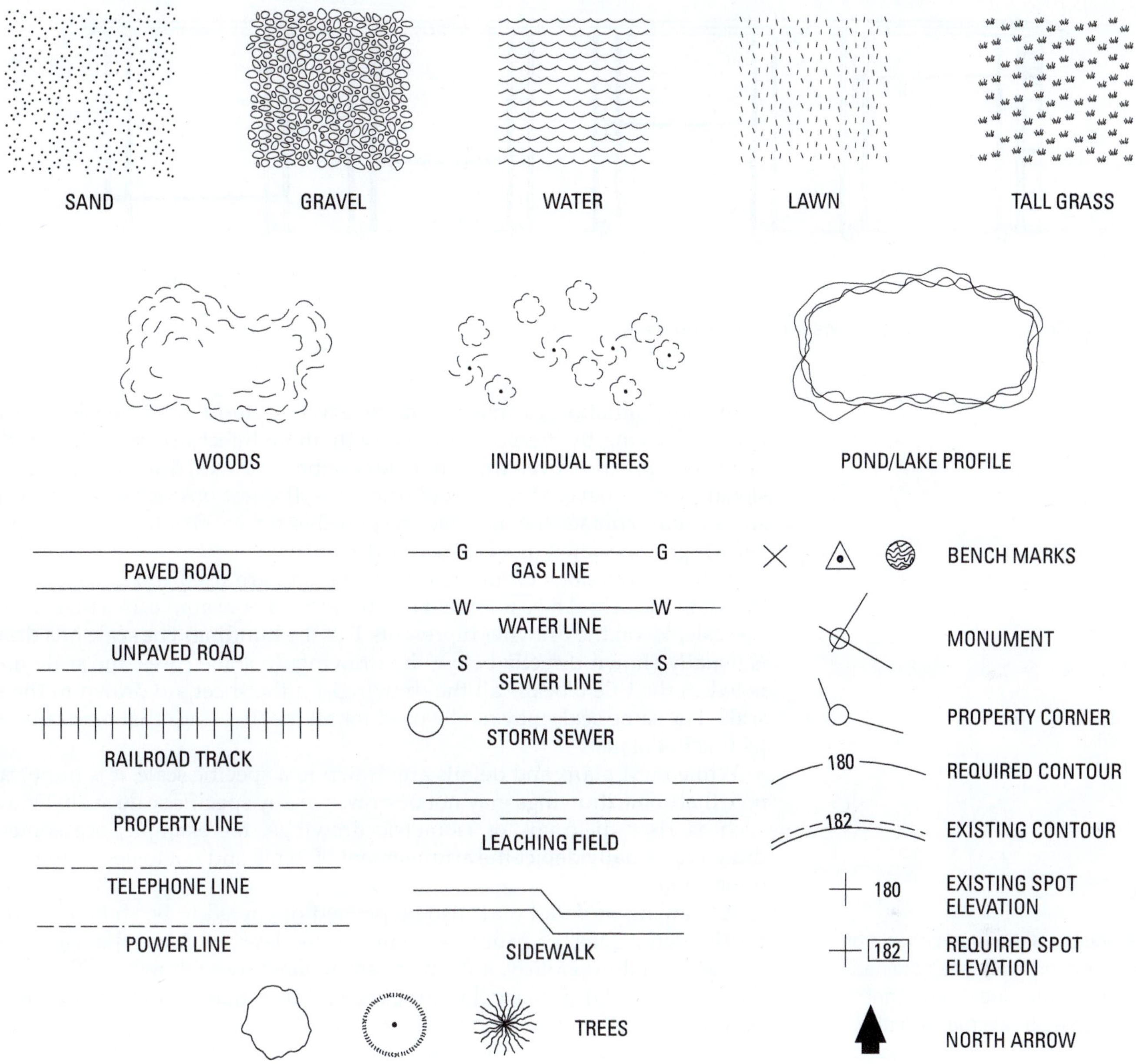

Figure 23 Common site plan symbols.

Architectural Terms Used in Drawings and Specifications

There are many architectural terms. Some of the most commonly used terms are listed in *Appendix 28208C*.

1.3.0 Dimensioning

Dimensions provided on drawings indicate actual sizes, distances, and measurements of the objects and spaces being represented. Dimensions may be indicated from outside to center, center to center, wall to wall, or outside to outside (*Figure 24*). In all cases, dimensions shown on drawings are noted in full scale. Section and detail views may use a **nominal size** in labeling. Nominal sizes or dimensions are approximate or rough sizes by which products such as lumber and concrete masonry units (CMUs) are commonly known and sold. For example, a 2 × 4 measures $1\frac{1}{2}$" × $3\frac{1}{2}$".

Nominal size: Approximate or rough size (commercial size) by which products such as lumber and CMUs are commonly known and sold. Nominal sizes are normally slightly larger than the actual sizes (e.g., 2 × 4s).

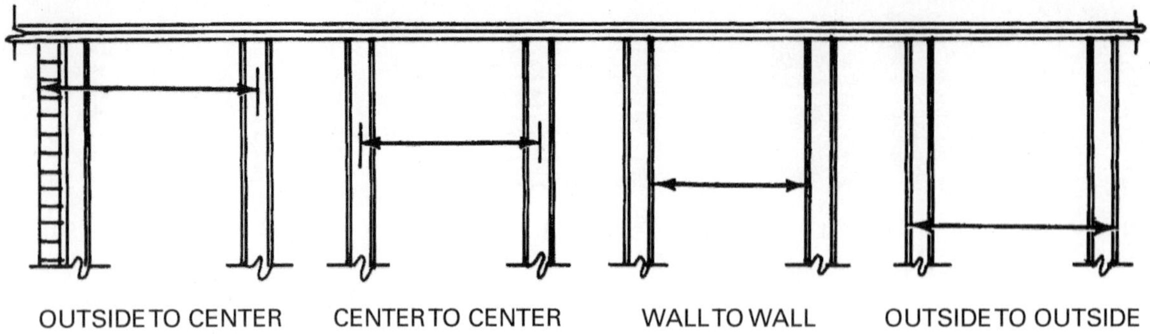

Figure 24 Common methods of indicating dimensions on drawings.

In some situations, it may be necessary to determine dimensions omitted from a drawing by measuring them with an architect's scale; however, this is not a good practice. Even though scales can be used with drawings, craft professionals should never scale a set of drawings. If questions regarding the dimensions occur, contact the architect responsible for confirming the dimensions. Drawings are created using a specified scale.

Inches or fractions of an inch on a drawing are used to represent feet and inches in the actual measurement of a building. For example, in a plan drawn to $\frac{1}{4}$" scale, $\frac{1}{4}$" on the drawing represents 1' of the building. The scale of a drawing is usually shown directly below the drawing. In some cases, the scale may be noted in the title block if all the drawings on the sheet are drawn to the same scale. The same scale may not be used for all the drawings that make up a complete set of plans.

While most plans and details are drawn to a specific scale, it is important to note that some drawings may not be drawn to any specific scale at all. Drawings such as riser diagrams or isometric drawings, for example, are sometimes drawn to visually depict the arrangement of items and not to depict their actual dimensions.

A drawing or detail may also be printed or copied to be shown as large as possible on a sheet of paper to maximize the level of detail that can be seen. As mentioned previously, it is important to label these drawings "Not to Scale (NTS)" (*Figure 25*). This label lets the reader know that an architectural or engineering scale cannot be used to determine dimensions on the drawing. If there is any uncertainty about the dimensions on a drawing, the architect should be contacted for clarification.

NOTE

Always use the dimensions shown on a drawing rather than ones obtained by scaling the drawing. Some print reproduction methods may slightly reduce or enlarge drawings, which can introduce errors if the print is scaled.

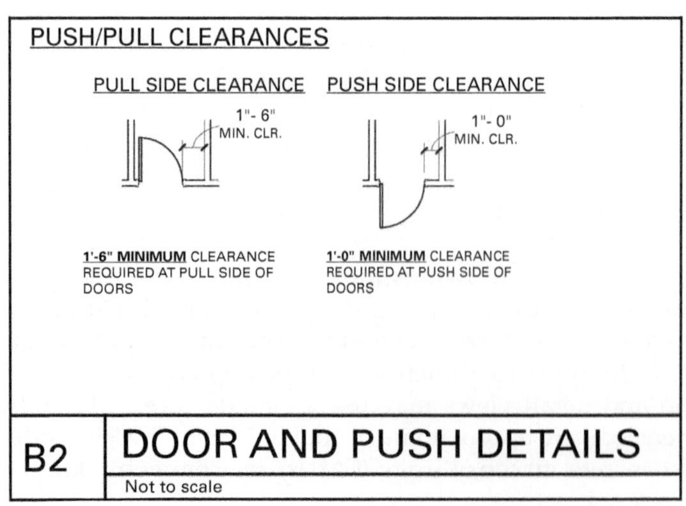

Figure 25 Not to Scale (NTS).

Architect's Scale

Measurements are usually made on construction drawings using an architect's scale rather than a standard ruler. Architect's scales like those shown here are divided into feet and inches and usually consist of several scales on one rule. Architect's scales are also available in other forms, such as tapes or with wheels.

(A) Architect's Scale Ruler

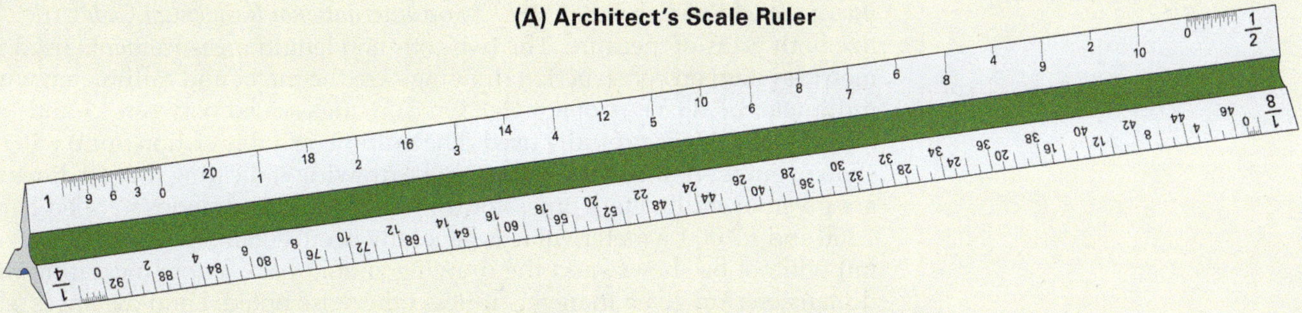

(B) The Three Sides of an Architect's Scale Ruler

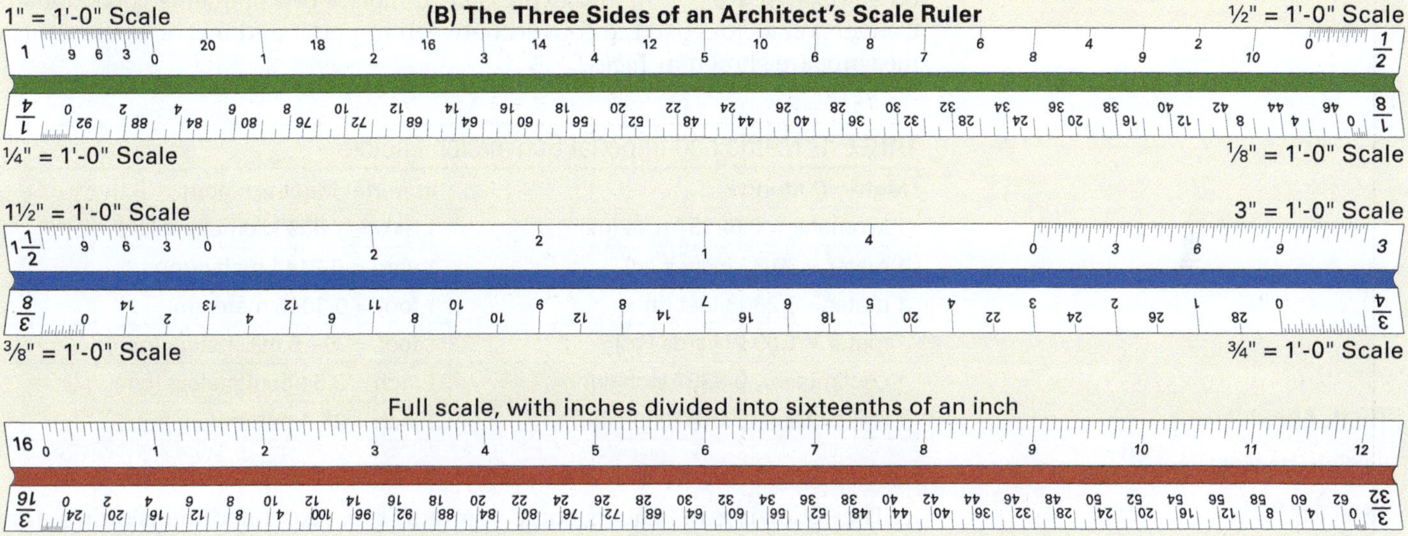

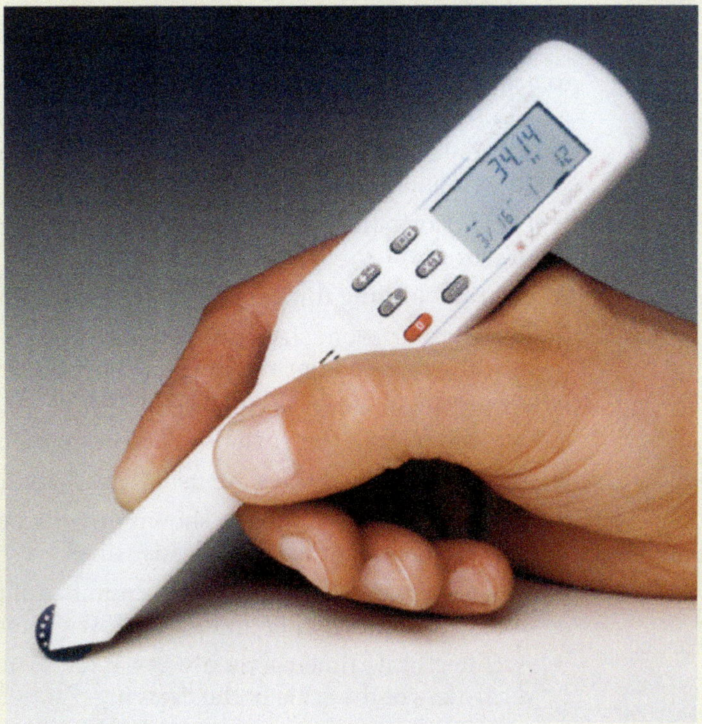

1.3.1 Metric Measurements on Drawings

In the United States, length dimensions shown on construction drawings are typically expressed in feet and inches using the imperial system of measurement. In Canada and many other parts of the world, the SI (International System of Units) metric system of measurement is used for dimensions. The *International Building Code*® (*IBC*®) and *International Residential Code*® (*IRC*®) utilize both units of measure. The two common length measurements used in the metric system on construction drawings are the meter and millimeter, with the millimeter being $\frac{1}{1,000}$ of a meter. On drawings scaled between 1:1 and 1:100, a millimeter value is typically used. The millimeter abbreviation (mm) will not be shown, but there should be a note on the drawing indicating that all dimensions are given in millimeters unless otherwise noted. On drawings scaled between 1:200 and 1:2,000, a meter value is generally used. Again, the meter abbreviation (m) will not be shown, and the drawing should have a note indicating that all dimensions are given in meters unless otherwise noted. Land distances (shown on site plans) are commonly expressed in meters or kilometers (1,000 meters). Conversion factors used to convert between imperial and metric linear units of measure are shown in *Table 1*.

TABLE 1 Metric and Imperial Conversion Factors

Metric Distance	Imperial Measurement
1 kilometer = 0.62137 mile (mi)	1 mile = 1.609 kilometers (km)
1 meter = 39.37 inches (in)	1 yard = 0.9144 meter (m)
1 meter = 3.2808 feet (ft)	1 foot = 0.3048 meter (m)
1 meter = 1.0936 yards (yd)	1 foot = 304.8 millimeters (mm)
1 centimeter = 0.3937 inch (in)	1 inch = 2.54 centimeters (cm)
1 millimeter = 0.03947 inch (in)	1 inch = 25.4 millimeters (mm)

Think About It

Scaling Drawings

Measuring the length of an object or line on a drawing and converting the measurement to the actual length using the given scale is called *scaling a drawing*. If a line on a floor plan measures 6" in length and the object being represented is drawn to a scale of $\frac{1}{4}$" = 1', what is the true length of the object?

Engineer's Scale

An engineer's scale is primarily used for land measurements, such as those included on site plans. Increments on an engineer's scale are shown in multiples of 10 and measurements are expressed in decimals. The most common scales used on an engineer's scale are 10, 20, 30, 40, 50, and 60, which can all be combined on a triangular scale.

Engineer's scales and architect's scales are not interchangeable. Always check the scale noted on the blueprint, and select the proper scale to use when scaling the drawing.

The following list shows some common practices for dimensioning drawings. Keep in mind that these are not rules and that the practices in your area may be different:

- Dimension lines are unbroken lines, with the dimensions placed above and near the center of the line.
- Dimensions over 1' are shown in feet and inches (not decimals). Dimensions less than 1' are shown in inches only. The common exception to this rule is the center-to-center distances for standard construction, such as for framing 16" on center (OC) or 24" OC.
- Dimensions are placed to be read from the right or from the bottom of the drawing.
- Overall building dimensions go to the outside of all other dimensions.
- Rectangular room sizes can be shown by stating width and length (usually near the room name), rather than using dimension and extension lines.
- Rooms are sometimes dimensioned from the center lines of partition walls, but wall-to-wall dimensions are more common.
- Window and door sizes are usually shown in window and door schedules.
- Dimensions that cannot be shown directly on the floor plan because of their size are placed at the end of leader lines.
- When stairs are dimensioned, the number of risers is placed on a line with an arrow indicating either an up or a down direction.
- Architectural dimensions always refer to the actual size of the building, regardless of the scale of the drawings.

1.0.0 Section Review

1. What is the primary drawing in a set of architectural plans?
   a. Floor plan
   b. Elevation drawing
   c. Foundation plan
   d. Layout plan

2. What type of drawing shows the interior of a wall or other structural feature?
   a. Detail drawing
   b. Foundation plan
   c. Section drawing
   d. Floor plan

3. What type of drawing shows the differences between what was originally on a plan and what was actually constructed?
   a. Shop drawing
   b. Structural drawing
   c. As-built drawing
   d. Detail drawing

4. What type of line indicates an object or area that is not being shown in its entirety?
   a. Object
   b. Break
   c. Lead
   d. Center

5. Who should craft professionals contact when they have questions about dimensions shown on a set of drawings?
   a. Engineer
   b. Contractor
   c. Architect
   d. Surveyor

2.0.0 Specifications

Objective

Describe the purpose of written specifications.

a. Summarize how specifications are organized.
b. Explain the importance of construction building codes.

Performance Task

3. Read and interpret written specifications.

Specifications are written instructions provided by architectural and engineering firms to the general contractor and, consequently, to the subcontractors.

Specifications are contractual documents used along with the construction drawings and are just as important as the drawings themselves. Specifications furnish what the drawings cannot, in that they define the quality of work to be done and the materials to be used. Specifications serve several important purposes, including the following:

- Clarify information that cannot be shown on the drawings
- Identify work standards, types of materials to be used, and responsibilities of various parties to the contract
- Provide information on details of construction

Specifications: Precisely written presentation of the details of a plan.

- Serve as a guide for contractors bidding on the construction project
- Serve as a standard of quality for materials and workmanship
- Serve as a guide for compliance with building codes and zoning ordinances
- Serve as the basis of agreement between the owner, architect, and contractors in settling any disputes

Specifications are legal documents. When there is a difference between construction drawings and specifications, the specifications usually take legal precedence over the construction drawings. However, it is often the case that the plans are more specific to the job than the specifications. Therefore, notes on the plans may be considered by the architect/owner to be the true intent. Carefully watch for discrepancies between the plans and specifications and report them to your supervisor immediately.

2.1.0 Specifications Organization

Specifications consist of various elements that may differ between construction jobs. For small projects, specifications may be simple; for large projects, they may be complex and detailed. Two types of information are typically included in a set of specifications: special and general conditions, and technical aspects of construction.

2.1.1 Special and General Conditions

Special and general conditions cover the nontechnical aspects of the contractual agreements. Special conditions cover topics such as safety and temporary construction. General conditions cover the following points of information:

- Contract terms
- Responsibilities for examining the construction site
- Types and limits of insurance
- Permits and payments of fees
- Use and installation of utilities
- Supervision of construction
- Other pertinent items

The general conditions section is the area of the construction contract where misunderstandings often occur. These conditions are usually much more explicit on large, complicated construction projects. Note that residential specifications often do not spell out general conditions and are basically material specifications only. An example of a typical material specifications document is shown in *Figure 26*.

Notes on Drawings

Sometimes the notes on drawings will contradict the specifications or other requirements. Even though the specifications usually take precedence, the notes often are closer to the true intent. If you notice any such discrepancy, clarify it with a supervisor right away.

Addenda and Change Orders

Addenda and change orders are contractual documents used to correct or change original construction drawings and specifications. The difference between the two documents is a matter of timing. An addendum is written before the contract is awarded, while a change order is drawn up after the contract has been awarded.

Jack DeMusse & Associates
East End Clinic Project

SECTION 061000 - ROUGH CARPENTRY

PART 1 - GENERAL

1.01 SECTION INCLUDES

A. Structural dimension lumber framing.

B. Exposed timber structural framing.

C. Roof-mounted curbs.

D. Roofing nailers.

E. Roofing cant strips.

F. Preservative treated wood materials.

G. Fire retardant treated wood materials.

H. Communications and electrical room mounting boards.

I. Concealed wood blocking, nailers, and supports.

J. Miscellaneous wood nailers, furring, and grounds.

1.02 RELATED REQUIREMENTS

A. Section 054000 - Cold Formed Metal Framing.

B. Section 061500 - Wood Decking.

C. Section 061800 - Structural Glued-Laminated Timber.

D. Section 076200 - Sheet Metal Flashing and Trim: Sill flashings.

1.03 SUBMITTALS

A. Product Data: Provide technical data on wood preservative materials, application instructions, and fasteners and adhesives.

B. Samples: For rough carpentry members that will be exposed to view, submit two samples, ____x____ inch (____x____ mm) in size illustrating wood grain, color, and general appearance.

C. Manufacturer's Certificate: Certify that wood products supplied for rough carpentry meet or exceed specified requirements.

1.04 DELIVERY, STORAGE, AND HANDLING

A. General: Cover wood products to protect against moisture. Support stacked products to prevent deformation and to allow air circulation.

B. Fire Retardant Treated Wood: Prevent exposure to precipitation during shipping, storage, or installation.

PART 2 - PRODUCTS

2.01 GENERAL REQUIREMENTS

A. Dimension Lumber: Comply with PS 20 and requirements of specified grading agencies.

ROUGH CARPENTRY 061000 - 1

Figure 26a Typical specification (1 of 5).

Jack DeMusse & Associates
East End Clinic Project

1. If no species is specified, provide any species graded by the agency specified; if no grading agency is specified, provide lumber graded by any grading agency meeting the specified requirements.
2. Grading Agency: Any grading agency whose rules are approved by the Board of Review, American Lumber Standard Committee (www.alsc.org) and who provides grading service for the species and grade specified; provide lumber stamped with grade mark unless otherwise indicated.

B. Lumber fabricated from old growth timber is not permitted.

2.02 DIMENSION LUMBER FOR CONCEALED APPLICATIONS

A. See Structural Notes for additional information.

B. Grading Agency: West Coast Lumber Inspection Bureau (WCLIB). — Dimension Lumber Requirements

C. Sizes: Nominal sizes as indicated on drawings, S4S.

D. Moisture Content: S-dry or MC19.

E. Miscellaneous Framing, Blocking, Nailers, Grounds, and Furring:
1. Lumber: S4S, No. 2 or Standard Grade.
2. Boards: Standard or No. 3.

2.03 EXPOSED DIMENSION LUMBER

A. Sizes: Nominal sizes as indicated on drawings, S4S.

B. Moisture Content: S-dry or MC19.

2.04 CONSTRUCTION PANELS

A. Communications and Electrical Room Mounting Boards: PS 1 C-D Plugged or better plywood, or medium density fiberboard; 3/4 inch (19 mm) thick; flame spread index of 25 or less, smoke developed index of 450 or less, when tested in accordance with ASTM E 84. — Panel Requirements

B. Other Applications:
1. Plywood Concealed From View But Located Within Exterior Enclosure: PS 1, C-C Plugged or better, Exterior grade.
2. Plywood Exposed to View But Not Exposed to Weather: PS 1, A-D, or better.
3. Other Locations: PS 1, C-D Plugged or better.

2.05 ACCESSORIES

A. Fasteners and Anchors:
1. Metal and Finish: Stainless steel for exterior, high humidity or preservative-treated wood locations, unfinished steel elsewhere.

2.06 FACTORY WOOD TREATMENT

A. Treated Lumber and Plywood: Comply with requirements of AWPA U1 - Use Category System for wood treatments determined by use categories, expected service conditions, and specific applications. — Treated Lumber
1. Fire-Retardant Treated Wood: Mark each piece of wood with producer's stamp indicating compliance with specified requirements.
2. Preservative-Treated Wood: Provide lumber and plywood marked or stamped by an ALSC-accredited testing agency, certifying level and type of treatment in accordance with AWPA standards.

B. Fire Retardant Treatment:
1. Manufacturers:

ROUGH CARPENTRY 061000 - 2

Figure 26b Typical specification (2 of 5).

Jack DeMusse & Associates
East End Clinic Project

a. Arch Wood Protection, Inc: www.wolmanizedwood.com.
b. Hoover Treated Wood Products, Inc: www.frtw.com.
c. Osmose, Inc: www.osmose.com.
d. Substitutions: See Section 016000 - Product Requirements.

2. Exterior Type: AWPA U1, Category UCFB, Commodity Specification H, chemically treated and pressure impregnated; capable of providing a maximum flame spread rating of 25 when tested in accordance with ASTM E 84, with no evidence of significant combustion when test is extended for an additional 20 minutes both before and after accelerated weathering test performed in accordance with ASTM D 2898.
   a. Kiln dry wood after treatment to a maximum moisture content of 19 percent for lumber and 15 percent for plywood.
   b. Treat exposed exterior rough carpentry items, including stairways, balconies, and covered walkways
   c. Do not use treated wood in direct contact with the ground.
3. Interior Type A: AWPA U1, Use Category UCFA, Commodity Specification H, low temperature (low hygroscopic) type, chemically treated and pressure impregnated; capable of providing a maximum flame spread rating of 25 when tested in accordance with ASTM E 84, with no evidence of significant combustion when test is extended for an additional 20 minutes.
   a. Kiln dry wood after treatment to a maximum moisture content of 19 percent for lumber and 15 percent for plywood at time of incorporation into the work.
   b. All interior rough carpentry items are to be fire retardant treated.
   c. Do not use treated wood in applications exposed to weather or where the wood may become wet.

C. Preservative Treatment:
1. Manufacturers:
   a. Arch Wood Protection, Inc; Product: www.wolmanizedwood.com.
   b. Chemical Specialties, Inc; Product: www.treatedwood.com.
   c. Osmose, Inc; Product: www.osmose.com.
   d. Conrad Forest Products: www.conradfp.com.
   e. Substitutions: See Section 016000 - Product Requirements.
2. Preservative Pressure Treatment of Lumber Above Grade: AWPA U1, Use Category UC3B, Commodity Specification A using waterborne preservative to 0.25 lb/cu ft (4.0 kg/cu m) retention.
   a. Kiln dry lumber after treatment to maximum moisture content of 19 percent.
   b. Treat lumber exposed to weather.
   c. Treat lumber in contact with roofing, flashing, or waterproofing.
   d. Treat lumber in contact with masonry or concrete.
   e. Treat lumber less than 18 inches (450 mm) above grade.
3. Preservative Pressure Treatment of Plywood Above Grade: AWPA U1, Use Category UC2 and UC3B, Commodity Specification F using waterborne preservative to 0.25 lb/cu ft (4.0 kg/cu m) retention.
   a. Kiln dry plywood after treatment to maximum moisture content of 19 percent.
   b. Treat plywood in contact with roofing, flashing, or waterproofing.
   c. Treat plywood in contact with masonry or concrete.
   d. Treat plywood less than 18 inches (450 mm) above grade.

ROUGH CARPENTRY 061000 - 3

Figure 26c Typical specification (3 of 5).

Jack DeMusse & Associates
East End Clinic Project

PART 3 - EXECUTION

3.01 PREPARATION

A. Coordinate installation of rough carpentry members specified in other sections.

3.02 INSTALLATION - GENERAL

A. Select material sizes to minimize waste.

B. Reuse scrap to the greatest extent possible; clearly separate scrap for use on site as accessory components, including: shims, bracing, and blocking.

C. Where treated wood is used on interior, provide temporary ventilation during and immediately after installation sufficient to remove indoor air contaminants.

3.03 BLOCKING, NAILERS, AND SUPPORTS

A. Provide framing and blocking members as indicated or as required to support finishes, fixtures, specialty items, and trim.

B. In framed assemblies that have concealed spaces, provide solid wood fireblocking as required by applicable local code, to close concealed draft openings between floors and between top story and roof/attic space; other material acceptable to code authorities may be used in lieu of solid wood blocking.

C. In metal stud walls, provide continuous blocking around door and window openings for anchorage of frames, securely attached to stud framing.

D. In walls, provide blocking attached to studs as backing and support for wall-mounted items, unless item can be securely fastened to two or more studs or other method of support is explicitly indicated.

E. Where ceiling-mounting is indicated, provide blocking and supplementary supports above ceiling, unless other method of support is explicitly indicated.

F. Specifically, provide the following non-structural framing and blocking:
1. Cabinets and shelf supports.
2. Wall brackets.
3. Handrails.
4. Grab bars.
5. Towel and bath accessories.
6. Wall-mounted door stops.
7. Chalkboards and marker boards.
8. Wall paneling and trim.
9. Joints of rigid wall coverings that occur between studs.

3.04 ROOF-RELATED CARPENTRY

A. Coordinate installation of roofing carpentry with deck construction, framing of roof openings, and roofing assembly installation.

B. Provide wood curb at all roof openings except where prefabricated curbs are specified and where specifically indicated otherwise. Form corners by alternating lapping side members.

3.05 INSTALLATION OF CONSTRUCTION PANELS

A. Communications and Electrical Room Mounting Boards: Secure with screws to studs with edges over firm bearing; space fasteners at maximum 24 inches (610 mm) on center on all edges and into studs in field of board.

ROUGH CARPENTRY 061000 - 4

Figure 26d Typical specification (4 of 5).

Jack DeMusse & Associates
East End Clinic Project

1. At fire-rated walls, install board over wall board indicated as part of the fire-rated assembly.
2. Where boards are indicated as full floor-to-ceiling height, install with long edge of board parallel to studs.
3. Install adjacent boards without gaps.
4. Size and Location: As indicated on drawings.
5. Paint all mounting boards. Leave one copy of firetreatment stamp visible (unpainted) for building inspector

3.06 SITE APPLIED WOOD TREATMENT

A. Apply preservative treatment compatible with factory applied treatment at site-sawn cuts, complying with manufacturer's instructions.

B. Allow preservative to dry prior to erecting members.

3.07 TOLERANCES

A. Framing Members: 1/4 inch (6 mm) from true position, maximum. — Construction Tolerances

B. Variation from Plane (Other than Floors): 1/4 inch in 10 feet (2 mm/m) maximum, and 1/4 inch in 30 feet (7 mm in 10 m) maximum.

3.08 CLEANING

A. Waste Disposal: Comply with the requirements of Section 017419.
1. Comply with applicable regulations.
2. Do not burn scrap on project site.
3. Do not burn scraps that have been pressure treated.
4. Do not send materials treated with pentachlorophenol, CCA, or ACA to co-generation facilities or "waste-to-energy" facilities.

B. Do not leave any wood, shavings, sawdust, etc. on the ground or buried in fill.

C. Prevent sawdust and wood shavings from entering the storm drainage system.

END OF SECTION

ROUGH CARPENTRY 061000 - 5

Figure 26e Typical specification (5 of 5).

2.1.2 Technical Aspects

The technical aspects section includes information on materials that are specified by standard numbers and by standard national testing organizations such as ASTM International. The technical data section of specifications can be of three types:

- *Outline specifications*—These specifications list the materials to be used in order of the basic parts of the job, such as foundation, floors, and walls.
- *Fill-in specifications*—This is a standard form that requires pertinent information to be filled in. It is typically used on smaller jobs.
- *Complete specifications*—For ease of use, most specifications written for large construction jobs are organized in the Construction Specifications Institute (CSI) format called *MasterFormat®*, which is discussed in the next section.

2.1.3 Specifications Format

For convenience in writing, speed in estimating, and ease of reference, the most suitable organization of specifications is a series of sections dealing with the construction requirements, products, and activities. Specifications must be easily understandable by the different trades. People who use the specifications must be able to find all information needed without spending too much time looking for it.

The most commonly used specification format in North America is MasterFormat. This standard was developed jointly by CSI and Construction Specifications Canada (CSC). Prior to 2004, the organization of construction specifications and suppliers catalogs was based on a standard with 16 sections known as *divisions*.

The divisions and their subsections were individually identified by a five-digit numbering system. The first two digits represented the division number, and the next three individual numbers represented successively lower levels of breakdown. For example, the number 13213 represented Division 13, subsection 2, sub-subsection 1, and sub-sub-subsection 3.

In 2004, the MasterFormat standard underwent a major change. What had been 16 divisions was expanded to 4 major groupings and 49 divisions, with some divisions reserved for future expansion. The first 14 divisions of the most recent version, MasterFormat 2020, are essentially the same as the previous format, with a few small changes. Subjects under the old Division 15—Mechanical have been relocated to Divisions 22 and 23. The basic subjects under Division 16—Electrical have been relocated to Division 26 (*Table 2*).

In addition, the numbering system for the new MasterFormat organization was changed to six digits preceding the decimal point to allow for more subsections in each division. In the new numbering system, the first two digits represent the division number. The next two digits represent subsections of the division, and the two remaining digits represent the third-level sub-subsection numbers. The fourth level, if required, is a decimal and number added to the end of the last two digits. For example, the number 132013.04 represents Division 13, subsection 20, sub-subsection 13, and sub-sub-subsection 04.

Specifications

Specifications supplement the related construction drawings because they contain details not shown on the drawings. Specifications define and clarify the scope of the job. They describe the specific types and characteristics of the components needed to do the job. They may also describe the installation methods for some of them. Many components are identified specifically by the manufacturer's model and part numbers. This information is used to purchase the hardware needed to accomplish the installation based on contractual requirements.

TABLE 2 Example of Divisions in the MasterFormat 2020 Standard

Division Numbers and Titles	
Division 00	Procurement and Contracting Requirements
Specification Group	
General Requirements Subgroup	
Division 01	General Requirements
Facility Construction Subgroup	
Division 02	Existing Conditions
Division 03	Concrete
Division 04	Masonry
Division 05	Metals
Division 06	Wood, Plastic, and Composites
Division 07	Thermal and Moisture Protection
Division 08	Openings
Division 09	Finishes
Division 10	Specialties
Division 11	Equipment
Division 12	Furnishings
Division 13	Special Construction
Division 14	Conveying Equipment
Division 22	Plumbing
Division 23	Heating, Ventilating, and Air Conditioning
Division 26	Electrical

2.2.0 Building Codes

Specifications are written to conform to building codes and the best practices of the construction trade. Model building codes provide minimum standards to guard the life, health, and safety of the public by regulating and controlling the design, construction, and quality of materials used in construction. Model building codes have also come to govern the use and occupancy of buildings, the locations of types of buildings, and the ongoing maintenance of buildings and facilities.

Model building codes are the legal instruments that enforce public safety in the construction of human habitation and assembly structures. They are used not only by the construction industry, but also by the insurance industry for compensation appraisals and claims adjustments and by the legal industry for court litigation.

Until 2000, there were three model building codes. The three code-writing groups—Building Officials and Code Administrators International (BOCA), International Conference of Building Officials (ICBO), and Southern Building Code Congress International (SBCCI)—combined their efforts under the umbrella of the International Code Council (ICC) with the purpose of writing one nationally accepted family of building and fire codes. The first editions of the *IRC* and *IBC* were published in 2000 and continue to be updated on a three-year cycle:

- The *IRC* addresses the design and construction of one- and two-family dwellings and townhouses.
- The *IBC* pertains to all other building construction not covered by the *IRC*.

Model building codes are written to be just that: models. Local jurisdictions, such as states, counties, or cities, model their building codes after the *IRC* and *IBC* and adopt them as their own. Once adopted by a local jurisdiction, the model building codes then become law. There are times when local jurisdictions change and add new requirements to adopted model code requirements to provide more stringent requirements and/or meet local needs. The provisions of the building codes apply to the construction, alteration, movement, demolition, repair, structural maintenance, and use of any building or structure within the local jurisdiction. An important general rule about codes is that in almost every case, the most stringent local code takes precedence.

The National Fire Protection Association (NFPA) publishes codes and standards designed to minimize the risk and effects of fire by establishing criteria for building, processing, design, service, and installation. Examples include *NFPA 72®*, the *National Fire Alarm and Signaling Code®*; and *NFPA 70®*, the *National Electrical Code®*. Craft professionals must be aware of the federal laws, building codes, and restrictions that affect the specific job being constructed. They should also have a basic understanding of some of the codes that pertain to other trades, such as the NFPA gas and electrical codes.

Project Organization

Many people are involved in the design, planning, and execution of a project. The following descriptions provide insight into the people involved in project organization and the roles they play:

- *Building owner*—The owner can be either a public entity or a private entity. Public projects are funded with public money from sources such as taxes, fees, and donations. Public owners can include federal, state, and local governments; school boards; park commissions; public universities; and transit authorities. Private projects are funded with private money from individuals, companies, and institutions.
- *Architect/engineer (A/E)*—The architect is the person or firm that does the design work for a project. Architects produce the drawings that govern every aspect of the project. In many cases, architectural firms also do engineering work because in commercial projects, a great deal of mechanical, structural, and electrical engineering work and other design work is required; hence the term *architect/engineer.* Some of the engineering activities include the following:
  - *Structural engineering*—The structural engineer designs the frame of the building.
  - *Site engineering*—The site engineer designs everything outside the footprint (perimeter) of the building.
  - *Geotechnical engineering*—The soils engineer tests the soil and may also perform concrete testing.
  - *Mechanical engineering*—The mechanical engineer designs the HVAC and plumbing systems.
  - *Electrical engineering*—The electrical engineer designs the lighting and electrical distribution systems.
- *General contractor (GC)*—In many cases, the GC manages the entire construction project, following the plans developed by the A/E. The GC schedules all the work on the project, buys the materials, and hires subcontractors to perform carpentry, plumbing, electrical, roofing, painting, and other work.
- *Construction manager (CM)*—The CM becomes involved earlier in the project than the GC. The CM provides construction expertise throughout the design phase of the project and acts as the owner's agent for the duration of the project. The GC often acts as the CM.
- *Building official*—The building official is the officer or designated authority responsible for the administration and enforcement of the building code.
- *Special inspector*—The special inspector is a qualified person employed or retained by an approved agency and approved by the building official as someone competent to inspect a specific type of construction.

2.0.0 Section Review

1. Which construction document takes legal precedence over construction drawings when there is conflicting information?
   a. Shop drawings
   b. Specifications
   c. CAD files

2. How many digits are in the numbering system for MasterFormat 2020?
   a. 4
   b. 5
   c. 6
   d. 7

3. Which building code relates to construction of one- and two-story family dwellings?
   a. *IRC*
   b. *IBC*
   c. MasterFormat Division 22
   d. ICC

Module 28208 Review Questions

1. What is the distance from the street to the front of a structure called?
   a. Front setback
   b. Right of way
   c. Foundation easement
   d. Greenway

2. Section drawings are used in both architectural and which other important plans?
   a. Structural plans
   b. Roof plans
   c. Floor plans
   d. Site plans

3. In a finish schedule, each room is identified by which of the following?
   a. Name and number
   b. Primary purpose
   c. Overall square feet
   d. Scheduled completion date

4. Who typically creates submittal drawings?
   a. Contractors
   b. Supervisors
   c. Manufacturers
   d. Architects

5. If a line on a drawing measures 4" in length and the drawing scale is $\frac{1}{4}$" = 1'-0", the line represents what length?
   a. 4'
   b. 8'
   c. 12'
   d. 16'

6. What is the most common specification format in North America?
   a. *IRC*
   b. *NFPA 72*
   c. MasterFormat
   d. ANSI specifications

7. Once adopted by a local jurisdiction, model building codes _____.
   a. become law
   b. are *highly* recommended
   c. replace all local building laws and ordinances
   d. must be approved by local residents

8. Which of the following is an important general rule about codes?
   a. Recently adopted codes take precedence.
   b. Detailed codes take precedence.
   c. Stringent local codes take precedence.
   d. Often-cited codes take precedence.

Answers to odd-numbered questions are found in the Review Question Answer Key in the back of the book.

Answers to Section Review Questions

Answer	Section	Objective
Section One		
1. a.	1.1.4	1a
2. c.	1.1.7	1a
3. c.	1.1.14	1a
4. b.	1.2.1	1b
5. c.	1.3.0	1c
Section Two		
1. b.	2.0.0	2a
2. c.	2.1.3	2a
3. a.	2.2.0	2b

MODULE 28203

Reinforced Masonry

Source: Shutterstock

Objectives

Successful completion of this module prepares you to do the following:

1. Name and describe the primary ingredients in grout and how it is prepared.
   a. Explain the characteristics of grout, including aggregate, water content, and compressive strength.
   b. Explain the procedures for mixing grout.
2. Describe how grout is placed.
   a. Explain low-lift grouting and how to place grout using this technique.
   b. Explain high-lift grouting and how to place grout using this technique.
   c. Explain why mortaring joints for grouted masonry is important.
   d. Explain how to use mechanical vibrators with grout.
3. Describe how to construct reinforced walls and masonry elements.
   a. Explain how to cut and bend rebar.
   b. Explain how to place rebar in reinforced walls.
   c. Explain how to install bond beams, bond-beam lintels, precast lintels, piers, pilasters, and columns.

Performance Tasks

Under supervision, you should be able to do the following:

1. Place grout in a hollow block wall and properly consolidate it.
2. Build a block pier, fill it solidly, and properly consolidate it by rodding or with a mechanical vibrator.

Overview

Reinforced masonry is masonry that has been reinforced with grout or with a combination of grout and rebar. Grout is a fluid cementitious mixture used to fill masonry cores or cavities to increase structural performance. Grout bonds wythes together, transfers lateral stress from masonry to rebar reinforcement, and helps support vertical loads. The grout bonds the steel and masonry together to carry loads as a single unit. This module presents information on the manufacture and use of grout in grouted and reinforced walls and reinforced masonry elements such as bond beams, lintels, piers, pilasters, and columns.

Digital Resources for Masonry

NCCER SCAN ME

Scan this code using the camera on your phone or mobile device to view the digital resources related to this craft.

NOTE

Codes vary among jurisdictions. Therefore, it is essential to consult the applicable code whenever regulations are in question. Referring to an incorrect set of codes can cause as much trouble as failing to reference codes altogether. Obtain, review, and familiarize yourself with your local adopted code.

1.0.0 Grout Ingredients and Preparation

Performance Tasks

There are no Performance Tasks in this section.

Objective

Name and describe the primary ingredients in grout and how it is prepared.

a. Explain the characteristics of grout, including aggregate, water content, and compressive strength.
b. Explain the procedures for mixing grout.

Grout is a fluid mix of cementitious material, water, admixtures, and aggregate that is used to reinforce load-bearing walls or enclosed spaces containing steel reinforcement (*Figure 1*). Grout is used in grouted walls, reinforced walls, and reinforced masonry elements.

A grouted wall is a hollow masonry wall in which the voids in the masonry units are filled with grout only. If reinforcing bar is added to the inside of the wall along with grout, the wall is classified as a reinforced wall. Other masonry structures filled with grout and reinforcing bar are called *reinforced masonry elements*.

Figure 1 Reinforced foundation wall.
Source: Shutterstock

1.1.0 Grout Characteristics

Grout must have enough water to be poured or pumped without piling up. This fluid consistency allows it to fill hollow masonry cores or voids. Grout flow is affected by the specified mix recipe of ingredients and the chosen mixing method. Failure to accurately follow mixing directions may cause voids or poor bonding.

1.1.1 Coarse and Fine Aggregates

Grout used in concrete masonry walls should comply with *ASTM C476-23, Standard Specification for Grout for Masonry*. The major difference between fine and coarse grout is not the consistency, but the aggregate used. For instance, the aggregate size in fine grout is 3⁄8" or less. Aggregates used in grout must comply with *ASTM C404-18, Standard Specification for Aggregates for Masonry Grout*. This specification also includes requirements for grading and cleanliness of the aggregate. Aggregate size is important because grout must fit into tight spaces.

1.1.2 Water Content in Grout

The amount of water in grout determines its consistency and relates to its compressive strength. Water affects grout in several key ways:

- It affects compressive strength.
- It allows grout to flow around reinforcement.
- It allows grout to fill corners and recesses without voids.

Grout water content and consistency will depend partly on external factors such as the air, the masonry unit, and the cavity size. Air and masonry units absorb the moisture from grout, effectively reducing its water-to-cement ratio and increasing its strength. For example, if the masonry unit is dry and highly absorptive, or if the air is hot and dry, masons must account for more water in the grout formula.

Another variable to note is surface area. If the cavity has a small surface area relative to its volume, the grout will need less water because larger surface areas will soak up more water.

1.1.3 Compressive Strength

Mix proportions produce grouts with compressive strengths ranging from 1,000 to 2,500 psi (pounds per square inch) after 28 days of curing. This wide range depends on the amount of mixing water used.

The grout's in-place compressive strength may exceed 2,500 psi before the masonry units absorb water. The water held in the masonry keeps the grout moist so that it can hydrate slowly. Slow hydration allows the grout to develop strength that increases beyond specifications over time.

1.2.0 Mixing Grout

Grout can be mixed at the jobsite or at a central batching plant designed to blend the materials in predetermined proportions. When grout is batched at a plant, continuous mixing is required to keep it from clumping. This is why grout is usually delivered in a truck-mounted mixer that continuously mixes the grout until it is applied. Grout is mixed at the jobsite only under unusual circumstances or in very small amounts that can be used quickly.

Typically, premixed grout is transferred from the truck mixer into a pump hopper (*Figure* 2). It is pumped from the hopper to its target location within 90 minutes of when the mixing initially started.

Figure 2 Grout pump and hopper.
Source: Shutterstock

Grout can be mixed at the jobsite in a two-speed mortar mixer, using the same general procedures as those used for mortar. The two-speed mixer (*Figure 3*) has a high mixing speed and a low agitating speed to prevent the grout from segregating without overmixing it.

Figure 3 Worker using a power mixer.
Source: SPEC MIX LLC

Before using any power mixer, review the equipment's operating manual and any general safety procedures for working with power equipment. However, the following procedures are common for mixing grout:

Step 1 Check the formula.
Step 2 Measure the ingredients.
Step 3 Add one-third of the amount of water.
Step 4 Add one-half of the sand.
Step 5 Start the mixer at high speed.
Step 6 Add the cement and lime.
Step 7 Add the remaining sand.
Step 8 Add one-third of the remaining water.
Step 9 Blend for one to two minutes.
Step 10 Add the remaining water.
Step 11 Blend for an additional three to four minutes, for a total of five minutes, minimum.
Step 12 Empty the grout into a container.
Step 13 If any grout remains in the mixer, switch the mixer to low (agitate) speed.
Step 14 Discard all grout not placed within 90 minutes after the water is added.
Step 15 Clean the mixer thoroughly after emptying the grout.

WARNING!

Grout is caustic. Always wear appropriate personal protective equipment (PPE) when mixing or placing grout.

1.0.0 Section Review

1. What is the *maximum* size of aggregate in fine grout?
   a. $\frac{1}{3}$"
   b. $\frac{3}{8}$"
   c. $\frac{1}{2}$"
   d. $\frac{5}{8}$"

2. Premixed grout should be delivered and applied to its target location within how long after initial mixing begins?
   a. 45 minutes
   b. 90 minutes
   c. Two hours
   d. Six hours

2.0.0 Grout Placement

Performance Task

1. Place grout in a hollow block wall and properly consolidate it.

Objective

Describe how grout is placed.

a. Explain low-lift grouting and how to place grout using this technique.
b. Explain high-lift grouting and how to place grout using this technique.
c. Explain why mortaring joints for grouted masonry is important.
d. Explain how to use mechanical vibrators with grout.

It is important to place grout slowly. Because of its viscosity and aggregates, grout can trap air bubbles as it fills a space. You must consolidate grout by vibrating or rodding to ensure that it fills all voids and adheres to reinforcement without air gaps. Vibrating involves the use of a mechanical vibrator to consolidate grout. Rodding involves poking the grout with a rod to consolidate it.

The amount of grout placed continuously without interruption is called a *lift*. Grout is applied in either a low lift or a high lift. Low-lift grout is poured to fill a space less than 5'-4" high, while high-lift grouting fills exceed this measurement. A lift of 12" or less can be consolidated by rodding, while a lift over 12" must be consolidated by mechanical vibration.

Before grouting, the mortar between the masonry units must be hardened and be strong enough to contain the wet grout. The hardening time can vary depending on various factors, such as the number of wythes and the weather conditions.

Blowout: The swelling or rupture of a cavity wall from too much pressure caused by pouring liquid grout into the cavity.

If the mortar is not firm enough, the internal pressure can cause a **blowout**. A blowout occurs when the structure buckles from the weight of the grout. A movement of even $\frac{1}{4}$" means the wall must be removed and rebuilt. The best practice is to proceed carefully and slowly.

Clean mortar joints are also vital because any mortar protrusions into the cavity prevent the grout from bonding properly. Keep mortar off any surface that will be grouted, and clean as you work.

WARNING!

Whether wet or dry, mortar and grout can be harmful because both contain cement. Dry cement dust can cause blood poisoning by entering open wounds and can cause chemical burns of the eyes, nose, mouth, throat, or lungs. Wet mortar, wet grout, and wet concrete can also cause chemical burns to the eyes and skin. Repeated contact with cement or wet concrete can also cause an allergic skin reaction known as *cement dermatitis*.

2.1.0 Low-Lift Grouting Techniques

Rebar: Reinforcing bar embedded in concrete, mortar, or grout to act together with the other components to resist loads.

The mortar's hardening can take from 3 to 24 hours, and grouting begins afterward. If the cores are to be reinforced, the steel reinforcement is positioned and secured. The reinforcing steel bars, called **rebar**, overlap and are secured in the foundation. They protrude above the grout level to provide the specified lap into the following masonry lift, as shown in *Figure 4*.

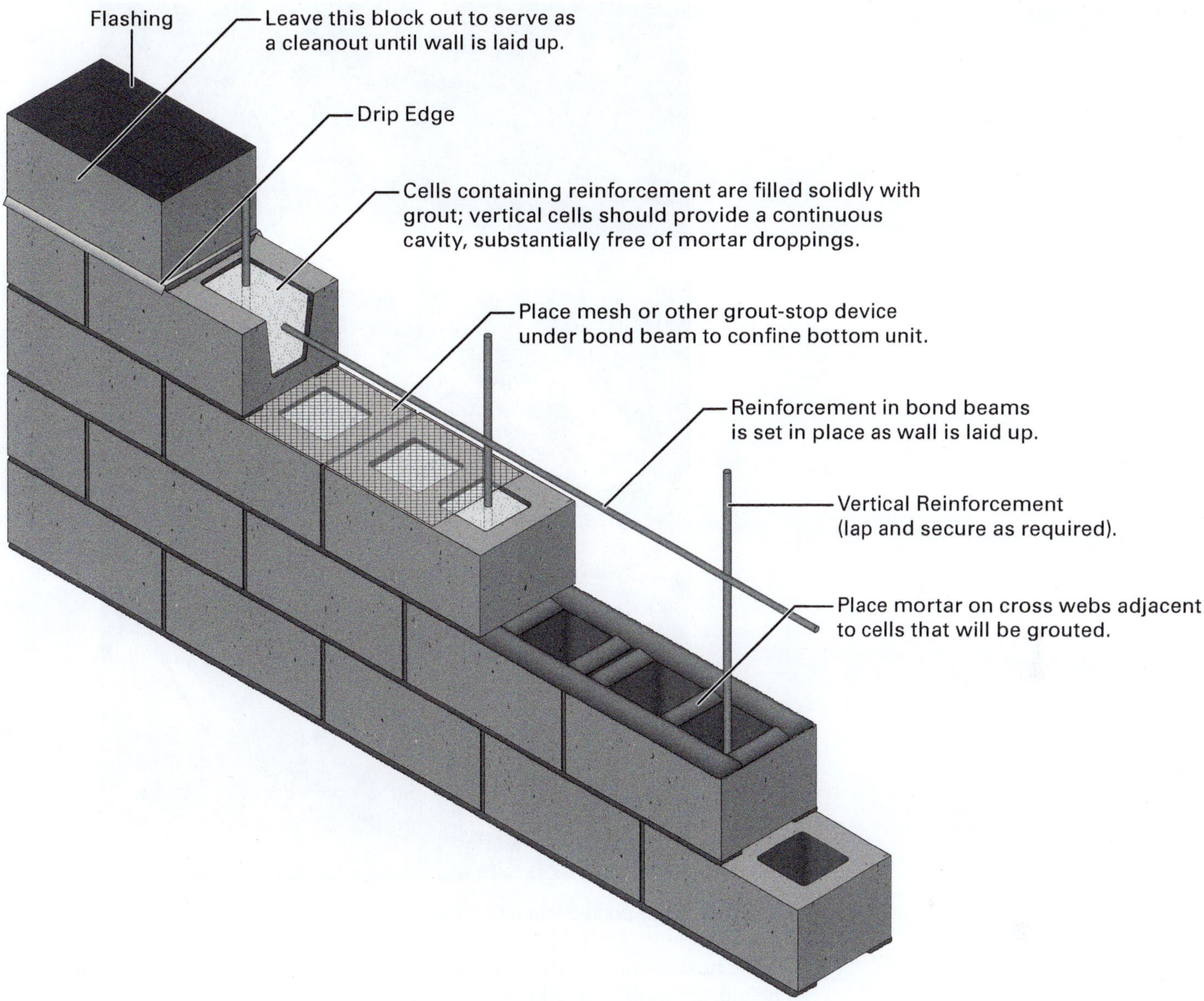

Figure 4 Typical reinforced concrete masonry construction using low-lift grouting.

Place wire mesh or another grout-stop device under **bond beams** to contain grout, or use a solid-bottom unit. Cells containing steel are filled solidly with grout. Vertical cells should provide a clean, continuous cavity.

The grout should not be poured to the very top of the block, but to within 1" of the top. This will form a **key** or keyway that permits the next lift of grout to bond with the block and the grout below it. The key allows the two lifts to lock together in a tongue-and-groove joint.

As the grout is placed, it should be applied carefully to reduce splatter (*Figure 5*). As one person places the grout, another should rod or vibrate it to ensure that no voids remain.

Bond beam: A course of masonry units with steel rebar inserted and held in place by a solid fill of grout or mortar; used as a lintel or reinforcement beam to distribute stress.

Key: A recess or groove in one placement of grout or concrete that is later filled with a new placement of grout or concrete so that the two lock together in a tongue-and-groove configuration.

Figure 5 Grout poured into a CMU wall.

The next section of the wall is then built, and reinforcements are installed. The mortar should be allowed to harden for a specified amount of time before the next lift of grout is poured. If there is another lift, the grout will be poured within 1" of the top of the last block. Pour the final wall section to the top of the last block.

2.2.0 High-Lift Grouting Techniques

High-lift grouting adds some complex features to wall designs. If nothing in the wall prevents the grout from flowing from top to bottom without interruption, the wall can be built to full height. However, cleanout openings must be made at the bottom of the wall or the bottom of every core containing reinforcement (*Figure 6*).

Cleanout openings allow masons and inspectors to guarantee that all inside surfaces of the cavity to be grouted are sufficiently clean and free of excess mortar protrusions, and that all rebar ties are secured properly.

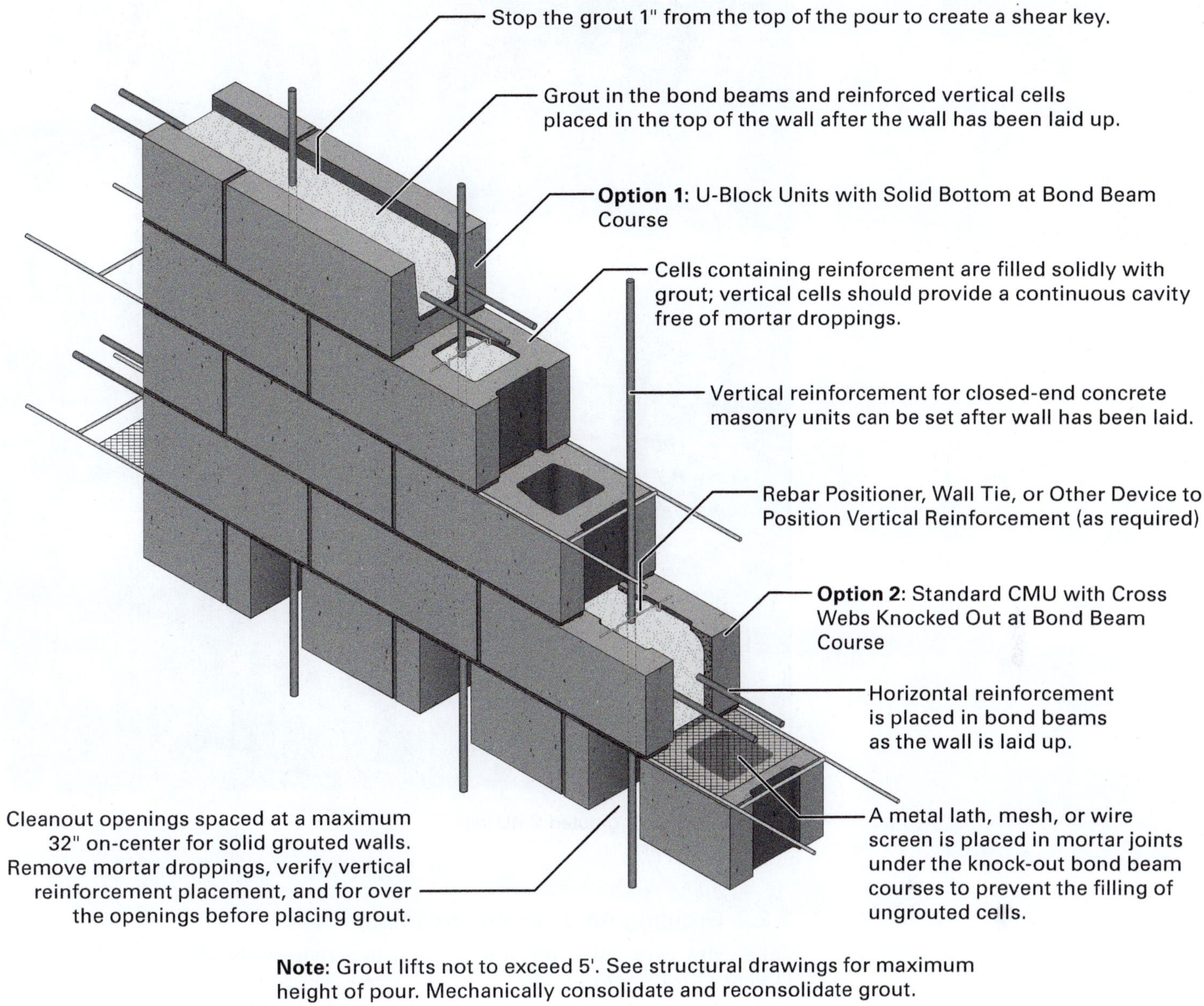

Figure 6 Spaced cleanout openings.

2.2.1 High-Lift Grouting for Cavity Walls

High-lift grouting for cavity walls requires building the wall to the specified height (*Figure 7*). The cavity is inspected and the cleanout openings are sealed. The grout is then poured into the cavity in lifts using a cement bucket or a pump. The grout is then vibrated and allowed to stiffen for about an hour before the remaining lift is poured. The average lift should be around 4' high. If the lift is higher, it will be difficult to fill potential voids.

The cavity should not be filled until the mortar has hardened enough to hold the weight of the grout. Project specifications or local building codes set the hardening time, which varies significantly depending on the work and locale.

High-lift grouting requires walls to be temporarily braced until the mortar and grout are fully set. Partially completed walls should also be braced against lateral loads from winds or other forces during construction. Dimension lumber can be used for bracing. In addition to any local building codes, all Occupational Safety and Health Administration (OSHA) requirements for strength, spacing, and height of bracing must be met.

Figure 7 Freshly grouted CMU wall.
Source: IMS Masonry, Inc.

2.2.2 Grouting Around Reinforcement Steel

Bridging: The mounding of grout or cement over an obstruction, creating a void under the obstruction.

Take extra care when pouring grout when there is hardware in the cavity. Do not allow the grout stream to separate over rods, spacers, rebars, or other embedded items. This leads to a condition called **bridging**, in which the grout mounds over the reinforcement and creates a void underneath. Further placing and consolidating may not fill that void.

To avoid bridging, fill the space underneath the obstruction before filling the space above. You may choose to rod or vibrate the grout before finishing the lift to reduce bridging.

2.3.0 Using Mortar Joints for Grouted Masonry

Mortaring joints becomes critical when the wall is grouted, because failure to do so leads to grout leaking, which will compromise the integrity and strength of the wall. To make sure head joints are filled with mortar, use a shoved joint. This technique consists of buttering the ends of each unit and shoving them forcefully into place to produce a solid joint with no voids in the mortar.

Mortaring block for grouted walls requires precision. If alternate cells are grouted, mortar must be spread for full bed joints on shells and webs. This will keep the grout from squishing out from between the cells.

Angle the inside of the mortar joint with the edge of the trowel. This bevel helps prevent mortar from protruding past the masonry unit. Carefully cut off mortar that pushes into the cavity, to prevent the grout from filling the space beneath.

2.4.0 Using Mechanical Vibrators for Grout

Mechanical vibration consolidates grout and concrete. Vibrating forces the air out of the grout and compacts it within the cavity and around the reinforcement. Air voids rise to the surface during vibration so that the cavity is solidly filled. Mechanical vibrators (also called *pencil vibrators*), as shown in *Figure 8*, come in various sizes. The most common models are fitted with backpack straps or are handheld devices.

Figure 8 Mechanical vibrator.
Source: Milwaukee Tool

The process of using a mechanical vibrator is straightforward. As with other power equipment, read the manufacturer's instructions and review safety procedures before operating any equipment. Here are some helpful tips:

- Turn the vibrator on before inserting it and turn it off before removing it from the grout, to avoid splatter.
- Lower the vibrator's head and shaft vertically into the full depth of the fresh grout layer, keeping the shaft as vertical as possible.
- Insert the head of the vibrator into the previous lift if it hasn't set yet.
- Manipulate the shaft up and down for roughly five to 15 seconds. Any longer, you risk segregation.
- Withdraw the vibrator slowly with a series of up-and-down motions.
- Keep the vibrator from hitting any reinforcing bars.

Vibration is complete when air bubbles rise to the surface and the grout settles into place.

2.0.0 Section Review

1. How long does mortar take to harden?
   a. From 3 to 24 hours
   b. From one to three hours
   c. From 24 to 36 hours
   d. One week

2. What is the *maximum* height of a low lift?
   a. 5'-4"
   b. 4'-5"
   c. 3'-6"
   d. 2'-0"

3. What is the *maximum* mortar joint protrusion on the inside face of a block wall?
   a. ½"
   b. ¾"
   c. 1"
   d. 1¼"

4. What is the *maximum* time a mechanical vibrator should be allowed to run in one location to avoid segregation?
   a. One minute
   b. 30 seconds
   c. 15 seconds
   d. Five seconds

3.0.0 Reinforced Walls and Masonry Elements

Performance Task

2. Build a block pier, fill it solidly, and properly consolidate it by rodding or with a mechanical vibrator.

Objective

Describe how to construct reinforced walls and masonry elements.

a. Explain how to cut and bend rebar.
b. Explain how to place rebar in reinforced walls.
c. Explain how to install bond beams, bond-beam lintels, precast lintels, piers, pilasters, and columns.

Using steel reinforcement in masonry walls allows the walls to withstand more stress and carry heavier loads than unreinforced masonry walls of the same size. The most commonly used reinforcement is steel bar (rebar). Rebar is used for vertical reinforcement and horizontal bond-beam reinforcement.

Each bar has identification stamped onto it, as shown in *Figure 9*.

ASTM specifications require that the manufacturer stamp each bar with the following information:

- A letter or symbol to indicate the manufacturer's mill
- A number corresponding to the size number of the bar (*Table 1*)
- A symbol or marking indicating the type of steel (*Table 2*)
- A marking to designate the grade (*Table 3*)

TABLE 1 Reinforcing Bar Properties

Bar Size, No.	Weight	Diameter
3	0.376 lb/ft	⅜"
4	0.668 lb/ft	½"
5	1.043 lb/ft	⅝"
6	1.502 lb/ft	¾"
7	2.044 lb/ft	⅞"
8	2.670 lb/ft	1"
9	3.400 lb/ft	1⅛"
10	4.303 lb/ft	1¼"
11	5.313 lb/ft	1⅜"

TABLE 2 Reinforcing Bar Types

Symbol/Marking	Type of Steel
A	Axle (ASTM A617)
S or N	Billet (ASTM A615)
I or IR	Rail (ASTM A616)
W	Low alloy (ASTM A706)

TABLE 3 Reinforcing Bar Grades

Grade	Identification	Minimum Yield Strength
40 and 50	None	40,000 to 50,000 psi (40 to 50 ksi)
60	One line or the number 60	60,000 psi (60 ksi)
75	Two lines or the number 75	75,000 psi (75 ksi)

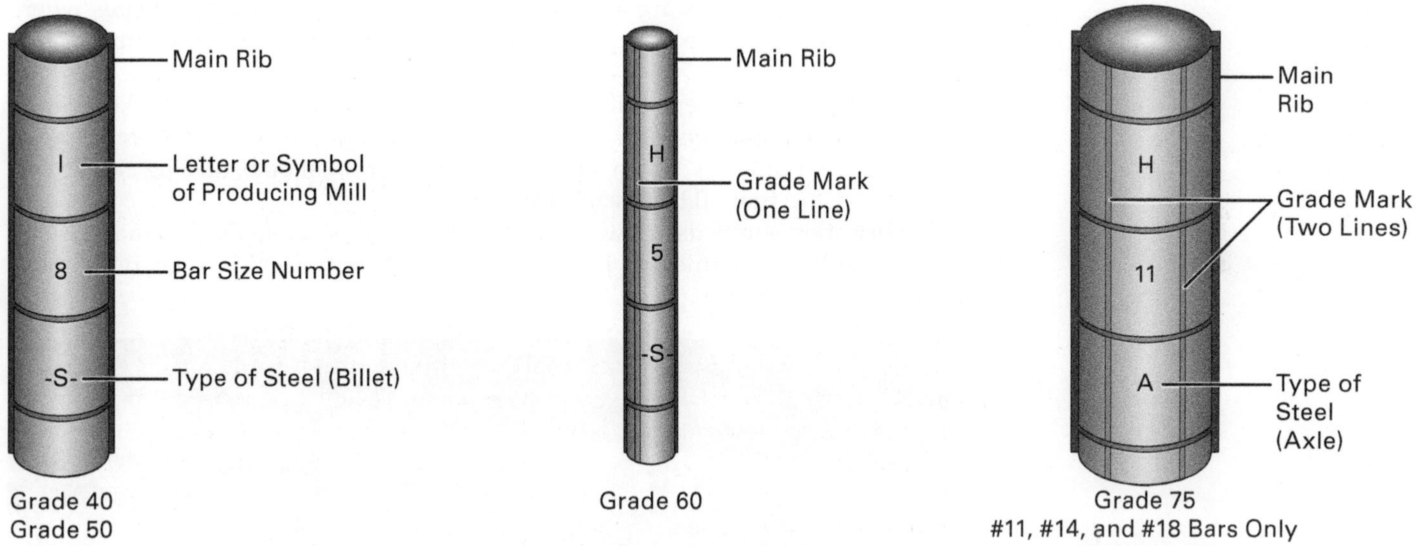

(A) Line System Grade Marks

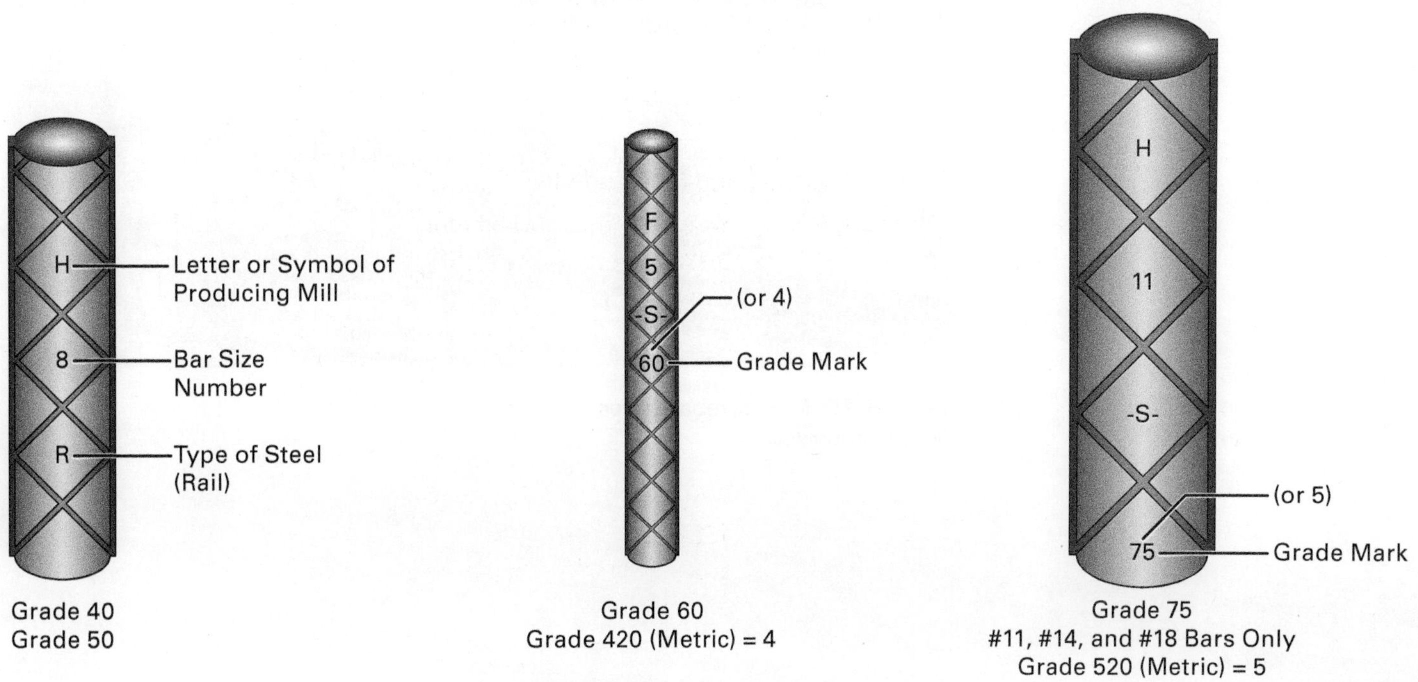

(B) Number System Grade Marks

Figure 9 Reinforcement bar identification.

Bar sizes are measured in eighths of an inch. The size number corresponds to the number of eighths of an inch of the nominal diameter of the bars. For example, a size 5 bar is $^5/_8$" in diameter.

Rebar may be coated with epoxy or plastic to increase corrosion resistance. Grade, deformation, size, coating, location, and spacing are all found in the project specifications.

Keeping rebar and all other reinforcement items clean and dry is important. Dirt and moisture will prevent a complete bond of the metal to the grout or mortar and will lead to long-term decay or corrosion inside the grouted cavity.

3.1.0 Cutting and Bending Rebar

Rebar can be cut using several manual and power tools. Wire mesh and rods up to size 4 ($^1/_2$" in diameter) can be cut quickly using a bolt cutter (*Figure 10*). Portable power shears are also used to cut rebar, especially on jobsites where lots of cutting must be done. Powered rebar cutters can be used to cut rebar up to size 7.

Rebar should be bent cold, and can be bent by hand using a hickey bar or other type of rebar bender (*Figure 11*). Hand benders can be used on rebar up to size 8 (1" in diameter). When a great deal of rebar must be bent, a jig can be created to ensure that all bends are uniform.

Fabrication shops use power benders; however, portable models may also be useful on large commercial jobsites. Only workers trained in their use should use a power bender.

Figure 10 Manual rebar cutter.
Source: Shutterstock

Figure 11 Hickey bar.
Source: Klein Tools

3.2.0 Placing and Tying Rebar in Reinforced Walls

When working with rods in a reinforced cavity wall, masons first place short steel dowel rods into the concrete footing before it hardens. Two wythes of masonry are then laid. Rebar is placed according to the plans or specifications. Rebar must be held in place so that it does not shift during normal construction or grout placement. The bar must be aligned, supported, and rigidly fastened before the grout is poured.

It is often necessary to secure rebar to another piece of rebar, such as when vertical and horizontal reinforcements cross. Soft-annealed iron wire is normally used for tying rebar. The ends of the wires are twisted together with pliers (*Figure 12*). It is important to bend the ends of the tie toward the rebar. Otherwise, air pockets may form during grouting.

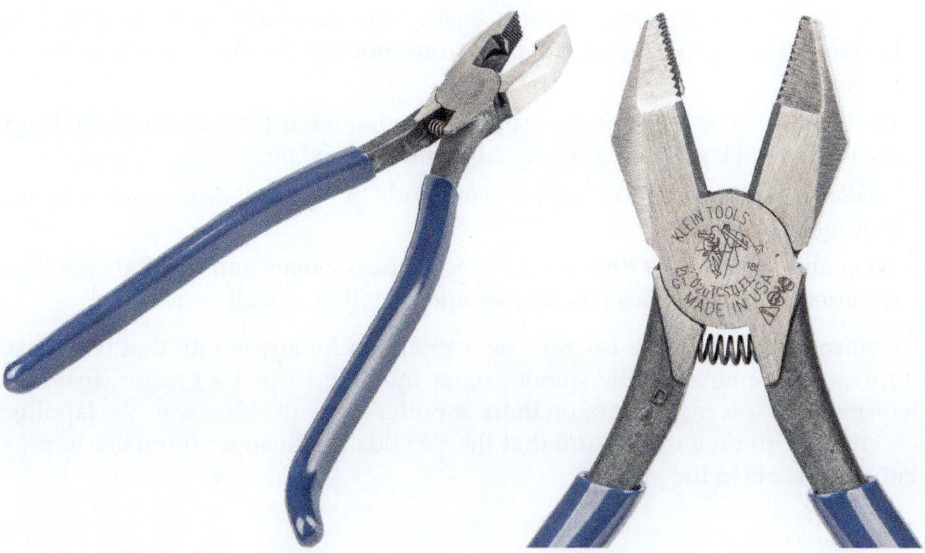

Figure 12 Pliers for tying rebar.
Source: Klein Tools

Several common ties are shown in *Figure 13*.

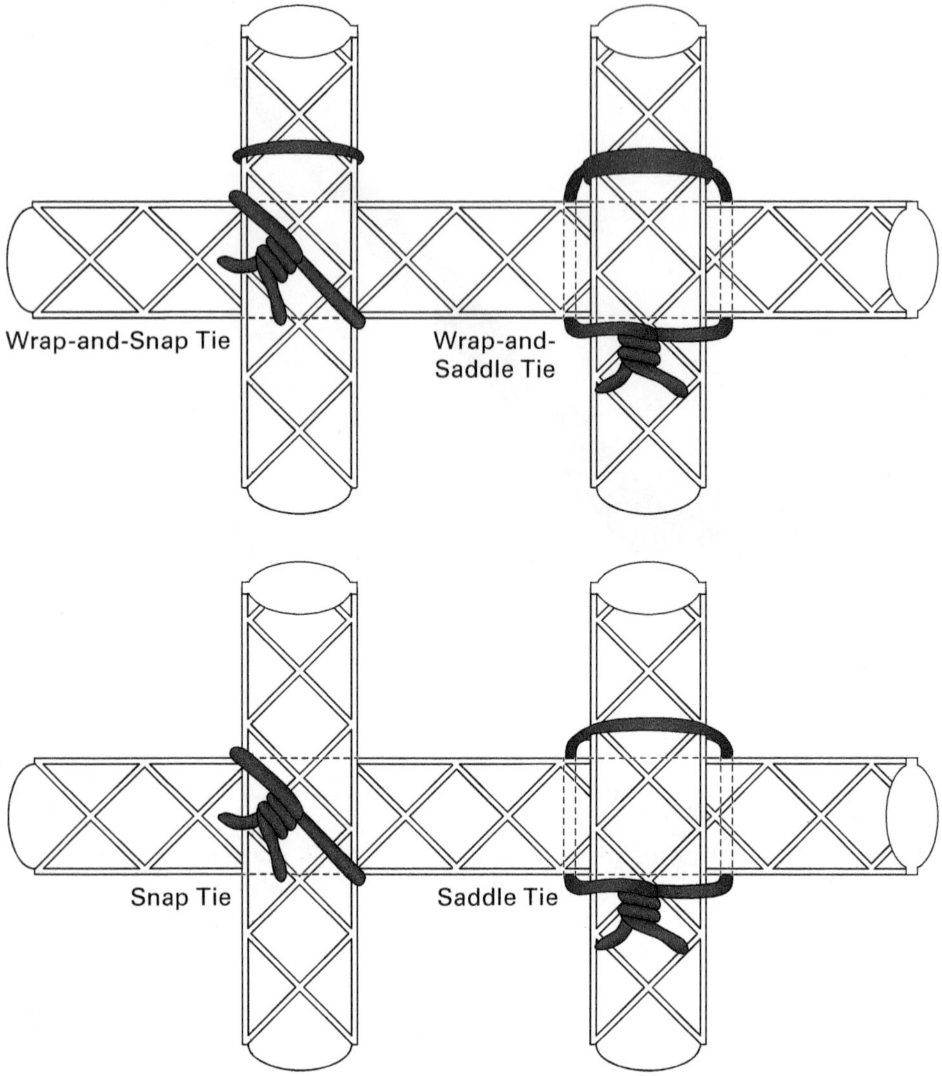

Figure 13 Types of ties.

Each type of tie has a particular application:

- *Snap tie*—The simplest and most basic. This tie is normally used in flat, horizontal work to prevent the bars from moving during concrete or grout placement.
- *Wrap-and-snap tie*—Normally used when tying rebar placed in walls to keep the horizontal bars from moving. Also called a *wall tie.*
- *Saddle tie*—Often used to hold hooked ends of bars in position when tying footing or other mats.
- *Wrap-and-saddle tie*—Often used to secure heavy mats and vertical bars that experience strong forces when lifted and installed in wall systems.

Deformed rebar (*Figure 14*) must be overlapped for any length that is at least 40 times its diameter, while smooth rebar must be lapped a greater distance. Deformed rebar is more common than smooth rebar, but either way, the lapping is secured with tie wire. Ensure that the tie ends are flush with the bar to prevent air pockets in the grout.

Figure 14 Deformed rebar.
Source: Shutterstock

Carbon fiber reinforcement (*Figure 15*) is also becoming a more popular option for some designs. Even though it is a relatively new product, some industry leaders see this alternative becoming the future reinforcement option because it is durable, is lightweight, requires less concrete cover, and can withstand roughly twice the tension of steel reinforcement.

Vertical rods are wired to horizontal rods to stabilize and brace the wall as it increases in height. In high-wind areas, a hook is added to the vertical reinforcement. Grout is poured between the rods as the courses are laid.

Figure 15 Carbon fiber reinforcement.
Source: Shutterstock

3.3.0 Installing Bond Beams and Bond-Beam Lintels

A bond beam is a grouted, reinforced horizontal or vertical collection of masonry units spaced through a structure to carry extra loads. A vertical bond beam or internal pilaster can be easily constructed in a hollow masonry wall to strengthen non-load-bearing structures such as garden walls.

This structure requires placing reinforcement before the sides of the pilaster are framed. The reinforcement can be metal lath, fine hardware cloth, or cut pieces of masonry. Wood should never be used for this purpose. The beam can be grouted using high-lift techniques. Temporary bracing may or may not be necessary, depending on the age and hardness of the mortar.

Bond beams can be spaced through an otherwise unreinforced wall to add lateral strength. When the core is underneath the bond beam, the grout stop is interrupted. If solid-bottomed bond beams are used, the cells beneath them must be grouted first.

The bond beam is constructed of a cut or specially shaped CMU, with rebar cradled horizontally in the course of CMUs. To prevent grout from leaking, wire lath or metal screening must be put over the cells that are not to be grouted, to keep the grout out. The rods are put into the beam block, as shown in *Figure 16*, and then grouted in place.

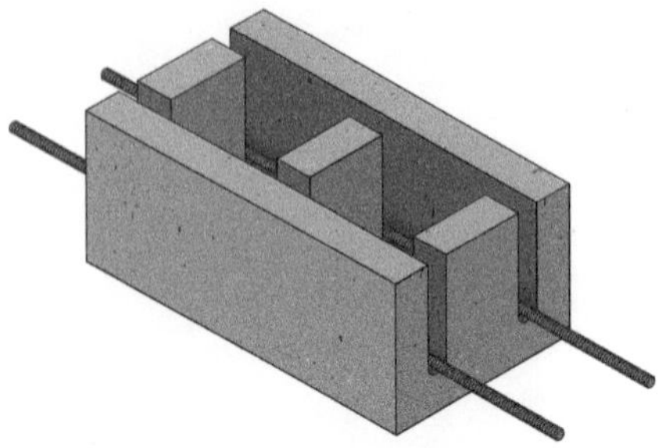

Figure 16 Placing rebar in a bond beam.

A horizontal bond beam can also be used as a lintel, which is the horizontal load-bearing member or beam over an opening. Bond-beam lintels are usually made of the same material as the wall. The inside of the brick or block is cut out, or a special precut block is used (*Figure 17*).

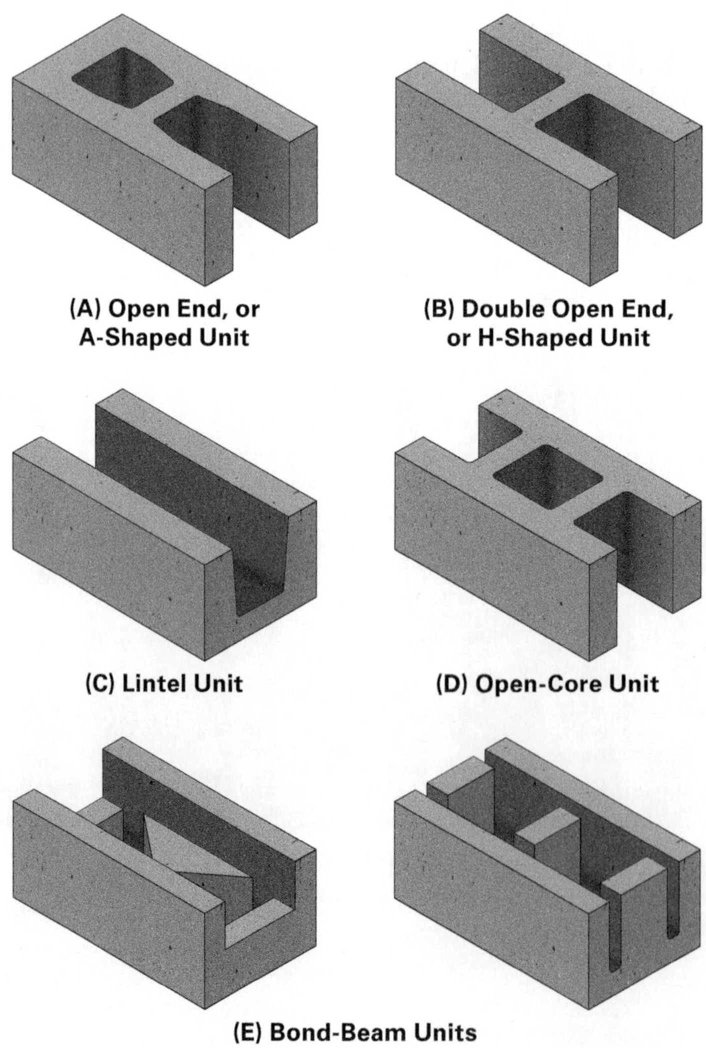

Figure 17 Bond-beam shapes.

If the bond-beam lintel may be built in place, use dimension lumber to build a brace to support the lintel on the bottom and sides. Then, lay out the cut masonry on the bracing and leave it in place until the lintel cures. A layer of concrete or grout is then poured into the cutout masonry. Steel reinforcement rods are laid in the cutouts and covered with another layer of concrete or grout.

3.3.1 Building a Bond-Beam Masonry Lintel

The following steps describe the proper sequence for building a bond-beam masonry lintel:

Step 1 Build the supporting wall to the proper height. Ensure that the supporting masonry jambs are level, plumb, ranged, and of the correct height.

Step 2 Install temporary support, called *shoring*, using wood, steel angles, or metal studs.

To allow easier shoring removal once the bond-beam masonry lintel is finished, the length of the shoring should be at least ½" less than the masonry opening and shimmed to the desired height.

Set the top of the support so that the first course of masonry units can be laid dry over the support with their tops aligning adjacent units in the supporting wall. Vertical supports can be fastened to the masonry jambs but should be placed at a height that will allow the header to be shimmed to the proper opening height.

Place vertical supports at each end of the lintel, or at regular intervals for larger openings. As a rule of thumb, place a vertical support every 24". The opening height should also be at least ½" less than the masonry opening.

Step 3 Set the first course of masonry over the shoring.

Units in this course should be buttered carefully, then set on the support and slid against adjacent units to avoid trapping mortar between the units and the temporary support.

A backer rod can be used in the bottom joint between these units to prevent mortar and grout from oozing out of the joint. Parge the inside of the head joints on all three sides.

Step 4 Install the lower reinforcing bars.

Ensure that the bars have a minimum clearance of ½" between the bar and the inside of the masonry unit. Use small pieces of block or tie wire to separate the bars and to secure them in place. Some lintels will require additional courses with horizontal reinforcing bars tied together by vertical reinforcing bars, commonly called *stirrups*.

Step 5 Once the lower reinforcing bars have been installed, the remaining masonry lintel courses can be installed.

Typically, all courses within a masonry lintel will require open-end units. Care must be taken to parge the inside joints to reduce the chance of a blowout. These courses must be installed with care so as not to disrupt the course bearing on the temporary support.

Step 6 When the mortar has set, grout the masonry lintel.

With the engineer's permission, lintels may be poured in two half pours. Do not allow excessive vibration of the lintel during the grouting process, as this can lead to a blowout.

Step 7 When the grout has achieved sufficient strength, remove the temporary shoring. Avoid damaging the finished masonry faces, especially if they are exposed.

Step 8 Fill and finish the joints on the underside of the lintel and any affected jamb joints.

This may require grinding or chiseling out mortar and grout if the inside of any head joint was not parged properly during the initial construction.

Brush and dampen the joints before filling them with mortar. Tool the joints to the proper finish, taking care to address each joint's vertical return.

3.4.0 Installing Precast Lintels

Precast reinforced concrete lintels are often used over openings in concrete block walls and foundations. They have rectangular cross sections, as shown in *Figure 18*, or one corner may have a recess to fit over a shaped block. Precast concrete lintels may be one piece or split into two thinner sections. This way, the lintel cross section that is needed can be made without requiring the manufacture of a special precast lintel.

Figure 18 Precast concrete lintel.
Source: Shutterstock

A solid, one-piece lintel is used to support only the weight of a block above the span of an opening (*Figure 19*). A one-piece lintel will typically be the same thickness as the masonry wall. Steel reinforcing rods may be cast into the lintel to provide increased tensile strength.

Figure 19 Masonry openings with lintels.
Source: Shutterstock

To place a precast concrete lintel, apply mortar to the masonry on either side of the frame. Carefully set the lintel into its proper position and press down until the mortar joint is the same thickness as all other joints. Then, check that the lintel is level and plumb. Remember that unless otherwise specified, the lintel needs a minimum bearing surface on each end (4" for brick, 8" for block).

3.5.0 Installing Piers, Pilasters, and Columns

Masons are also responsible for installing other types of reinforced masonry elements and lintels. Three common reinforced masonry elements a mason will encounter on the job are **piers**, pilasters, and **columns**.

Piers: A vertical reinforced masonry element typically shorter than a column or pilaster. Piers may be designed to carry loads or may be purely ornamental.

Columns: A vertical reinforced masonry element designed to support a load, the width of which never exceeds three times its thickness and the height of which always exceeds four times its thickness.

3.5.1 Installing Piers

Piers (*Figure 20*) are typically shorter than pilasters or columns. The *International Building Code®* requires that the nominal heights of solidly grouted foundational piers are limited to no more than 10 times their nominal thickness; the nominal height of a pier that is not solidly grouted may not exceed four times the pier's nominal thickness.

Structural engineers design solid, reinforced piers to carry concentrated loads. Piers require reinforcement when carrying flexural loads (loads that can cause the pier to bend), such as the thrust of an arch, a steel beam, a roof truss, or a floor joist. Some piers are designed with bearing plates set in a thin mortar

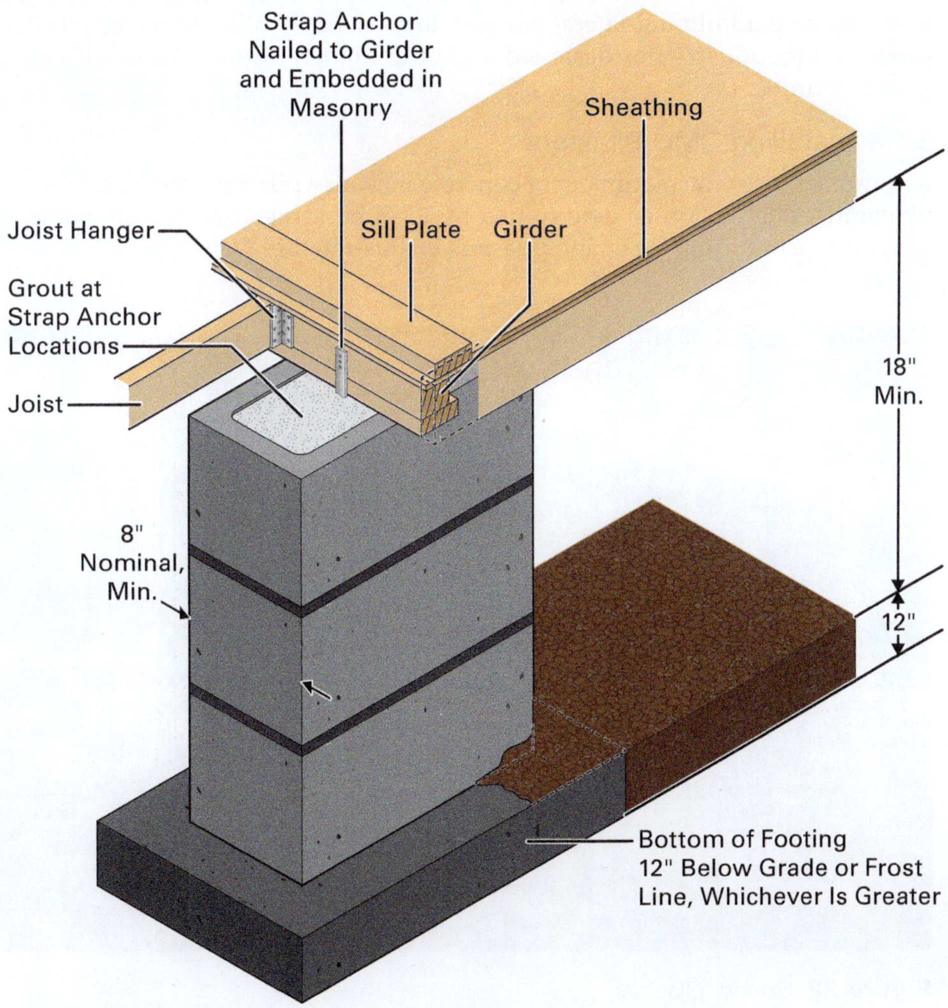

Figure 20 Example of a concrete masonry pier.

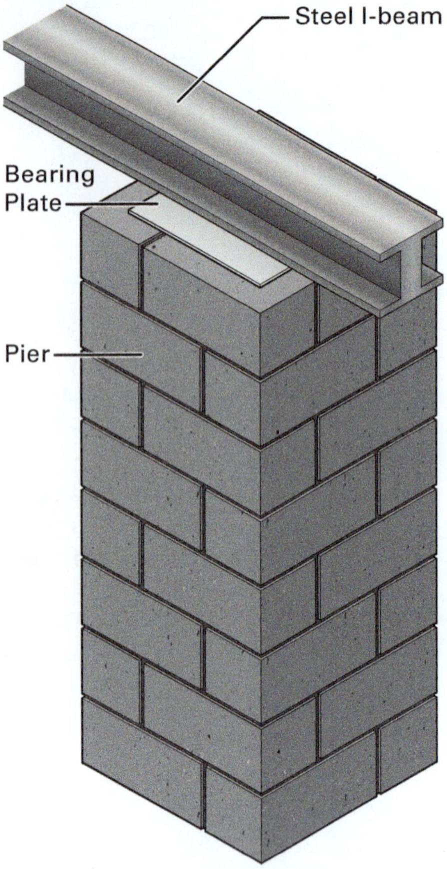

Figure 21 Pier with bearing plate.

bed to help the pier bear the load placed on it (*Figure 21*). When installing a narrow pier, carefully select brick to ensure that they do not vary in length. Otherwise, both sides of the pier will not be plumb for the pier's full length.

3.5.2 Installing Pilasters

A pilaster is a vertical section of a thickened wall. This vertical section of masonry is tied directly to the wall with wall ties or by bonding the masonry units with mortar.

Pilasters project from the wall on one or both sides. If the pilaster projects on a single side, it is called a *flush pilaster.* When they appear as ornaments, pilasters are often called *buttresses*, and may be finished at the top with a sloping face. The portion of the double-sided pilaster that shows on the outside of a structure is also typically referred to as the *buttress*.

If the pilaster also projects inside the wall, the inside projection is called a *counterfort*. According to the side of the wall that shows the projection, pilasters may also be called *interior pilasters* or *exterior pilasters*. *Figure 22* shows typical pilasters protruding from the face of a brick structure.

Although most pilasters reinforce or support a wall, they can serve several other functions. They may support heavy loads, such as vertical roof or floor loads, or they may just provide a decorative effect.

In reinforced masonry construction, pilasters are sometimes used as recesses (more commonly called *chases*) to hide heating or air-conditioning ducts. Pilasters also hide vertical steel beams and offer protection from fire damage.

In the case of fire resistance, the pilaster wall is usually only 4" thick, because it acts as a fire screen instead of primarily supporting loads. Finally, pilasters may give additional lateral support to masonry walls. By using pilasters properly, the industry has designed and built higher, thinner walls with plain masonry units.

3.5.3 Installing CMU Pilasters

Requirements for the placement of concrete masonry pilasters can be calculated by engineering design or determined by local building code requirements. A general rule of thumb for sizing pilasters is illustrated in *Figure 23*.

Figure 22 Exterior pilasters.
Source: Shutterstock

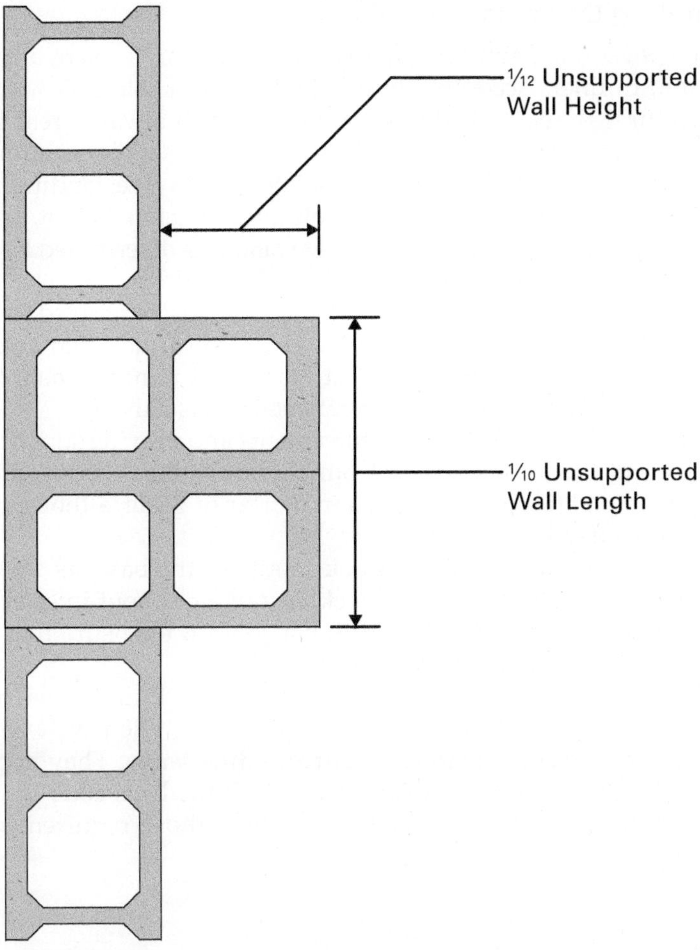

Figure 23 General requirements for pilaster size.

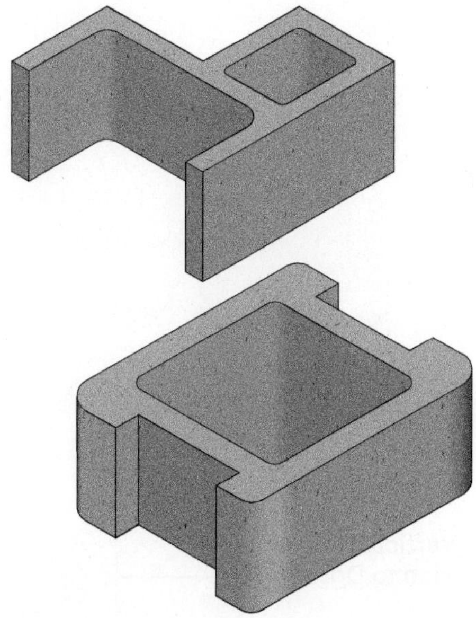

Figure 24 Pilaster units.

The following guidelines can be used to judge the sizes of pilasters and projections:

- The pilaster projection should be no less than $^{1}/_{12}$ the pilaster height between supports.
- The pilaster width should be no less than $^{1}/_{10}$ the wall length between pilasters or other vertical supports.

Special concrete masonry units can be used when laying pilasters (*Figure 24*), or plain units similar to those used for the wall itself can be used. If hollow units are used, they can be grouted and may contain embedded reinforcements. Grouted pilasters have all vertical joints fully mortared to keep the grout from flowing into adjacent cells.

Pilasters appear rectangular or square and are classified according to their connection to the wall. Bonded pilasters are constructed as an integral part of the wall, while unbonded pilasters are used where control joints are needed for crack control. Lateral support is provided using suitable connections.

NOTE

Always check local building codes to determine the size and spacing of pilasters.

3.5.4 Installing Columns

Columns (*Figure 25*) are similar in design and function to piers but are generally taller and thinner. According to *ACI CODE-530*, a column's width is never greater than three times its thickness, and its height is always greater than four times its thickness. Columns, unlike piers, are always solid or reinforced.

Columns may be built inside a wall (isolated) or as freestanding structures. Structural engineers must consider the column's effective height, which is the clear height between supports. The effective height is determined by the risk of movement or twisting when the column is under load.

Codes require columns to be vertically and laterally reinforced. A minimum of four vertical bars is required for column reinforcement. When installing vertical ties at the base or below the lowest horizontal reinforcement in a beam, girder, slab, or drop panel, the spacing between ties is halved.

Lateral ties enclose the vertical reinforcements and provide support. They are placed to prevent the reinforcement from buckling when the column is under compression. Lateral ties may be placed in mortar or grout, although placement in grout is more effective.

Columns are sometimes built with cleanouts in the base, as was shown in *Figure 25*. When using full concrete block, install a cleanout in each cell of the base course. Use a vibrator when grouting a column to ensure that all air has been evacuated and that all cavities have been filled.

As their name suggests, wall columns have walls built between them but are not structural elements of the wall (*Figure 26*). When wall columns are used, they are constructed before the surrounding walls. They may incorporate a vertical steel pipe or I-beam at their core. If the local code provides seismic requirements for columns, be sure to follow those requirements during construction.

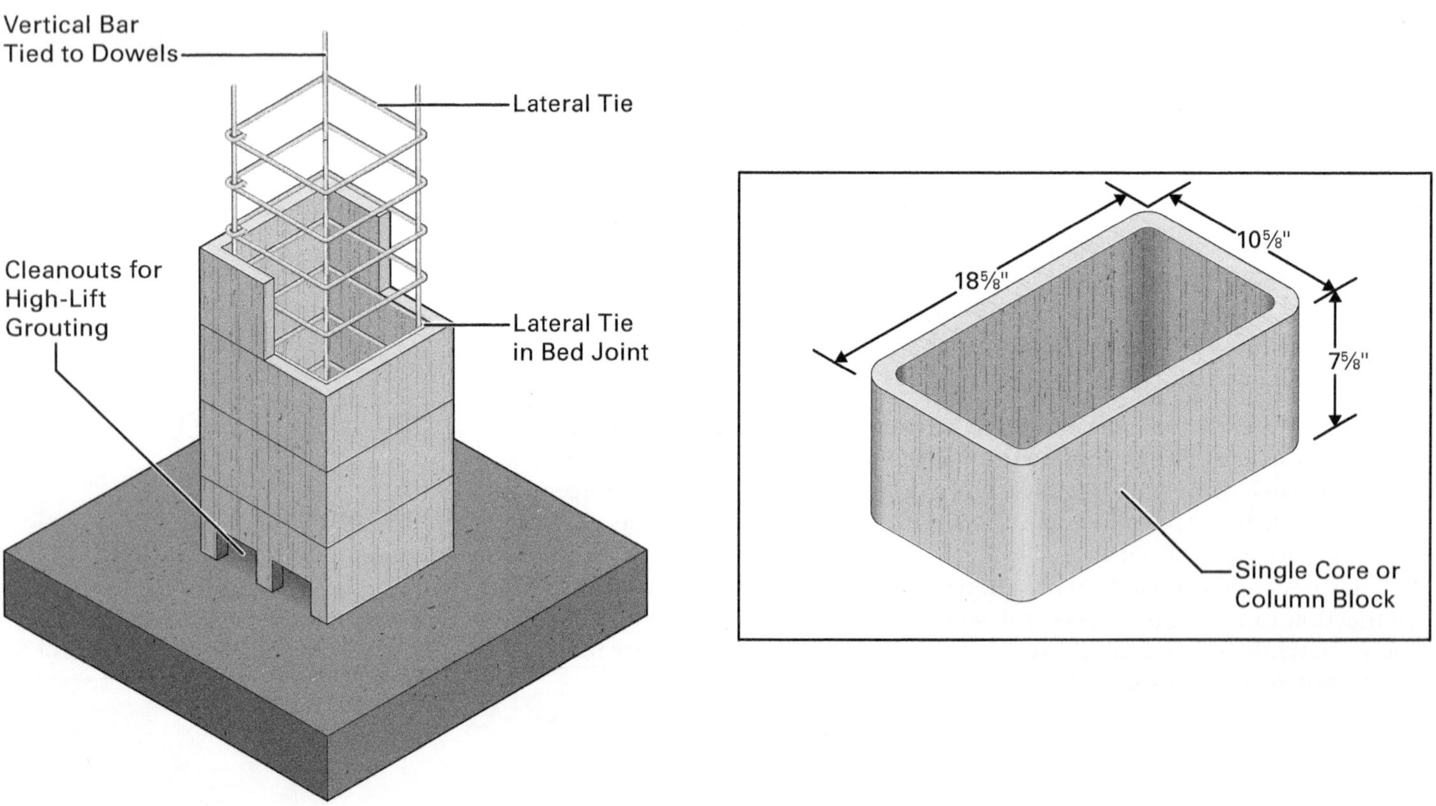

Figure 25 Masonry column.

Figure 26 Wall columns.
Source: Shutterstock

3.0.0 Section Review

1. What is the *most common* hand tool for bending rebar in the field?
   a. Hickey bar
   b. Pliers
   c. Mechanical vibrator
   d. Bolt benders

2. A bar spacer is used to align, support, and rigidly fasten rebar in place _____.
   a. once grout has been poured
   b. before grout is poured
   c. only when using low-lift grouting
   d. only when using high-lift grouting

3. When installing bond beams, prevent grout from leaking into cells that are not to be grouted by _____.
   a. installing a bearing plate over the cells with a thin layer of mortar
   b. using CMUs that do not have holes
   c. placing wire lath or metal screening over the cells
   d. using grout with coarse aggregate

4. How thick will a one-piece lintel typically be?
   a. Twice the nominal height of the masonry units used in the wall
   b. Thicker than the masonry wall
   c. Thinner than the masonry wall
   d. The same thickness as the masonry wall

5. What is a buttress?
   a. A pilaster that does not project from the wall
   b. A double-sided pilaster
   c. A portion of a double-sided pilaster that shows on the outside of a structure
   d. A portion of a double-sided pilaster that shows on the inside of a structure

Module 28203 Review Questions

1. A reinforced wall contains _____.
   a. only grout
   b. both grout and steel reinforcement
   c. voids filled with rubble
   d. only steel reinforcement

2. Masons should account for additional water when mixing grout in what type of weather?
   a. Cool and dry
   b. Hot and damp
   c. Cool and damp
   d. Hot and dry

3. The amount of grout placed in one uninterrupted operation is called a _____.
   a. layer
   b. level
   c. lift
   d. flight

4. Rodding techniques can be used to consolidate grout in what lift measurement?
   a. 36" or more
   b. 18" or more
   c. 36" or less
   d. 12" or less

5. How many lifts would it take to complete high-lift grouting for a cavity wall with a height of 8'?
   a. One
   b. Two
   c. Three
   d. Four

6. When grout mounds over a reinforcement element and creates a void, this is called _____.
   a. bridging
   b. a separation
   c. voiding
   d. a blowout

7. Mortar protruding into a cavity will prevent the grout from _____.
   a. curing properly
   b. consolidating
   c. forming a complete bond
   d. segregating

8. Turn off mechanical vibrators _____.
   a. after removing the vibrator from the grout
   b. as the vibrator is being removed from the grout
   c. as a surface skin begins to form over the grout
   d. when the grout level rises slightly

9. Reinforcing bar (rebar) is made from new or recycled _____.
   a. cast iron
   b. aluminum
   c. steel
   d. fiberglass

10. What is the diameter of size 6 rebar?
    a. ⅛"
    b. ⅜"
    c. ½"
    d. ¾"

11. What material is most commonly used to secure pieces of rebar together?
    a. Tie wire
    b. Rebar nylon
    c. Brass wire
    d. 550 cord

12. A horizontal load-bearing member placed over a masonry opening is called a _____.
    a. stile
    b. spreader beam
    c. strongback
    d. lintel

Answers to odd-numbered questions are found in the Review Question Answer Key in the back of the book.

Answers to Section Review Questions

Answer	Section	Objective
Section One		
1. b.	1.2.0	1a
2. b.	1.2.0	1b
Section Two		
1. a.	2.1.0	2a
2. a.	2.2.1	2b
3. a.	2.3.0	2c
4. c.	2.4.0	2d
Section Three		
1. a.	3.1.0	3a
2. b.	3.2.0	3b
3. c.	3.3.0	3c
4. d.	3.4.0	3c
5. c.	3.5.2	3c

MODULE 28204

Masonry Openings and Metalwork

Source: iStock/Getty Images

Objectives

Successful completion of this module prepares you to do the following:

1. Describe the methods and materials used to install masonry openings.
   a. Describe how to use and install doorframes and window frames.
   b. Describe how to use and install windowsills.
   c. Describe how to use and install steel lintels.
   d. Describe how to use and install chases and recesses.
2. Describe the methods and materials used to tie a single masonry wythe together.
   a. Describe how to use and install ladder and truss joint reinforcement.
3. Describe the methods and materials used to tie two masonry wythes together.
   a. Describe how to use and install flexible anchors.
   b. Describe how to use and install horizontal anchors.
4. Describe the methods and materials used to tie a masonry wythe to structural elements.
   a. Describe how to use and install rigid ties and bolts.
   b. Describe how to use and install bearing plates.

Performance Tasks

Under supervision, you should be able to do the following:

1. Assemble or install a hollow metal doorframe.
2. Describe lintel applications.
3. Explain the purpose and placement of a bearing plate.

Overview

Many masonry structures require the installation of reinforcements to increase the structure's strength and resistance to external forces. Reinforcements are installed around openings such as doorframes, windowsills, lintels, chases, and recesses because these are common structural weak points. They are also embedded into the mortar of masonry wythes to tie wythes together and connect wythes to structural elements. This module discusses standards for the various types of openings and reinforcements, and the steps and procedures for installation.

Digital Resources for Masonry

SCAN ME

Scan this code using the camera on your phone or mobile device to view the digital resources related to this craft.

NOTE

Codes vary among jurisdictions. Therefore, it is essential to consult the applicable code whenever regulations are in question. Referring to an incorrect set of codes can cause as much trouble as failing to reference codes altogether. Obtain, review, and familiarize yourself with your local adopted code.

1.0.0 Masonry Openings

Performance Tasks

1. Assemble or install a hollow metal doorframe.
2. Describe lintel applications.

Objective

Describe the methods and materials used to install masonry openings.

a. Describe how to use and install doorframes and window frames.
b. Describe how to use and install windowsills.
c. Describe how to use and install steel lintels.
d. Describe how to use and install chases and recesses.

Every structure requires openings for lighting, ventilation, and egress. With accurate planning and design, these various opening types shouldn't create any particular difficulties for the mason. However, masons must be fully aware of the exact positions of all openings and be ready to apply special techniques if they are called out in the approved construction documents. Improper layout and misreading of the plans could result in expensive and time-consuming errors that must be fixed to ensure structural integrity in these key areas.

Masons must accurately build masonry spans, such as placing **sills**, lintels, and **jambs** to fit windows and doors, ensuring that these components are strong enough to carry the load above them. In placing **chases** or recesses, masons must be aware of the effect that these openings have on the strength of the wall, as well as how they fit into a room's floor plan. Several specialized block types can accommodate many of these specific situations (*Figure 1*).

Sills: Horizontal members under doors or windows. Slip sills fit inside the doorframe or window frame; lug sills extend beyond the frame and into the masonry on the jamb sides of the frame.

Jambs: The side of an opening, or the vertical framing member on the side of the opening, usually for doorframes and window frames.

Chases: Continuous recesses built into walls to receive pipes, wires, or heating ducts.

The four parts of any window opening are the sill, side jambs, and header or lintel. For doors, sills may be set directly on the foundation or on a few courses of masonry.

The most efficient openings are sized to accommodate the dimensions of the chosen masonry material. Doing so reduces material waste and labor costs from cutting operations.

Window frames and doorframes are generally available in sizes that will fit in brick openings without the need for excessive cutting of brick. These standard sizes permit the brick coursing to fit evenly, with the opening height and width based on the modular dimensions of the brick.

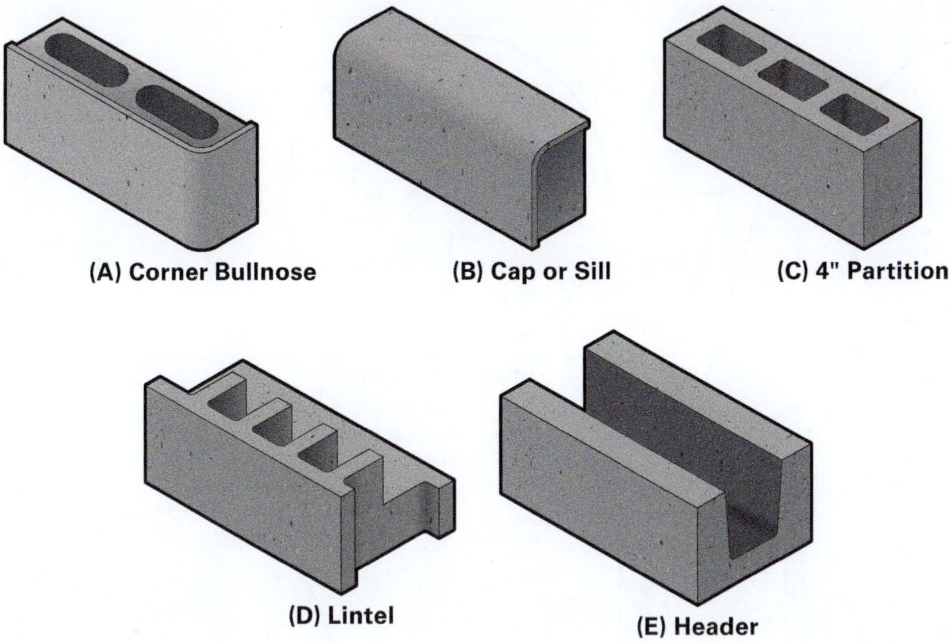

Figure 1 Specialized block.

If cutting is necessary, it is most efficient to precut all required units in batches of the same size. Adjusting the head joint thickness can compensate for small differences in cut dimensions.

Setting the windows is generally within the scope of a carpenter or window installer. Always check windows for level, plumb, and elevation before laying masonry units.

1.1.0 Installing Doorframes and Window Frames

Figure 2 shows a typical hollow metal doorframe with a welded and tab-locked corner. Metal frames must be set in place before the opening is laid up with a square-cut jamb block. Remember that all joints around the jamb must be filled with mortar to avoid air or moisture leaks. The cores closest to the jamb should be grouted to provide additional strength around the doorframes.

Different steel gauges are used for specific applications due to durability variances. Hollow metal steel frames come with at least three steel ties or fasteners for each jamb. Doorframes are sometimes equipped with floor knees for anchorage to the floor.

Doorframes are reinforced at all critical points, including the corners, hinges, strikes (the plate surrounding the frame recess that receives the door's latch or bolt), and closer locations.

Hollow metal aluminum frames are popular for both windows and doors (*Figure 3*). They come in a variety of finishes and are quite durable. All screws, nuts, washers, and other installation materials are often shipped with frames from the manufacturer, along with weatherproofing or soundproofing seals.

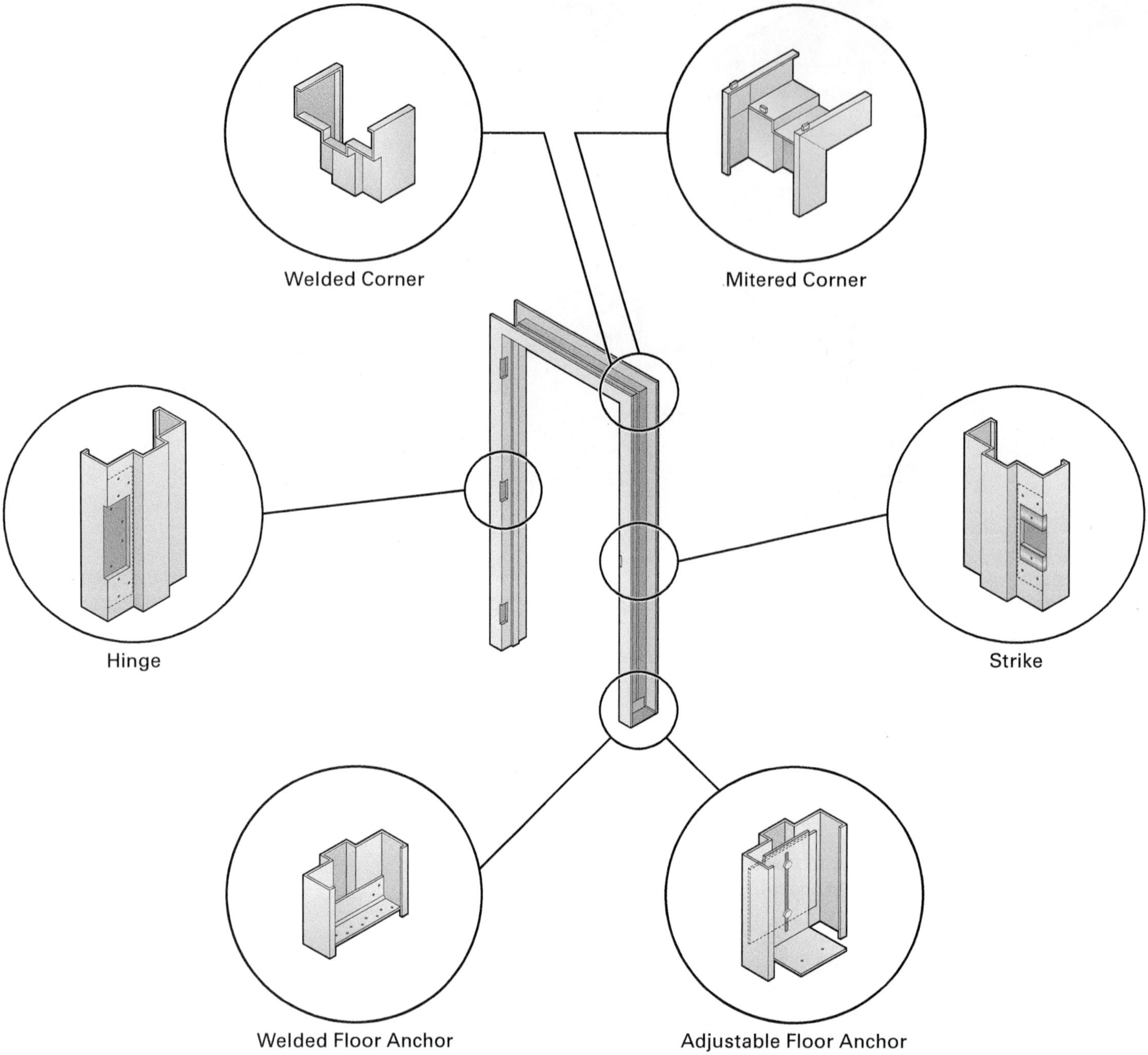

Figure 2 Standard hollow metal doorframe.

Both steel and aluminum frames can be ordered assembled (setup) or unassembled (knockdown). Unassembled frames offer savings in transportation and handling costs. These knockdown jambs and stretchers are typically wrapped during transport to prevent dents and scratches. The corners are mitered and reinforced with lock tabs for rigid joining. A factory-assembled frame is typically formed with welded corners that are ground smooth at the factory.

Figure 3 Aluminum windows on a commercial building.
Source: Shutterstock

Most manufacturers offer standard, special, and drywall options for hollow metal frames:

- A standard frame simply uses a standard design and dimensions. As such, it will fit most wall conditions. A standard frame may be ordered as either knockdown or setup.
- Special frames are made-to-order and usually delivered assembled. Unlike standard frames, they may have fasteners welded in place. Special frames often require special wall conditions, so check the drawings before installing them.
- Drywall frames are designed for installation after the drywall partition is in place. The drywall partitions can be placed, and even finished, before installing this type of frame. Drywall frames are always furnished knockdown and delivered with simple installation instructions.

The distance from the face of the frame to the face of the masonry wall is known as the **reveal**. Carpenters are generally responsible for setting the frame in a level and plumb position to the mason's jamb block. The mason should always check the frame to guarantee that it is level and plumb before finishing the wall.

Reveal: The side of an opening in a wall for a window or door. This is the part of the masonry jamb around a window or doorframe that can be seen from the frame to the face of the masonry wall.

Fasteners or ties provide design flexibility for standard frames. Ties for stud walls and for existing structures normally come welded in place on the jambs. Most fasteners are snap-in styles for masonry walls that can be ordered to accommodate a wide range of wall sizes. Doorframes may have adjustable floor ties that are set flush or that extend below the finish floor. *Figure 4* shows an adjustable frame tie that snaps onto the jamb.

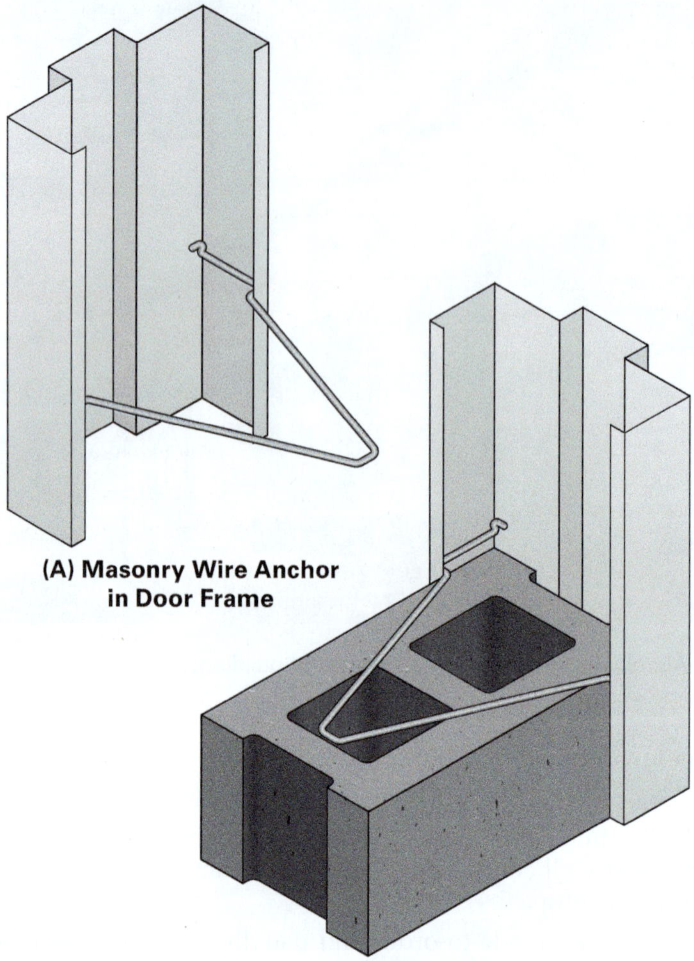

Figure 4 Adjustable frame tie.

1.1.1 Knockdown-Frame Installation

Knockdown frames come from the manufacturer unassembled. The components must be bolted or welded together before they are installed in the wall. The following steps are general guidelines for installing a knockdown frame:

Step 1 Check the frame to ensure that it is for the correct opening and that all parts are present.

Step 2 Push the top of the hinge jamb against the wall. Hold the top of the jamb in place, and push the bottom against the wall at the base.

Step 3 Position the frame head against the wall. Align the head tabs with the jamb slots, slide the head toward the jamb, and insert the tabs into the slots.

Step 4 Push the top of the strike jamb against the wall, aligning the jamb slots and head tabs as before. When the tabs engage, push the bottom of the frame against the wall at the base.

Step 5 Check the frame for level and reveal. Adjust as necessary.

Step 6 Adjust all jamb fasteners until they contact the supporting structure.

Step 7 Plumb the hinge jamb and secure it. Check the frame for level again. Plumb the strike jamb and secure it. Fasten at least the top, bottom, and middle of each jamb.

Step 8 Check the square, **skew**, and plumb of the frame before securing it permanently according to the manufacturer's recommendations.

Step 9 If the wall is not complete around the frame, continue laying masonry. Completely fill the void between the masonry and the frame with mortar or grout. If you are laying block, fill the end cells of all block against the jamb with grout.

Step 10 Clean any smeared mortar from the exposed frame.

Skew: The condition when two parts come together at an angle that is not 90° or perpendicular to each other.

1.1.2 Setup Doorframe Installation

The following steps apply to setup frames. This type of hollow metal frame is most commonly found on masonry jobs. *Figure 5* shows this type of frame set in concrete masonry units (CMUs), commonly called *block*.

Step 1 Ensure that the frame is free of manufacturing defects and damage, and that it matches the correct opening.

Step 2 Make sure the door will open on the correct side and in the correct direction.

Step 3 Mark the frame location, set the frame, and secure it into position with braces.

Step 4 Check the reveal to ensure that the door is set back the proper distance.

Step 5 Check the frame for plumb, square, and skew. Adjust as necessary.

Step 6 Brace the frame both vertically and horizontally. Brace it horizontally at the bottom and in the middle; this is depicted in Figure 7. Scrap dimension lumber can be used for braces.

Step 7 Begin laying the masonry units adjacent to the frame. Completely fill the void between the frame and the masonry unit with mortar or grout as the laying progresses. Ensure that both jambs are filled with mortar equally as courses are laid to keep frames from shifting. If laying block, fill the end cells of all block against the frame jamb with grout.

Step 8 Install at least three fasteners at each jamb as the work progresses. Follow the manufacturer's recommendations on the types and locations of the fasteners. At a minimum, fasten the top, bottom, and middle of each jamb close to the hinges.

Constantly check the frame location and plumb as the wall is built.

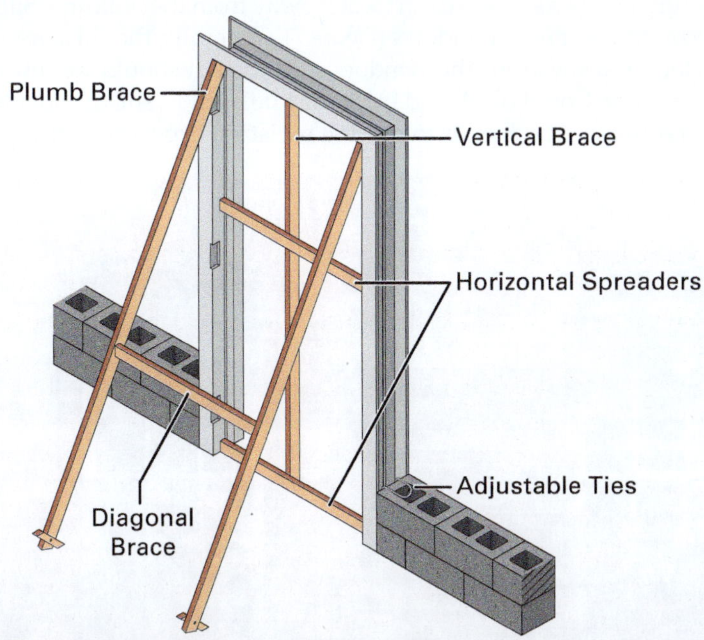

Figure 5 Brace-assembled doorframe.

1.1.3 Bracing Jigs

Setting assembled doorframes and window frames usually requires lumber to be cut for bracing. Several manufacturers offer ready-made proprietary bracing jigs for this work. Some consist of telescoping poles with universal clamps and adjustment tensioners on each end (*Figure 6*).

These jigs are designed so that one end clamps to the frame and the other end clamps to other fixed structural members already in place.

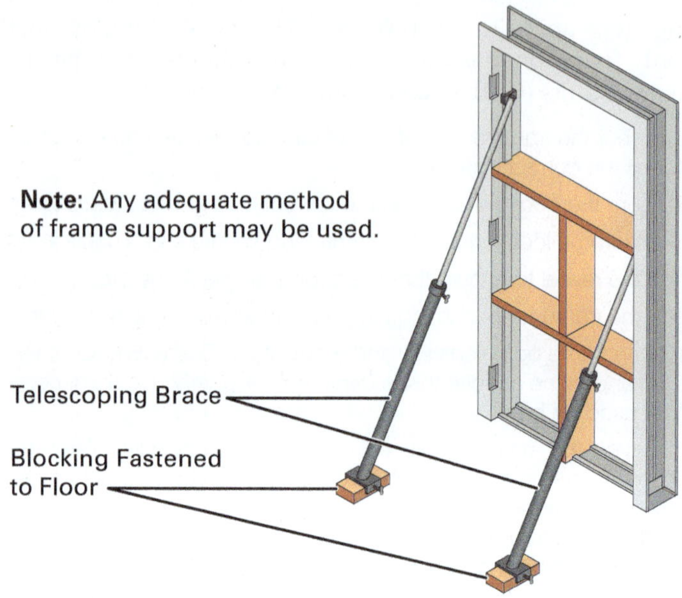

Figure 6 Telescoping brace.

1.2.0 Installing Sills

The primary function of a sill is to divert water away from the building. Sills can also provide a decorative accent to windows (*Figure 7*). Generally, the sill slope should be at least 15°, sloping away from the window, and the sill should extend at least 1" beyond the wall face. Cored block and brick may be used in sills, although uncored units may be required at both ends to prevent moisture exposure and dirt buildup.

Figure 7 Decorative sills.
Source: Shutterstock

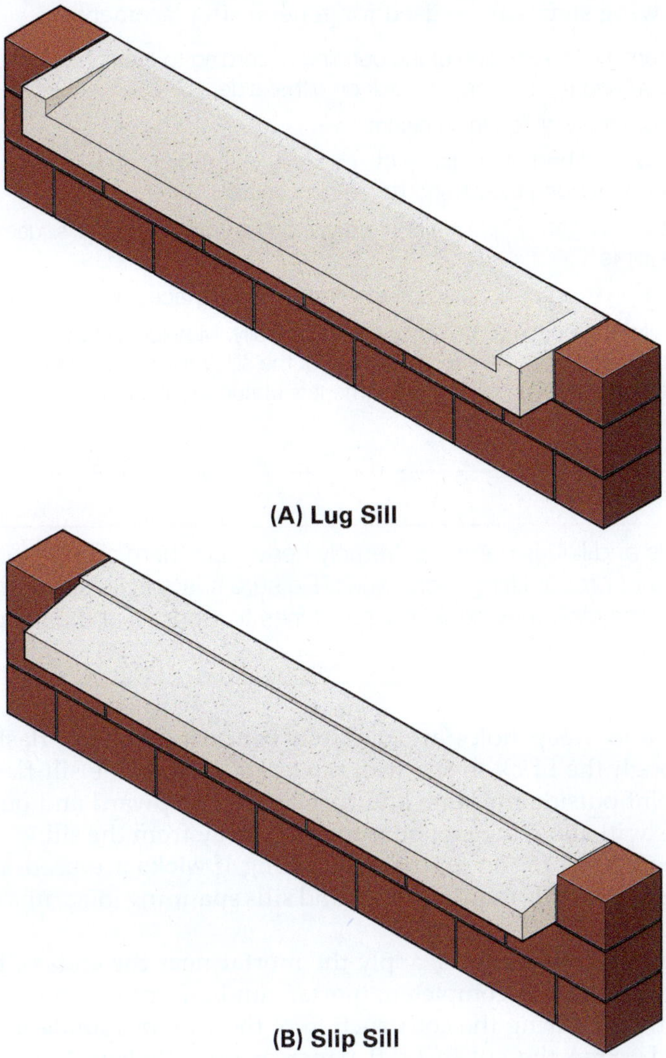

Figure 8 Lug and slip sills.

Sills can be built in place or prefabricated. Common materials include CMUs, brick, stone, wood, metal, and concrete. Typically, single-unit sills are classified as either slip sills or lug sills, as shown in *Figure 8*. Slip sills are the same length as the window or door opening, while lug sills extend into the masonry on either side of the opening. The slip sill may be left out when the opening is laid, and set later. The lug sill, however, must be set when the masonry reaches the bottom of the opening.

Sills support window frames and doorframes so that they will not sag, crack, or tilt. They also prevent moisture penetration and add a finished appearance to the opening. One-piece sills are designed with a slight downward slope from the edges to the center of the sill for water runoff.

Sills cover the full widths of openings and are supported by either the block, slab, or foundation directly below the opening. The use of precast sills, which can be placed in openings after the wall or foundation has been laid up or cast in place, is a common practice.

Sills may have slight variations that affect their method of installation. Masons should always check the plans and the sill construction shop drawings to determine the proper installation method before beginning.

The following steps can be used for general sill placement:

Step 1 Determine the position of the opening according to the number of courses under the sill and the number of block on either side.

Step 2 Lay up the wall to the sill height.

Step 3 Apply a full bed of mortar to the masonry unit directly under the sill. The width of all joints around the sill should remain constant.

Step 4 Set the sill gently in place and press it down until the mortar squeezes between the joints.

Step 5 Remove excess mortar and smooth the joint surfaces on both sides of the wall.

Step 6 As always, check all measurements carefully. Measure joint thickness and check the sill level along the horizontal line of the sill. If the sill is not level, press down at either end. If this slight adjustment is unsuccessful, remove the sill, apply fresh mortar, and reset.

NOTE

The term *drip* refers to a cut or casting on a masonry unit designed to let water drip without running down the wall. The term *drip edge* refers to the bent edge of a piece of flashing that serves the same function.

WARNING!

Concrete sills and lintels can be extremely heavy. Lift them properly, using two people and proper lifting techniques. Be sure sills and lintels are properly supported. If they fall, they could cause injuries to workers and damage to materials.

Wicks: Bundles of loosely twisted or braided fibers woven together to form cords that can carry water away from an area by capillary action, which occurs as long as the drip end is lower than the absorption end.

Flashing and weep holes are required beneath sills. The flashing should extend through the brick to the wall's exterior. Extend the sill flashing to the first head joint outside the jamb and turn the ends upward and outward for at least 1". This will allow the moisture to drain away from the sill.

Place weep holes 16" on top of the flashing if **wicks** are used and 24" if no wicks are used. All sills require a drip, and sills spanning long runs may require support anchors.

When laying block lug sills, apply the mortar near the ends of the sill only. After the wall has been completed, mortar can be forced under the sill. When sills are installed during the construction of the wall or foundation, they must be protected. Wrap the sill in kraft paper or polyethylene to prevent mortar from falling on the sill.

After the block sill has been properly laid, stretch a taut line and lay the block course by course, including the jamb block. Keep all block level, plumb, and aligned. Fill all joints around the jamb to prevent moisture penetration.

Brick for sills should be laid on edge, rowlock style, and sloped at a recommended 15° or $^3/_4$" to shed water and produce a finished look (*Figure 9*).

Brick windows begin a full number of courses above the beginning of the wall. Using the 4" modular system makes the coursing easier, but a common mistake is forgetting that three courses equal 8". At sill height, lay the sloped rowlock course to occupy the height of two normal courses.

When the mortar is firmly set, place the window frame and brace it to remain level, plumb, and square. Bracing remains in place until the wall is built over the top of the frame. The difference can be adjusted with mortar if the top course is not within $^1/_4$" of the top framing piece.

On brick veneer walls, the brick is butted flush against the window jambs, and the frame is nailed into the wall. Regardless of how frames are fastened to the masonry wall, remember that the frame should be set the proper distance back from the face of the wall. The position depends upon the thickness of the wall.

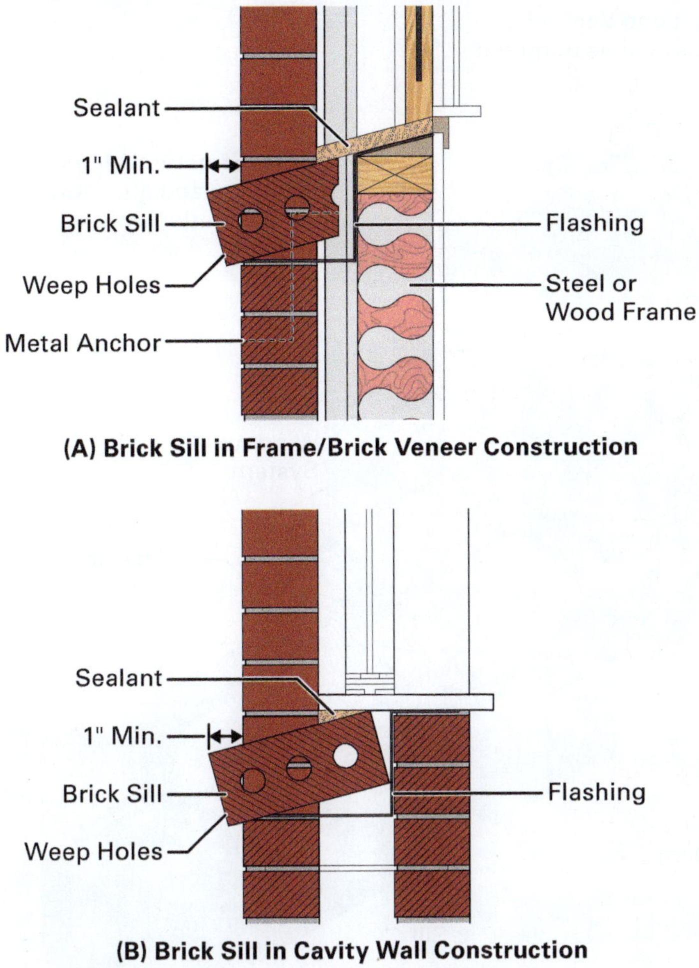

Figure 9 Brick windowsills.

1.2.1 One-Piece Slip and Lug Sills

For block construction, sills are usually one piece of concrete or cut stone. They support window frames and doorframes so that they won't sag, crack, or tilt. They also prevent the entry of water, wind, and dirt.

Sills cover the full widths of openings, supported either by the masonry under the opening or by the slab or foundation directly beneath. The usual practice is to place precast slip sills in an opening after the wall or foundation has been laid up. *Figure 10* shows a detailed cross section of sill placement.

Slip sills are the same width as the opening, while lug sills extend into the masonry on either side of the opening. The slip sill can be omitted when the opening is laid and set later. The lug sill, however, must be set when the masonry is up to the bottom of the opening.

Checking Plumb and Level on Doorframes

Take care to ensure that a metal frame is absolutely plumb and square. An improper installation will affect the door's performance and make fitting the door difficult or impossible.

Use a level to plumb a door jamb. One way to check squareness is by measuring diagonally from top to bottom at the corners. Equal measurements indicate that the frame is square.

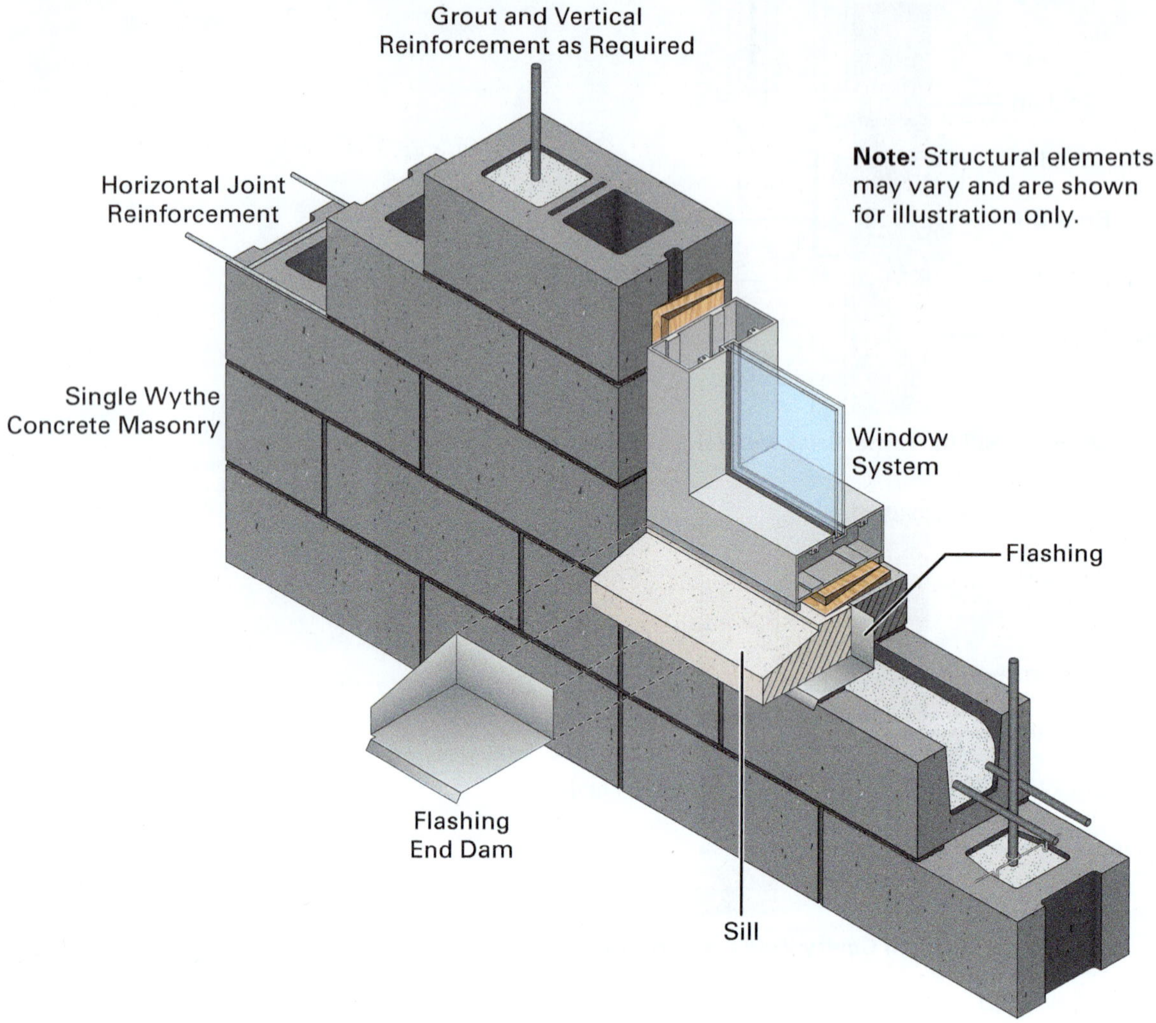

Figure 10 Typical detail for a slip sill.

1.2.2 Laying One-Piece Sills

Sills of different types or from different manufacturers may need different installation methods. Always check the plans and manufacturer's documents before installing a sill. The following is a general outline for placing a one-piece sill:

Step 1 Use the plans to determine the position of the opening according to the number of courses under the sill and the number of masonry units on either side of it.

Step 2 Lay up the wall to the height of the masonry under the sill.

Step 3 Apply a full bed of mortar to the masonry directly under the sill. The width of all joints around the sill should remain constant.

Step 4 Set the sill gently in place and press it down until the mortar squeezes from the joints.

Step 5 Remove excess mortar and smooth the joint surfaces on both sides of the wall.

Step 6 Measure joint thickness and check for level along the horizontal line of the sill. If the sill is not level, press down at either end. If this slight adjustment cannot level the sill, remove the sill, apply fresh mortar, and reset.

When lug sills are laid, mortar should be applied only near the ends of the sill. Mortar can be forced under the sill after the wall is completed.

1.2.3 Brick Sills

Brick for sills are usually laid on edge, rowlock style, and sloped about 15° or ¾" to shed water. The slope is maintained by thickening the mortar bed, and the laying angle is maintained by a line.

Figure 11 shows this type of sill, which is flush with the edges of the window opening. Note that raising the edge of the rowlock produces a slight lip. The window frame can be set flush against this lip to hold it tightly in place and minimize dirt collection.

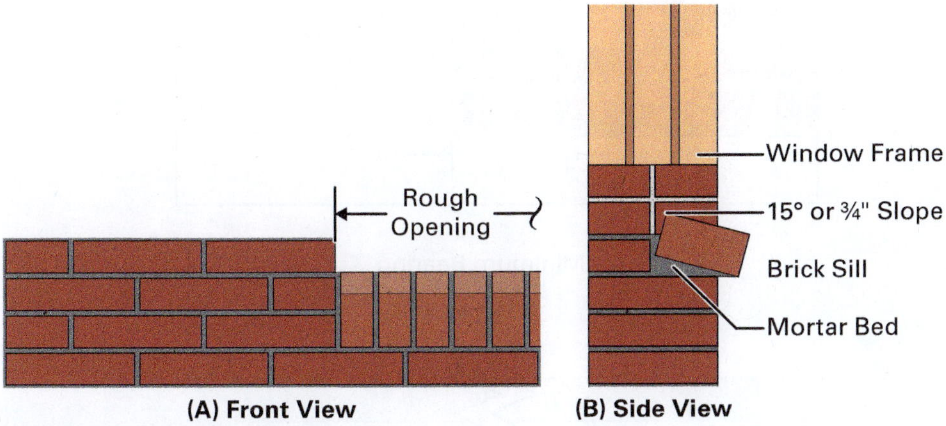

Figure 11 Rowlock brick windowsill.

Most brick sills are slip sills that do not extend past the window frame. Sometimes specifications call for lug sills that must be set once the masonry reaches the bottom of the opening.

1.2.4 Lintels

When the masonry courses have been completed to the height of the jamb, the lintel can be set. This is the final step in completing the construction of any opening. A lintel supports windows, doors, or other openings, ensuring that no loads are placed underneath the frame. The building design and the supported load determine the proper lintel type.

Steel, precast concrete, lintel block, reinforced brick, or metal straps can be used as lintel material. Be sure the masonry underneath that supports the lintel has been set up before placing the lintel.

The ends of the lintel transfer the load to the underlying masonry. To distribute the weight evenly, these ends should extend into the masonry on either side a minimum of 8" for block and 4" for brick. *Figure 12* shows a properly installed brick lintel. Insets vary according to project specifications.

Slight movement often occurs at the location of lintels. For this reason, control joints are often located at the ends of the lintel to allow it to move slightly in response to temperature changes without cracking the underlying joint.

A noncorrosive metal plate is set under the supported end of the lintel at the side of the opening. A full mortar bed is added to the plate and then raked out to a depth of ¾" when it is still soft enough to tool. The opening in the mortar joint is later filled with caulking. The lintel can then move on the plate without causing movement in the masonry.

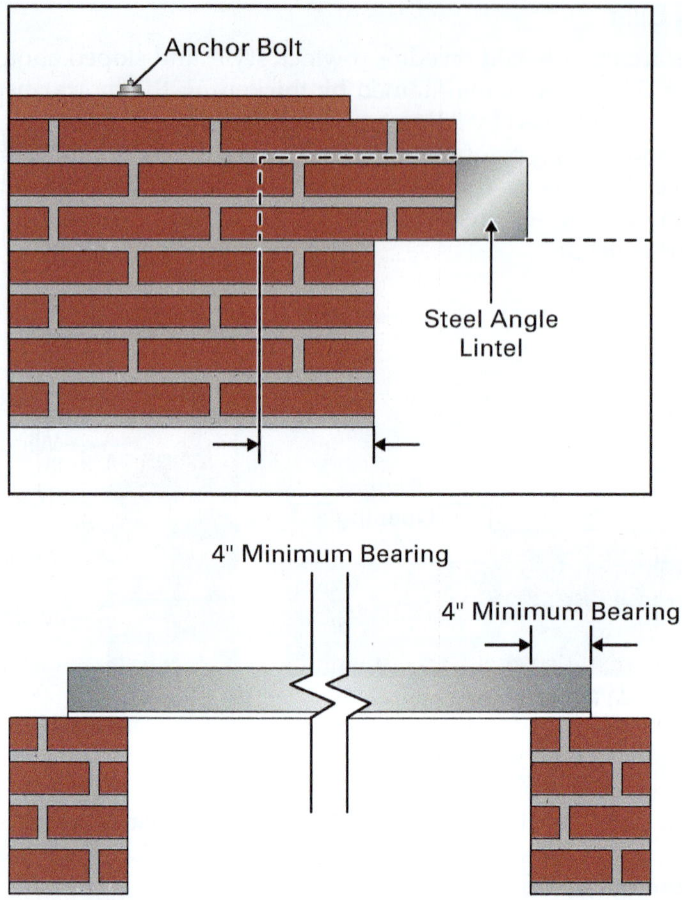

Figure 12 Minimum bearing is required for each end of the brick lintel.

1.3.0 Installing Steel Lintels

Masons use two types of steel lintels: angle irons and I-beams. Angle irons, shaped like the capital letter *L* when viewed in cross section, are one of the most common types of steel lintels (*Figure 13*).

I-beam lintels, resembling the capital letter *I*, support longer spans or concentrated loads (*Figure 14*).

Corrugated ties are welded to the faces of I-beams to attach masonry veneers or face shells that completely encase the beam. A continuous plate is welded to the bottom of the beam to support the weight of the masonry. These plates are then connected to supporting masonry jambs with steel bearing plates, anchor bolts, or rebar.

To install I-beams, follow these steps:

Step 1 Install the required connector. Ensure that the connectors are centered on the wall, set to the proper elevation, and placed the proper distance apart.

Step 2 Place the I-beam lintel, ensuring that the welded or bolted connections are made as specified. It is common for the steel erection contractor to perform this step. Place at least one temporary vertical support midspan along the lintel.

Step 3 Set the first course of masonry over the steel lintel. Units in this course should be buttered carefully, set on the support, and slid against adjacent units to avoid trapping mortar between the units and the lintel. It is common for the first few courses over an I-beam lintel to be reinforced bond beams to provide additional strength.

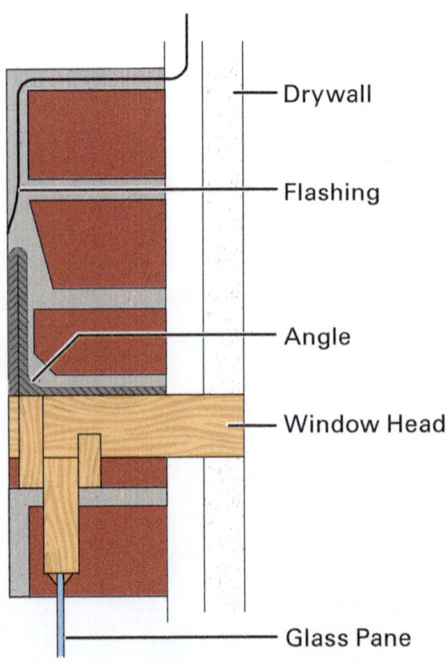

Figure 13 Angle iron used as a lintel.

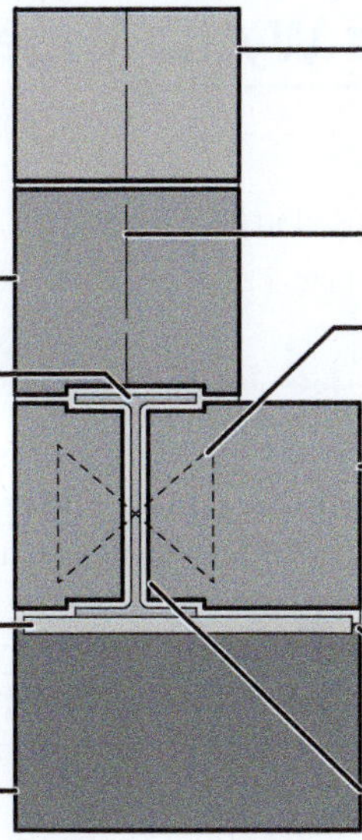

Figure 14 I-beam used as a lintel.

Steel lintels installed over large openings or carrying heavier-than-normal loads should have a larger load-bearing surface at each end and be thicker than those used over narrower openings or to carry normal loads. Larger spans require the lintel to extend even deeper into the wall. The lintel should be strong enough to support the loads and resist excessive bending.

Chases and Recesses

Chases and recesses are horizontal or vertical spaces in a wall that contain plumbing, heating ducts, electrical wiring, or other equipment. Often, specifications require these mechanical items to be placed behind the finished wall surface. Plans that call for chases or recesses should be formed as the wall is built. Otherwise, the other trades must chisel them out after the block is in place, which is expensive and time-consuming.

1.0.0 Section Review

1. Which of the following is *not* a masonry opening element?
   a. Reveal
   b. Sill
   c. Jamb
   d. Lintel

2. The primary function of a sill is to _____.
   a. support the weight of the load over the masonry opening
   b. divert water away from the building
   c. serve as a step or platform
   d. be a decorative element

3. The minimum thickness of a steel lintel should be _____.
   a. $\frac{1}{8}$"
   b. $\frac{1}{4}$"
   c. $\frac{1}{3}$"
   d. $\frac{1}{2}$"

2.0.0 Tying a Single Wythe Together

Performance Tasks

There are no Performance Tasks in this section.

Objective

Describe the methods and materials used to tie a single masonry wythe together.

a. Describe how to use and install ladder and truss joint reinforcement.

Moisture and temperature changes can cause masonry materials to expand and contract, which can cause them to crack. Horizontal joint reinforcement can maintain the structural integrity of a single-wythe masonry structure.

Horizontal joint reinforcement contributes to structural integrity by evenly distributing lateral loads and tying the masonry elements together mechanically. In addition to tying the wythe together, horizontal joint reinforcement reinforces masonry structures in the following ways:

- Bonding intersecting walls
- Limiting movement between different materials in a wall
- Giving horizontal structural reinforcement

Horizontal joint reinforcement is usually made of galvanized or stainless steel wire, and like vertical rebar, it comes in different gauges.

2.1.0 Installing Horizontal Joint Reinforcement

Continuous horizontal joint reinforcement for single-wythe walls is manufactured in ladder (rectangular) and truss (triangular) styles (*Figure 15*). Ladder is more common as it doesn't interfere with reinforcement or grout application. It is made of galvanized steel to prevent corrosion and is available in different wire sizes.

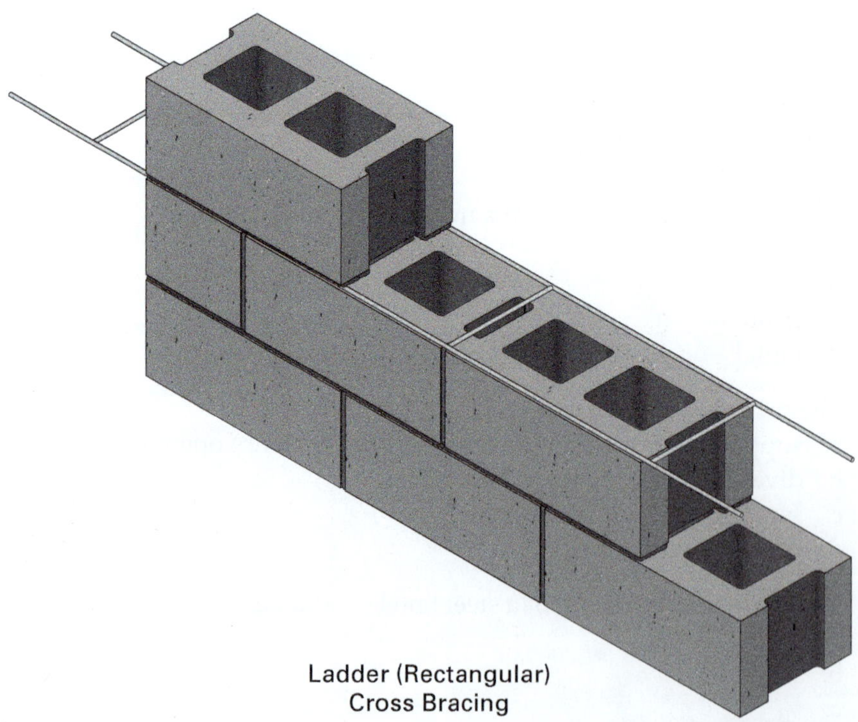

Figure 15 Continuous joint reinforcement for tying single-wythe walls together.
Source: Hohmann & Barnard

Standard widths are approximately $1\frac{5}{8}$" less than the wall thickness. This sizing ensures good mortar coverage on both faces of the wall. Cross wires are welded diagonally or perpendicular to the longitudinal wires.

Common joint reinforcement vertical spacing is every 8", 16", or 24". The distance between wall ends or control joints is known as a **panel**.

Panel: A section of wall between control joints, wall ends, or a control joint and wall end.

Joint reinforcement should be placed in the first and second bed joints immediately above and below wall openings. It should extend at least 24" or to the end of the panel. In addition, reinforcement should be placed two or three courses immediately below the top of a wall. Continuous joint reinforcement should be cut at control joints and expansion joints.

Continuous joint reinforcement can be shaped, cut, and bent to fit corners. It must be overlapped at least 6" to provide proper load transfer. When cutting reinforcement, remember to leave enough space for a good mortar joint to bury the ends. Prefabricated corners and tees are available for virtually any wall configuration.

WARNING!

Continuous joint reinforcement can cause a safety hazard. If steel wire protrudes past the end of the course, it can injure anyone passing by. Avoid accidents by bending down any surplus continuous reinforcement, and bend the wire up carefully when preparing to lay the intersecting block.

2.0.0 Section Review

1. The typical spacing for cross wires in continuous joint reinforcement is _____.
   a. 12"
   b. 14"
   c. 16"
   d. 18"

3.0.0 Tying Two Masonry Wythes Together

Objective

Describe the methods and materials used to tie two masonry wythes together.

a. Describe how to use and install flexible anchors.

b. Describe how to use and install horizontal anchors.

Performance Tasks

There are no Performance Tasks in this section.

Anchors and fasteners connect masonry veneer to an interior masonry wythe within a cavity wall. Anchors may be continuous (also called *three-wire*) joint reinforcements or individual anchors attached to the backing wall. Continuous joint reinforcement is used for multi-wythe walls of like-masonry materials.

Tab joint reinforcement is used for tying block and brick wythes together and is particularly well suited for transferring lateral and vertical loads.

3.1.0 Installing Flexible Anchors

Some special types of anchors are adjustable to allow some movement to reduce cracks. A common type of flexible anchor is the veneer anchor (*Figure 16*).

Figure 16 Veneer anchor.
Source: Hohmann & Barnard

Seismic anchors (*Figure 17*) are a special type of flexible anchor. They serve the same purpose as seismic reinforcements, but instead of being connected to each other as part of a continuous joint reinforcement system, they are connected individually to the structural backing.

Figure 17 Seismic anchors.
Source: Hohmann & Barnard

3.2.0 Installing Horizontal Anchors

Horizontal anchors, also called *unit ties*, do not transfer as much vertical load as continuous ties but are useful for connecting composite wythes in a solid wall. Rigid unit ties are available in box, Z-type (*Figure 18*), and corrugated versions. Brick manufacturers often recommend specific types of wall ties for use with their products.

Some rectangular or Z-type ties may have crimps or dips in them that are designed to allow condensation inside a masonry cavity to collect and drip down. However, because crimps can weaken the tie's load transfer strength, some building codes do not permit their use. Newer rectangular or Z-type ties have plastic drip rings on the wire instead of a crimp.

Shaped metal ties and hardware cloth are also used for joint reinforcement or for tying materials together. *Figure 19*A shows some common tie shapes. Some have dovetail anchors, which fit into special slots. On some jobs, the shaped metal bars may be called *anchors*. *Figure 19*B shows a piece of wire mesh wall tie. The rectangular mesh strips are used alone or with a metal tie bar to tie non-load-bearing walls to intersecting walls.

Because they allow some limited movement, adjustable anchors are often installed as individual unit ties. *Figure 20* shows some adjustable unit ties for attaching masonry to a backup concrete, wood, metal stud, or block structure. These units consist of two parts, with movement allowed between them.

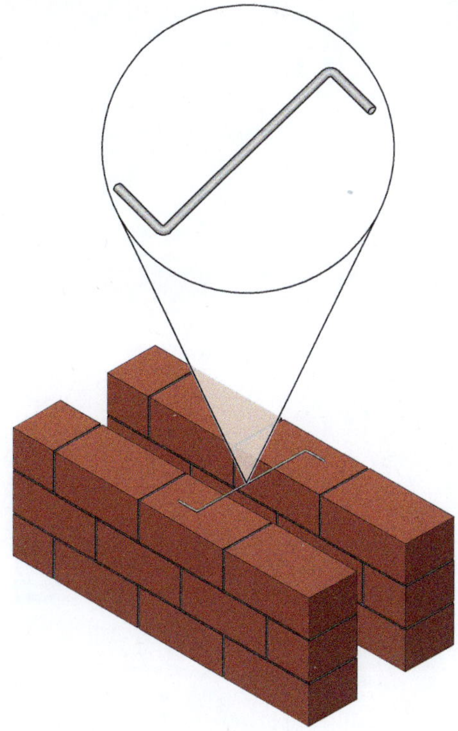

Figure 18 Z-type unit tie.

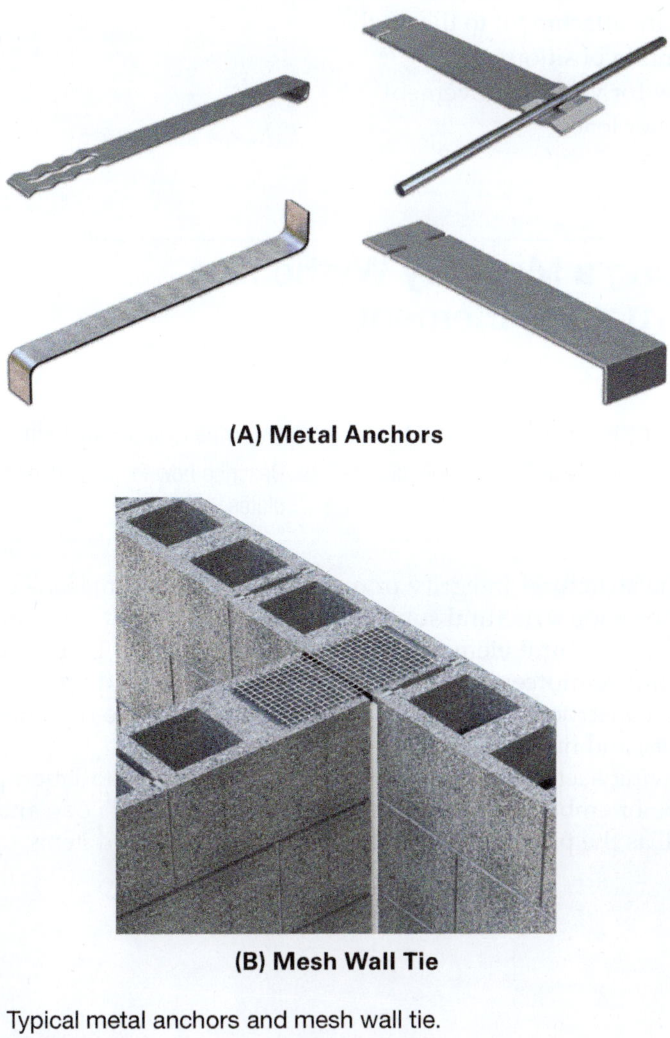

(A) Metal Anchors

(B) Mesh Wall Tie

Figure 19 Typical metal anchors and mesh wall tie.
Source: Hohmann & Barnard

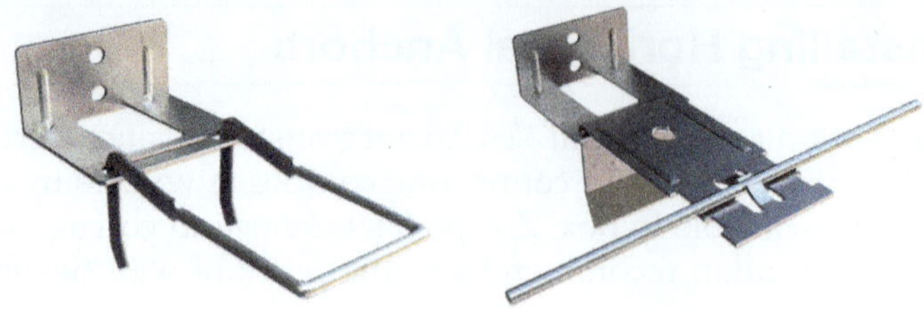

Figure 20 Adjustable unit ties.
Source: Hohmann & Barnard

The plate is attached to the backup structure and the wire part is mortared to the masonry unit. Adjustable assemblies are used to tie masonry face wythes to a masonry backup. The two parts of these assemblies allow contraction and expansion between the two wythes. Both of these adjustable horizontal ties maintain position but do not transfer load.

3.0.0 Section Review

1. A common type of flexible anchor is the _____.
   a. corrugated anchor
   b. veneer anchor
   c. three-wire truss
   d. tab tie

2. Individual adjustable horizontal ties do not _____.
   a. require attachment to the wythe
   b. maintain position
   c. allow for vertical movement
   d. transfer load

4.0.0 Tying a Masonry Wythe to a Structural Element

Performance Task

3. Explain the purpose and placement of a bearing plate.

Objective

Describe the methods and materials used to tie a masonry wythe to structural elements.

a. Describe how to use and install rigid ties and bolts.
b. Describe how to use and install bearing plates.

To ensure the structural integrity of masonry structures, masons often install devices that provide structural support to masonry elements by tying masonry wythes to the structural elements of the wall or building. Like other forms of horizontal joint reinforcement, these devices are embedded into the mortar or grout during construction. Commonly used devices include rigid ties and bolts, bearing plates, and hurricane ties.

The following sections discuss the applications and installation procedures for each type of embedded item. Always refer to the local code and specifications, as well as the project plans when installing embedded items.

4.1.0 Installing Rigid Ties and Bolts

Rigid ties include galvanized metal strips and galvanized bolts. These masonry wall ties resist compression and shear stresses, while flexible ties do not. For example, the strip tie is commonly used for attaching load-bearing partition walls.

Tie-bolt devices can be simple threaded rods, as shown in *Figure 21*, or complex sleeve-and-tie devices with various head types. These heavy-gauge connectors usually fasten a beam, ledge, sill plate, or other structural member to a masonry wall. Different ties may be set in mortar or epoxy, while others are power driven to meet engineering specifications.

(A) "L" Bolt

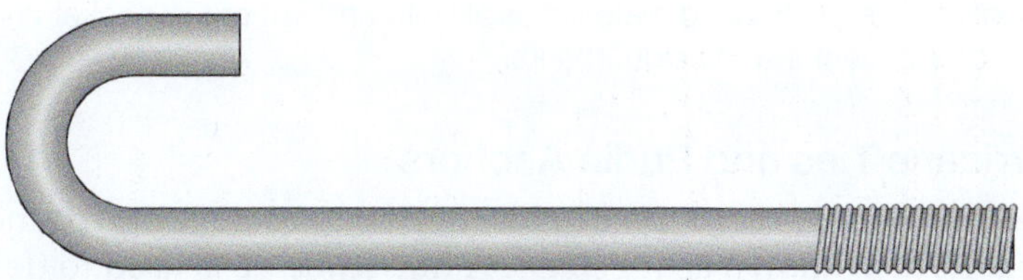

(B) "J" Bolt

Figure 21 Simple rod tie bolts.

When the plans call for bolts to anchor other structure components, they must be mortared in place when tightened to avoid loosening later. They are spaced according to the plans or drawings, and are generally set in the center of a wall unless otherwise specified.

Study the drawings before starting the wall, because the brick may have to be notched so that the angled ends of the bolts fit snugly. Use a line to ensure that all the bolts are set the same distance from the face of the wall, and check the drawings for the height at which the bolts must protrude above or beyond the wall.

Once the locations are set, form a base under each bolt with mortar. If the bolt protrudes into a core, use masonry scraps as a base. Fill the mortar around the bolt tightly, then use a trowel or a stick to pack the mortar.

Depending on the location, make sure the bolt shank is plumb or level, and keep the threaded part of the bolt free of mortar. Lastly, clean the threads so that the nut will easily thread onto the bolt. Installing bolts for steel beams or plates requires precise spacing, as the steel is predrilled and there is no room for adjustments.

4.2.0 Installing Bearing Plates

Another type of embedded item is a bearing plate, which serves as a connection point between masonry walls and steel structural members, such as beams or joists. The bearing plate is welded or bolted to the structural member and embedded into the masonry using welded studs, anchor bolts, welded rebar, or a combination.

Project specifications may call for bearing plates under the ends of lintels, arches, beams, or other concentrated vertical loads. Bearing plates are also used on top of columns, pilasters, or other load concentrators.

To install a bearing plate, follow these steps:

Step 1 Refer to the design drawings to determine the lateral location and the top of the plate's elevation. Mark the plate's center on the wall.

Step 2 Include any required bond-beam units and reinforcement. Determine if the vertical reinforcement will conflict with the plate, and cut as needed to avoid conflict.

Step 3 Plug the sides of the **beam pocket** to contain the grout. This may require the use of small, cut pieces of masonry and mortar to separate the pocket from the remainder of the wall.

Step 4 Apply double-bond-beam mesh to ensure that the grout does not hang into the pocket and interfere with installation.

Step 5 Confirm that the bearing plate fits easily into and out of the pocket and matches the proper center and height marks.

Beam pocket: An opening in a vertical masonry wall that allows an intersecting structural beam to bear on or pass through the wall.

4.2.1 Hurricane Ties and Purlin Anchors

Masons use two kinds of straps in masonry construction: hurricane ties and purlin anchors, also known as *PA straps*. A hurricane tie is used to tie the frame to the foundation. It provides vertical reinforcement and reduces the effects of lift in high winds.

PA straps provide horizontal reinforcement for masonry structures. They are used to fasten or connect wooden framing structures to masonry support walls. These straps are flat galvanized metal straps with bends designed to engage masonry construction on one end and with holes designed to allow for bolted and/or nailed connections to structural elements on the other end.

Anchoring Systems

Masonry anchoring systems are designed to retrofit veneer anchors for concrete, block, and wooden backing. They work with mechanical expansion, screws, or epoxy adhesive. These anchoring systems are used either in repairs where ties were not originally installed, to replace a failed system, or in new construction.

4.0.0 Section Review

1. Load-bearing partition walls are often attached using _____.
   a. strip ties
   b. seismic ties
   c. bearing plates
   d. corrugated ties

2. To distribute the vertical load over an arch, the project specifications may call for the installation of _____.
   a. a seismic anchor
   b. bearing plates
   c. continuous joint reinforcement
   d. a saddle

Module 28204 Review Questions

1. Before laying masonry units, the locations of wall openings should be marked on the _____.
   a. sill
   b. floor plan
   c. footing
   d. door schedule

2. The horizontal top element of a window is the _____.
   a. lintel
   b. jamb
   c. sill
   d. muntin

3. To accommodate openings in a brick wall, the bond must be worked out by _____.
   a. the mason
   b. the architect
   c. an engineer
   d. the window installer

4. On each jamb, hollow steel doorframes are usually provided with a minimum of three _____.
   a. floor knees
   b. tab locks
   c. steel ties
   d. bearing plates

5. Most fasteners used for standard frames in masonry walls are _____.
   a. welded in place on jambs
   b. installed after the drywall partition
   c. setups
   d. snap-ins

6. A setup doorframe installation should be braced horizontally _____.
   a. at the bottom and in the middle
   b. in the middle only
   c. at the bottom only
   d. at the top and at the bottom

7. For proper drainage, what is the typical brick sill slope?
   a. 5°
   b. 7.5°
   c. 15°
   d. 33°

8. If wicks are not used with sill flashing, a spacing of 24" should be used for _____.
   a. rigid ties
   b. weep holes
   c. bearing plates
   d. perimeter drains

9. For windowsills in brick buildings, the preferred choice of material is _____.
   a. stone
   b. metal
   c. precast concrete
   d. brick

10. Control joints are often installed at each end of a lintel to _____.
    a. create a stronger bond
    b. allow slight movement
    c. provide tensile strength
    d. level the lintel

11. The width of an angle-iron lintel that is used to support 4" masonry units should not be less than _____.
    a. 3½"
    b. 4"
    c. 4½"
    d. 5"

12. One problem with using a truss-type joint reinforcement is that it _____.
    a. is difficult to cut to size
    b. is more expensive than ladder-type joint reinforcement
    c. interferes with easy placement of vertical rebar
    d. is hard to keep flat

13. Compared to the thickness of the wall in which it is installed, the standard width of continuous horizontal joint reinforcement is approximately _____.
    a. 1⅝" greater
    b. 1⅝" less
    c. 1¾" greater
    d. 1¾" less

14. Ties must be bedded in mortar at least ⅝" from the exposed face of a wall to prevent _____.
    a. corrosion
    b. cracking
    c. condensation
    d. lateral movement

15. Tab-type reinforcements transfer _____.
    a. only vertical loads
    b. both lateral and vertical loads
    c. only lateral loads
    d. only lateral and diagonal loads

16. Flexible anchors connect materials with different rates of heat expansion to prevent _____.
    a. veneer separation
    b. windstorm damage
    c. formation of cracks
    d. lateral movement

17. Corrugated, Z-type, and box ties are examples of _____.
    a. continuous ties
    b. horizontal anchors
    c. flexible ties
    d. seismic reinforcements

18. Slots made in concrete receive the dovetail anchors of _____.
    a. shaped metal ties
    b. tab-type reinforcements
    c. J-bolts
    d. slip lugs

19. Bearing plates are installed to handle concentrations of _____.
    a. lateral pressure
    b. vertical load
    c. horizontal load
    d. shear stress

Answers to odd-numbered questions are found in the Review Question Answer Key in the back of the book.

Answers to Section Review Questions

Answer	Section	Objective
Section One		
1. a.	1.0.0	1
2. b.	1.2.0	1a
3. b.	1.3.0	1c
Section Two		
1. c.	2.1.0	2a
Section Three		
1. b.	3.1.0	3a
2. d.	3.2.0	3b
Section Four		
1. a.	4.1.0	4a
2. b.	4.2.0	4b

Advanced Laying Techniques

Source: Shutterstock

Objectives

Successful completion of this module prepares you to do the following:

1. Identify the structural principles and fundamental uses of basic wall types.
   a. Describe the common characteristics of solid masonry walls.
   b. Identify the structural principles and fundamental uses of hollow masonry walls.
   c. Explain the common characteristics of cavity walls.
   d. Identify the common characteristics of composite walls.
   e. Discuss the fundamental uses of anchored veneer walls.
   f. Describe the primary function of retaining walls.
   g. Explain the common characteristics of freestanding walls.
2. Identify the requirement for and function of control joints and expansion joints.
   a. Identify the effects of temperature and moisture on control joints and expansion joints.
   b. Identify the uses of control joints.
   c. Identify the uses of expansion joints.
3. Lay out and construct various corners and intersections.
   a. Lay out and construct toothing.
   b. Lay out and construct corbeling.
   c. Lay out and construct intersecting walls.
   d. Lay out and construct angled corners.

Performance Tasks

Under supervision, you should be able to do the following:

1. Lay out and construct a composite wall with control joints and expansion joints.
2. Lay out and construct intersections.
3. Lay out and construct angled corners.

Overview

After mastering the basic techniques of laying concrete masonry units and brick in courses and wythes, masons need to develop the knowledge and skills required to build more complex structures. This module explains the construction of walls, control and expansion joints, and corners and intersections.

Typically, the architect or engineer will determine specific construction procedures based on various factors. These include wind loading, lateral force, soil condition, and the structure's function. However, on smaller projects such as retaining walls, masons may be required to recommend the best approach to the client or owner. Masons must be prepared to determine safe and proper practices and know where to look for essential design and structural details.

Digital Resources for Masonry

SCAN ME

Scan this code using the camera on your phone or mobile device to view the digital resources related to this craft.

NOTE

Codes vary among jurisdictions. Therefore, it is essential to consult the applicable code whenever regulations are in question. Referring to an incorrect set of codes can cause as much trouble as failing to reference codes altogether. Obtain, review, and familiarize yourself with your local adopted code.

1.0.0 Structural Principles and Uses of Walls

Performance Tasks

There are no Performance Tasks in this section.

Objective

Identify the structural principles and fundamental uses of basic wall types.

a. Describe the common characteristics of solid masonry walls.
b. Identify the structural principles and fundamental uses of hollow masonry walls.
c. Explain the common characteristics of cavity walls.
d. Identify the common characteristics of composite walls.
e. Discuss the fundamental uses of anchored veneer walls.
f. Describe the primary function of retaining walls.
g. Explain the common characteristics of freestanding walls.

Masonry walls offer sustainability, durability, ease of maintenance, design flexibility, and an attractive appearance at a competitive price point. Masonry walls range from simple, single-wythe walls with no reinforcement to highly reinforced multi-wythe walls. They can be plain foundation walls or highly ornate and decorative walls (*Figure 1*). These various wall types are classified into seven basic categories:

- Solid wall
- Hollow wall
- Cavity wall
- Composite wall
- Anchored veneer wall
- Retaining wall
- Freestanding wall

Retaining and freestanding walls are two common types that are built to provide something other than an enclosure. Although freestanding walls are not subjected to the same pressures as retaining walls, reinforcement and bonding are important for both wall types. Proper foundation construction is also an important part of both types of structures.

Figure 1 Ornate decorative brick wall.
Source: Alamy Stock Photo

1.1.0 Common Characteristics of Solid Masonry Walls

Solid masonry walls are commonly used for either load-bearing or non-load-bearing construction. These walls are typically built of solid masonry units laid up with full mortar joints. Masonry units are considered solid if they have a void area of less than 25%.

A single-wythe wall is the most straightforward design. These can be built using brick or uncored concrete masonry units (CMUs), commonly known as *block*. The single-wythe brick wall shown in *Figure 2* can be used for low-perimeter walls, partition walls, or other non-load-bearing applications.

Figure 2 Single-wythe solid brick wall.

A single-wythe solid block wall, shown in *Figure 3*, can be either load-bearing or non-load-bearing. It can structurally support floor or roof loads. When used as a non-load-bearing partition wall, it offers excellent fire resistance and sound insulation.

Multi-wythe solid walls of brick or block, such as those shown in *Figure 4*, are used in load-bearing construction, or when the design requires a finished wall surface on both sides.

NOTE

Always check to make sure which building codes are used in your area. Codes are updated periodically; make sure you are using the latest version.

The two wythes of masonry are bonded together, and headers, horizontal anchors, or continuous joint reinforcement secure the wythes to each other. The wall system is masonry bonded, and the units overlap in alternate courses. It is common to fill the collar joint with mortar to completely bond the two wythes.

In 8" walls, header courses can be used for the full width of the wall section. However, masonry headers are often replaced by metal horizontal anchors or continuous joint reinforcement in modern techniques. They are grouted solidly in the bed joints to tie multiple wythes together. Today, solid masonry walls do not typically exceed 12" in thickness, except under special conditions.

Figure 3 Single-wythe solid block wall.

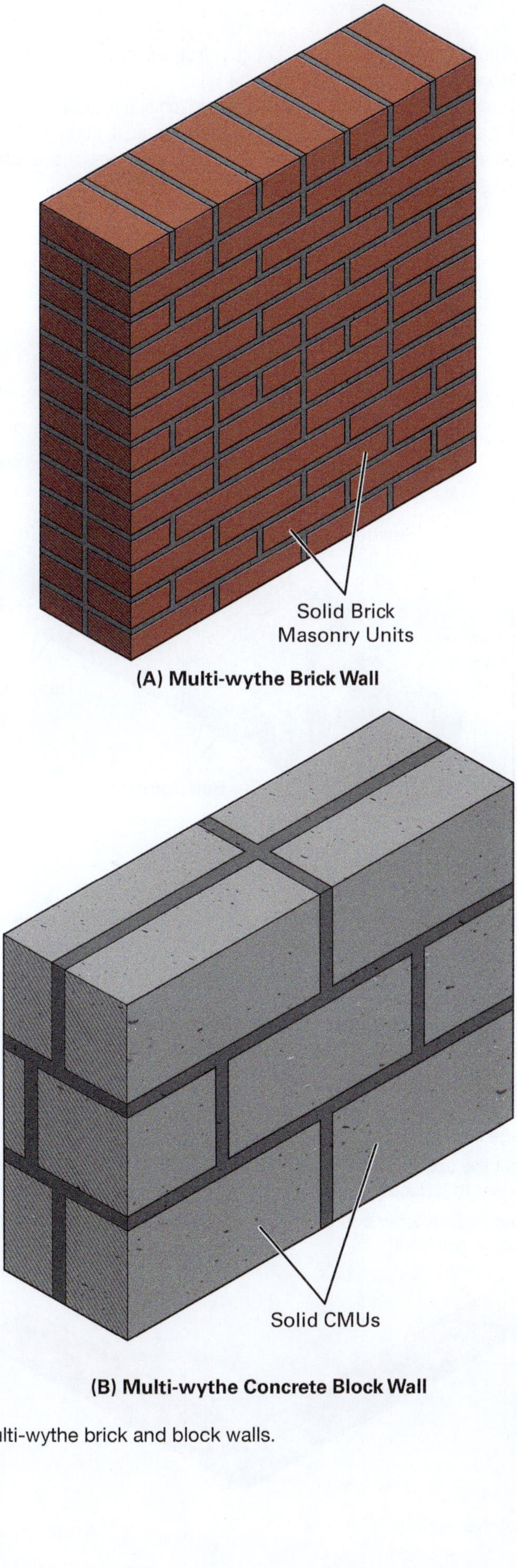

Figure 4 Multi-wythe brick and block walls.

Terminology

Masons need to understand terms for basic laying techniques related to bonding directions, the position of the masonry unit, and the type of mortar joint.

Stretcher, *header*, and *rowlock* refer to the basic position of a masonry unit within a wall. The stretcher unit is in the normal position, with its side exposed in the wall's face. The header unit is laid flat on its widest side; its end is exposed in the wall's face. The rowlock is laid on the side with the end exposed in the wall's face.

Collar, *head*, and *bed joints* refer to the position or type of mortar joint. Head and bed joints refer to the vertical and horizontal mortar joints that are seen in the exposed face of a wall. A collar joint is the mortar joint between the wythes of masonry in a multi-wythe wall system.

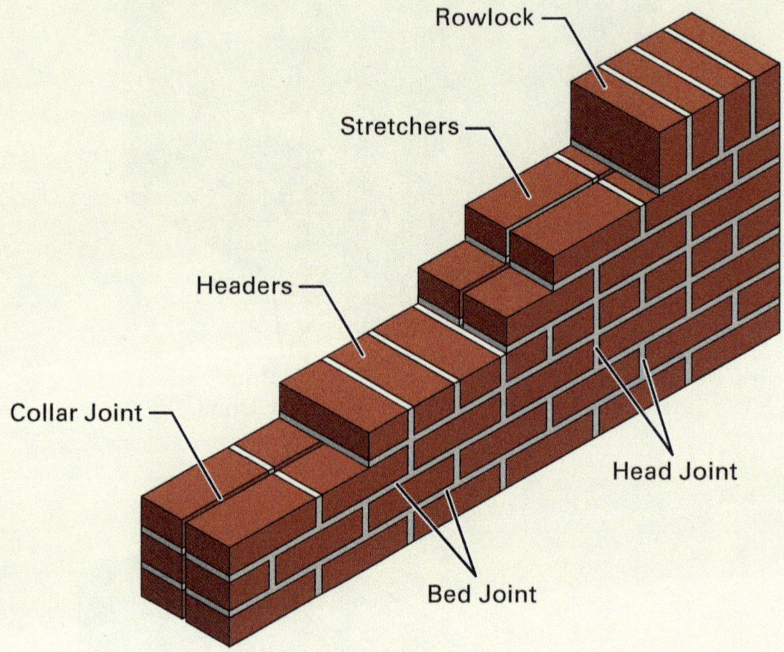

A typical solid masonry wall uses an 8" common bond. Headers are placed every seven courses for a double-wythe wall. The heading course serves as a bond between the inside and outside 4" tiers. This design can be laid up quickly and produces an effective bond.

The recommended procedure for constructing an 8" common-bond wall with leads is as follows:

Step 1 Lay out the wall location as with any single-wythe wall. In addition, locate a second line inside the first wall line. The distance between lines is equal to the average brick length used.

Step 2 Lay out the bond to eliminate excess cutting for any windows or doors. This is important to reduce waste and maintain patterns.

Step 3 Lay the first stretcher course, as shown in *Figure 5*. Four or five stretchers should be sufficient.

Figure 5 Stretcher course.

Figure 6 Header course.

Step 4 Lay the header course, as shown in *Figure 6*. The header course generally begins at the second course if not specified.

Step 5 Lay successive courses of stretchers until the next specified header course is reached. Level, plumb, and range each course as it is laid.

Step 6 Lay the header course, **breaking the bond** as usual.

Step 7 Finish laying the first corner, and complete the second corner similarly.

Step 8 Build the first stretcher course of the wall by laying the outside first, and then the inside. These should be laid from the leads to the center. Level, plumb, and range each course before beginning the header course when working with a partner, or from lead to lead when working alone. A line should be used on all backing courses to ensure a good face.

Step 9 Lay the header course from each lead toward the center. Level, plumb, and range the header course as you did in Step 8.

Step 10 Lay the outside wythe up to the next header course, keeping it level, plumb, and ranged.

Step 11 Lay the inside tier up to the same height. Be careful to keep it level with the outside tier as it is laid up. This is important to ensure a level surface for the header course.

Step 12 Continue the work until the desired height is reached. Tool the joints and brush the wall.

Breaking the bond: Starting a course with a cut so as to center the header unit over the head joints in the course below.

1.2.0 Structural Principles and Fundamental Uses of Hollow Masonry Walls

Hollow masonry walls are built with hollow masonry units. These include brick, block, or structural tile. Typically, all units are laid with full mortar joints.

Hollow masonry walls may be either load-bearing or non-load-bearing and single wythe or multi-wythe. Construction of hollow masonry walls is like construction of solid masonry walls.

Bonding requirements will vary depending on the local conditions and building codes. Pattern bond is used on some types of multi-wythe walls. Alternatively horizontal anchors (*Figure 7*) and continuous joint reinforcement (*Figure 8*) are used to bond wythes. These are made from corrosion-resistant rods or wire that is embedded in the horizontal mortar joints and across all wythes.

Anchor type and spacing are determined by the design team and local code requirements. Horizontal anchors are usually placed in alternate courses and staggered so that no two anchors form a continuous vertical line.

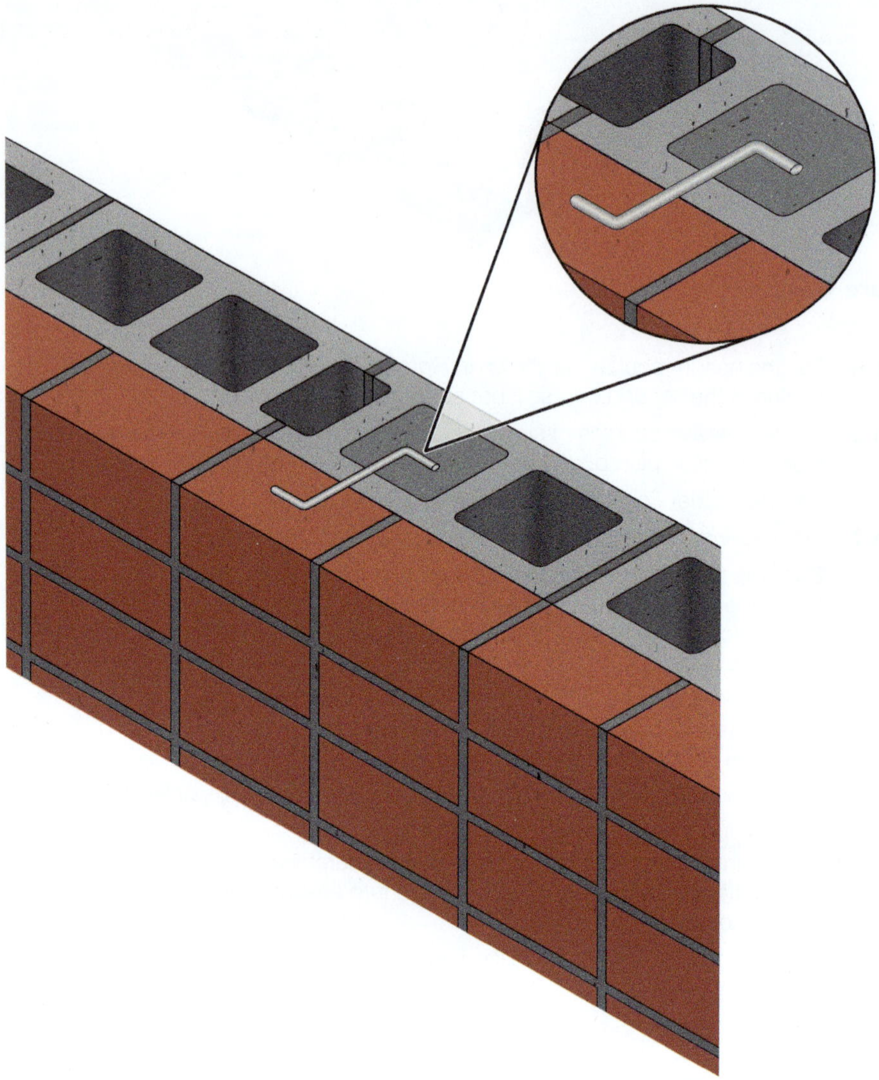

Figure 7 Example of a horizontal anchor.

Figure 8 Examples of truss-type continuous joint reinforcement.
Source: Hohmann & Barnard

Figure 9 shows adjustable horizontal anchors for veneer walls. Using adjustable anchors speeds up the construction of multi-wythe walls because the mason can lay each wythe independently instead of laying up both wythes at the same time.

Continuous joint reinforcement serves three functions:

- It acts as horizontal reinforcement for single-wythe masonry walls, usually with vertical spacing of no more than 16".
- It provides reinforcement to control cracking.
- It bonds multi-wythe walls without using individual metal anchors.

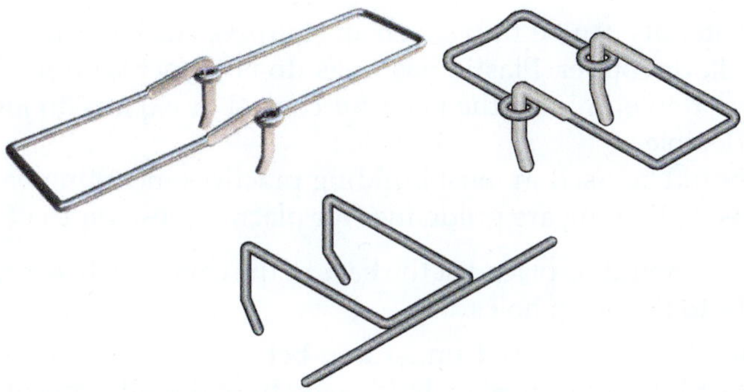

Figure 9 Adjustable horizontal anchors.
Source: Hohmann & Barnard

1.3.0 Common Characteristics of Cavity Walls

Cavity walls are built in two wythes that are completely separated by an unbroken air space. The inner and outer walls are tied together by rigid metal anchors or continuous joint reinforcement embedded in the mortar joints. The face wythe and the backup wythe may be of similar materials.

The thickness of the face wythe is generally 4" (nominal dimension), while the backup wythe can be thicker, depending on the wall's structural requirements. The general rule is that the thickness of the interior wythe should be equal to or greater than the thickness of the outer wythe.

In most cases, the inner wythe carries the structure's vertical load, while the outer wythe acts as a weather barrier and visual design feature.

The cavity wall is designed to offer three major advantages over solid or hollow masonry wall systems:

- It increases the insulating value of the wall.
- When properly constructed, it prohibits the passage of water or moisture across the wall, especially if proper flashing and weep holes are used. Any moisture that comes through the first wythe of a cavity wall flows to the bottom and out through the weep holes. Flashing is installed at the base of the cavity wall and any other necessary locations in the wall as required by the design to direct water to the weep holes.
- It prevents condensation on interior surfaces. If the cavity is left open, a vapor barrier or damp proofing is required to prevent interior condensation.

In the past, masonry wythes of cavity walls were held together with masonry headers or bonders. However, the advantage of a cavity wall system is to have an air space that is completely free of obstructions, for insulation and drainage purposes. To accomplish this, corrosion-resistant horizontal anchors or continuous joint reinforcement is used to bond the wythes of masonry.

1.3.1 Mortar and Mortar Joints

The bond formed between the mortar, the metal anchors or joint reinforcement, and the masonry units is an important single property in cavity wall construction.

Because the cavity wall is separated by an unbroken air space, the mortar joints must be completely filled if the wythes are to act as a unified wall system.

Make certain that the cavity remains as free of mortar droppings as possible. If mortar droppings remain in the cavity, they may create dams that clog the weep holes and prevent moisture from escaping. Excess mortar in the cavity can also create bridges between wall wythes where condensation can cross.

1.3.2 Flashing

Any moisture that penetrates the outer wythe of a cavity wall should be free to flow to the bottom of the cavity, where it is channeled by flashing materials to the weep holes.

Flashing materials can be thin sheets of stainless steel, thermal plastic, rubberized asphalt, or copper. Plastic flashings do not react with mortar ingredients, and they often eliminate the need for control or expansion joints because the plastic is flexible.

Flashing should be used in most building practices including base, sill, roof, and head areas. Following are guidelines for placing these types of flashing:

- Base flashing should be placed at the base of the cavity to direct any moisture in the cavity to the weep holes.
- Sill flashing should be placed under and behind all sills. These sills should extend at least 1" from the wall to drain water from the sill, prevent water stains on the masonry units, and provide a drip rail for the underside of the sill.
- Roof flashing should be placed where roof lines connect to masonry walls. This consists of a base flashing placed under the roofing and turned up against the masonry units, and counterflashing overlapped against the base flashing and extended securely into the mortar joints.
- Head flashing should be placed over all openings except those completely protected by an overhang.

Examples of these types of flashing are shown in *Figure 10*.

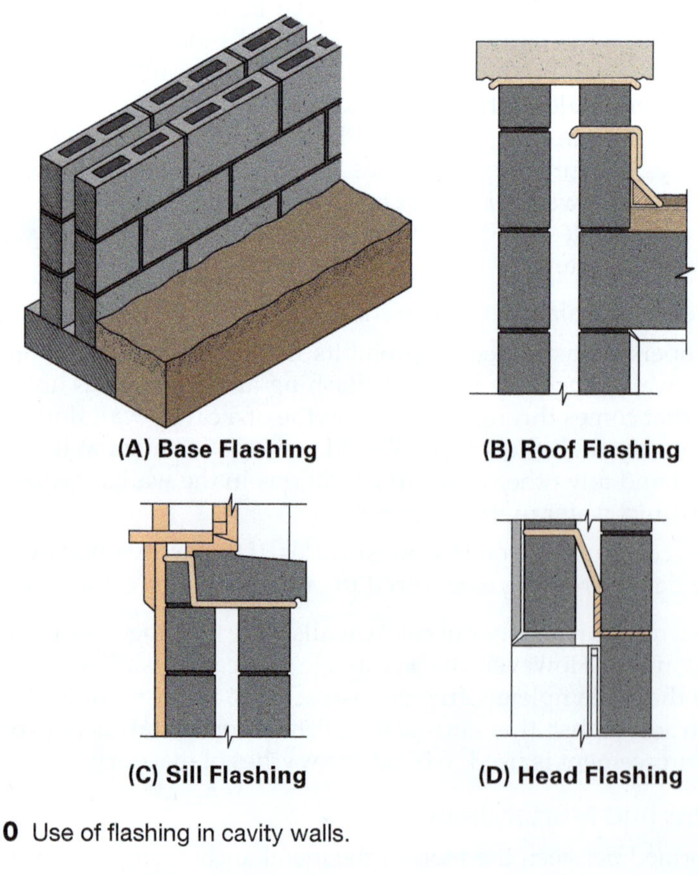

Figure 10 Use of flashing in cavity walls.

Through-wall flashing should be placed on a thin bed of sealant, then covered with a thin mortar bed to create the proper bond for the next masonry course. The masonry on which the flashing is bedded should be smooth enough that the flashing is not punctured. Any joints in the flashing must be completely sealed.

Remember that the effect of flashing on wall strength depends on the mortar placement and the mortar–flashing–masonry bond. Flashing does not affect a wall's compressive strength unless the flashing is placed directly in contact with the masonry units.

1.3.3 Weep Holes and Cell Vents

Flashing concealed in tooled mortar joints is not self-draining. Weep holes or cell vents must be provided to allow the moisture to escape through the outer wythe, as shown in *Figure 11*. Without these holes, water can be trapped within the wall cavity, causing moisture problems that may weaken the structure.

Weep holes should be provided in head joints immediately above flashing sheets. Do not place weep holes where they will be covered with soil after the construction is complete.

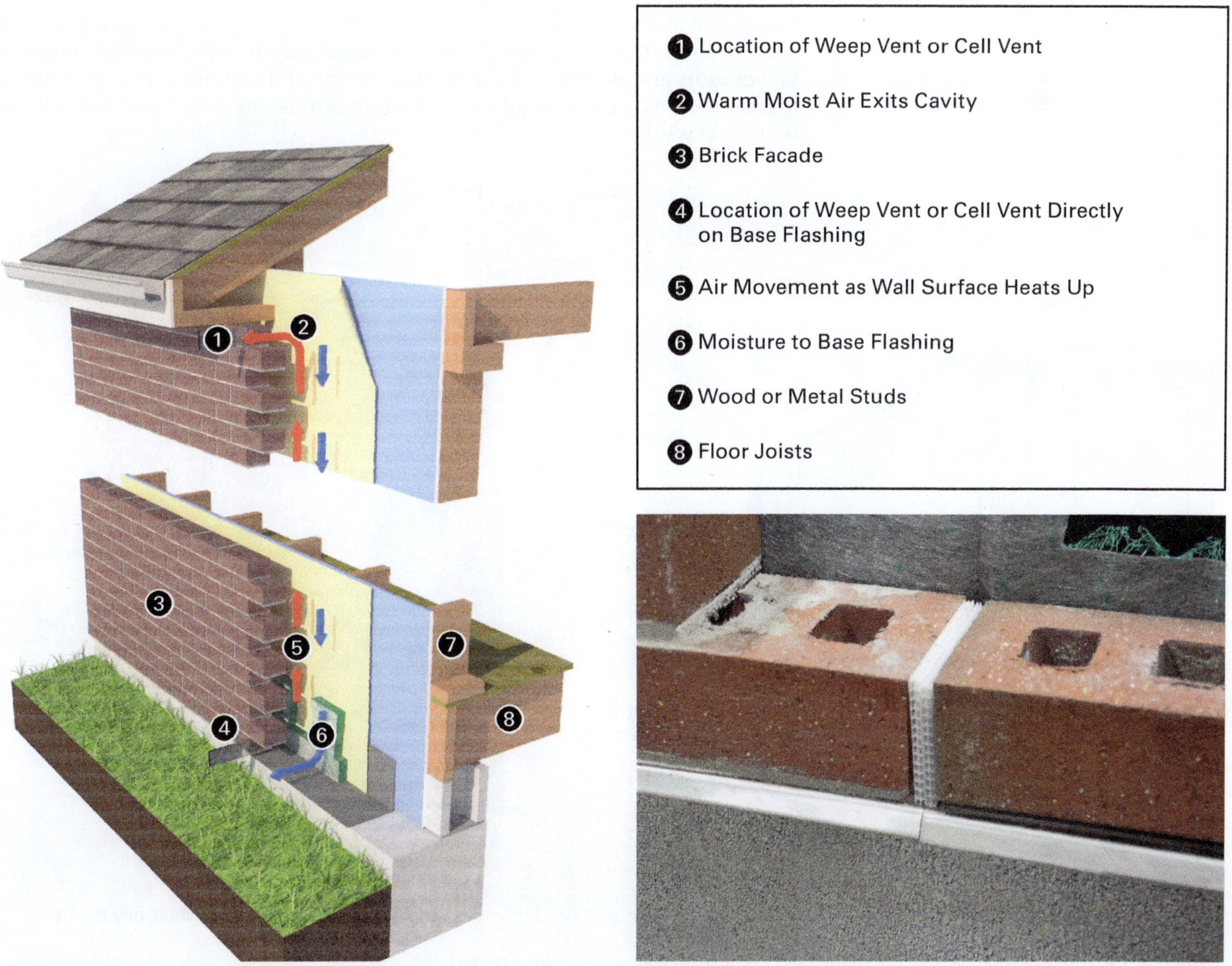

Figure 11 Weep holes and cell vents.
Source: Mortar Net Solutions – Masonry Division of Keene Building Products

1.4.0 Common Characteristics of Composite Walls

A composite wall is a multi-wythe wall system consisting of wythes of the same or different materials bonded together to distribute loads evenly across the entire system.

The wythes of masonry are bonded together in several ways:

- By using a full vertical collar joint
- By using masonry headers
- By using corrosion-resistant horizontal anchors or continuous joint reinforcement

Composite designs are typically used for exterior walls consisting of a brick- or stone-face wythe bonded to a concrete masonry or structural clay tile backup wythe. In some parts of the country, this type of construction is incorrectly called a *brick veneer*. *Figure* 12A shows a typical brick/block composite wall using a ladder-type continuous joint reinforcement. *Figure* 12B shows a different arrangement for bonding using headers to bond the two wythes every six courses. Because of the continuous collar joint, composite systems are particularly effective in preventing the passage of rain through the wall.

1.4.1 Bonding for Composite Walls

Composite walls present two problems that the mason must consider. The load-bearing strength of the wall is limited by the strength of the weakest combination of units and mortar and the strength of one of the wythes. This means that to ensure even distribution of the load, masons must properly bond the anchors, units, and wythes to one another.

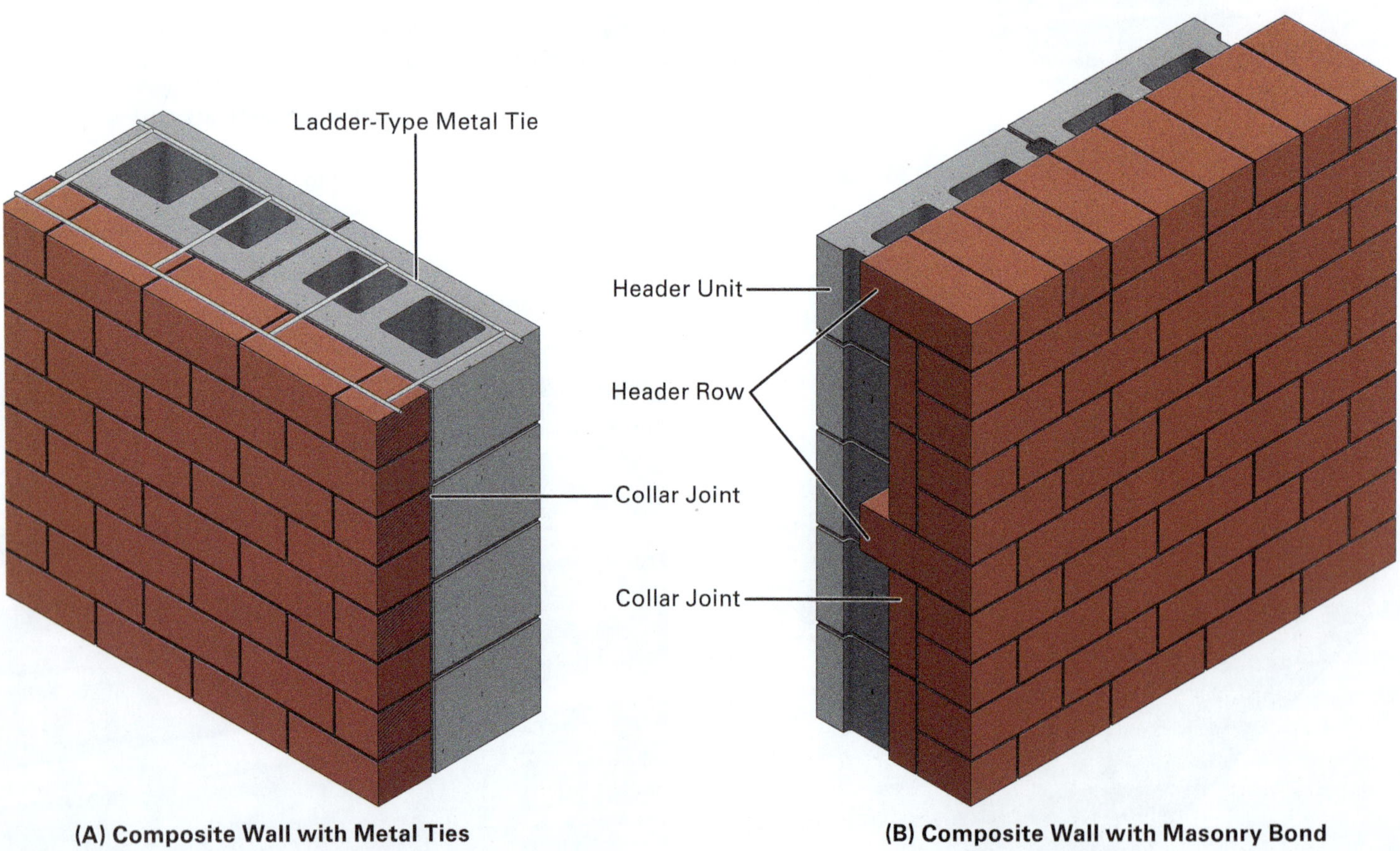

Figure 12 Composite wall with metal ties and composite wall with masonry bond.

Because the wall is composed of different materials, its parts will expand and contract at different rates with changes in temperature and moisture. However, such movements are generally insufficient to endanger the wall's stability if all mortar joints are properly filled and tooled.

The work should have full, solid, and compact head and collar joints, along with smooth bed joints. Although it is difficult to solidly fill collar joints and joints between brick headers, extra effort should be made to do this so that the wall will act as a unified structure that can resist moisture penetration and cracking.

Proper bonding is easiest to achieve when the brick courses are level with the block, particularly at bonding height. For this reason, it is best to carry up the facing and backing together.

Unless necessary, the facing should not be carried up more than five courses ahead of the backing. Carrying up the backing and inserting the wall horizontal anchors or continuous joint reinforcement so that the facing can be laid up later may cause problems if the brick is not level with the block at the height of the horizontal anchors or continuous joint reinforcement.

1.4.2 Using Brick Headers

Using brick headers may present special problems if the headers are not laid correctly. Masons must take extra care with any cross joint (a mortar joint spread across the full length of the brick header), or moisture penetration could occur. Single head joints applied only to the end of the header are not acceptable.

Another common problem presented by headers is the difficulty of ensuring that the set of the mortar is not broken when tying between brick and heavy block. This occurs when the block backup unit is placed immediately after a header course of brick. The block is placed on the lower course of block as well as the back half of the brick header. If the block is laid before the mortar under the header stiffens, the header brick will settle unevenly.

In this case, a crack may develop in the mortar joint under the header, leading to moisture penetration. Also, a twisted header brick will ruin the appearance of the wall face.

To avoid this problem, build to the height of the header course before laying any backup block. Blockwork can be **humored** to remain in line with the brickwork.

Humored: A slang term for gradually adjusting a section of masonry to correct alignment issues.

The block must also be laid exactly level in height to match the outside brickwork, or the header will not lie level across the two wythes. The steps in this procedure are as follows:

Step 1 Lay two courses of backup block to the proper height.

Step 2 Lay five courses of brick facing using full mortar joints. Cut joints flush at the back of the wall.

Step 3 Lay the brick header with a full mortar cross joint.

Step 4 Continue with the remaining courses in the same manner.

1.4.3 Using Continuous Joint Reinforcement and Horizontal Anchors

Continuous joint reinforcement is the most popular type of metal tie used for composite walls. Continuous joint reinforcement not only ties the masonry work together, but also reduces shrinkage in the mortar joints and prevents cracking from settlement.

Continuous joint reinforcement should overlap when two pieces are used together. This ensures that the tensile strength will not be interrupted throughout the length of the bed joint. It is extremely important that the reinforcement is embedded and covered completely with mortar before any masonry unit is laid on top of it.

Horizontal anchors, also called *unit ties,* can also be used for composite wall construction. The Z-type unit tie and the rectangular, or box, unit tie are the two most commonly used. Both are made from steel, which may be coated to prevent corrosion.

To build a block-and-brick composite wall using metal anchors, the following steps are recommended:

Step 1 Lay two courses of backup block to the proper height.

Step 2 Lay five courses of brick facing using full mortar joints. Cut joints flush at the back of the wall.

Step 3 Place the metal reinforcement and spread the mortar bed joint. Make sure the reinforcement is fully embedded in the mortar joint.

Step 4 Continue with the remaining courses in the same manner.

It should be noted that Steps 1 and 2 may be reversed. With either sequence, the facing and backup wythes are laid up together at each spacing of the reinforcement.

1.5.0 Fundamental Uses of Anchored Veneer Walls

A veneered wall system consists of a single wythe of masonry (as a facing) attached to a structural system. The masonry veneer acts as a protective coating for the structure—as a barrier to wind and rain, and as an insulator of sound and heat. The veneer may be brick, stone, tile, or uncored concrete masonry materials. The masonry facing is not a load-bearing wythe (designed to carry only its own weight). Although these masonry units are fastened to the framework, they are not bonded to backing with masonry or mortar. *Figure 13* shows brick veneer on a frame surface, and *Figure 14* shows brick veneer on a masonry surface.

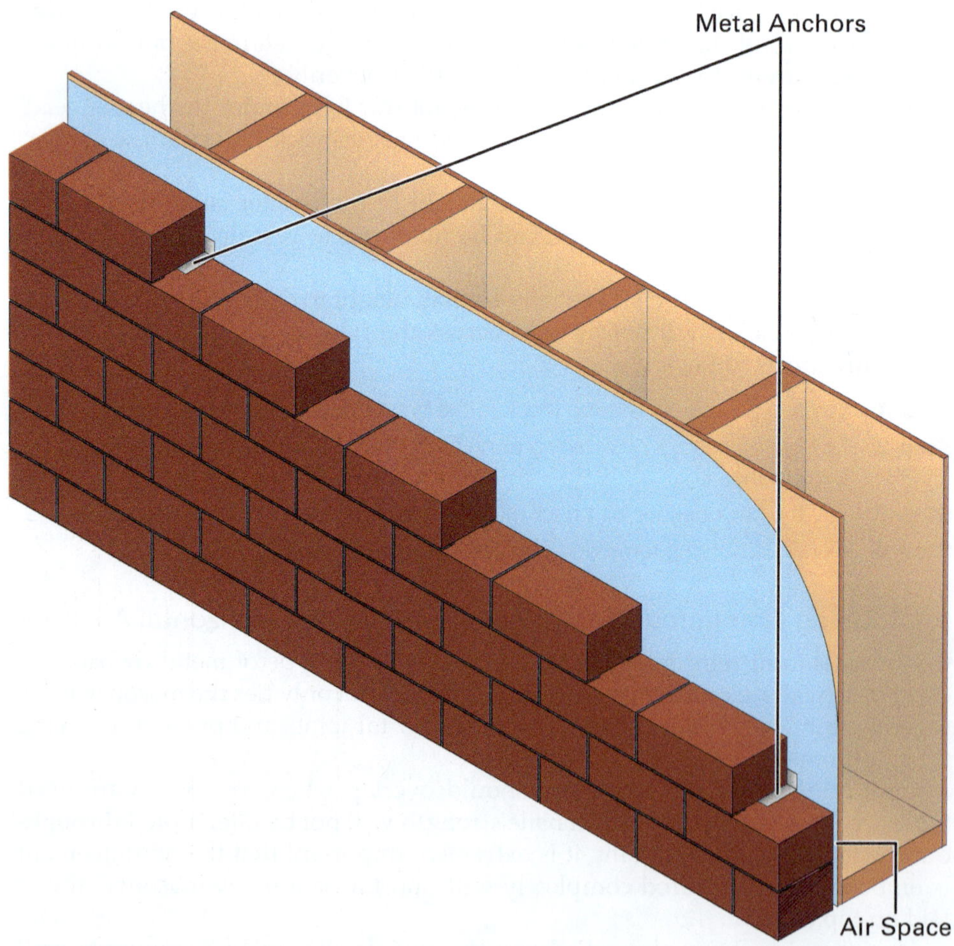

Figure 13 Brick veneer on a frame surface.

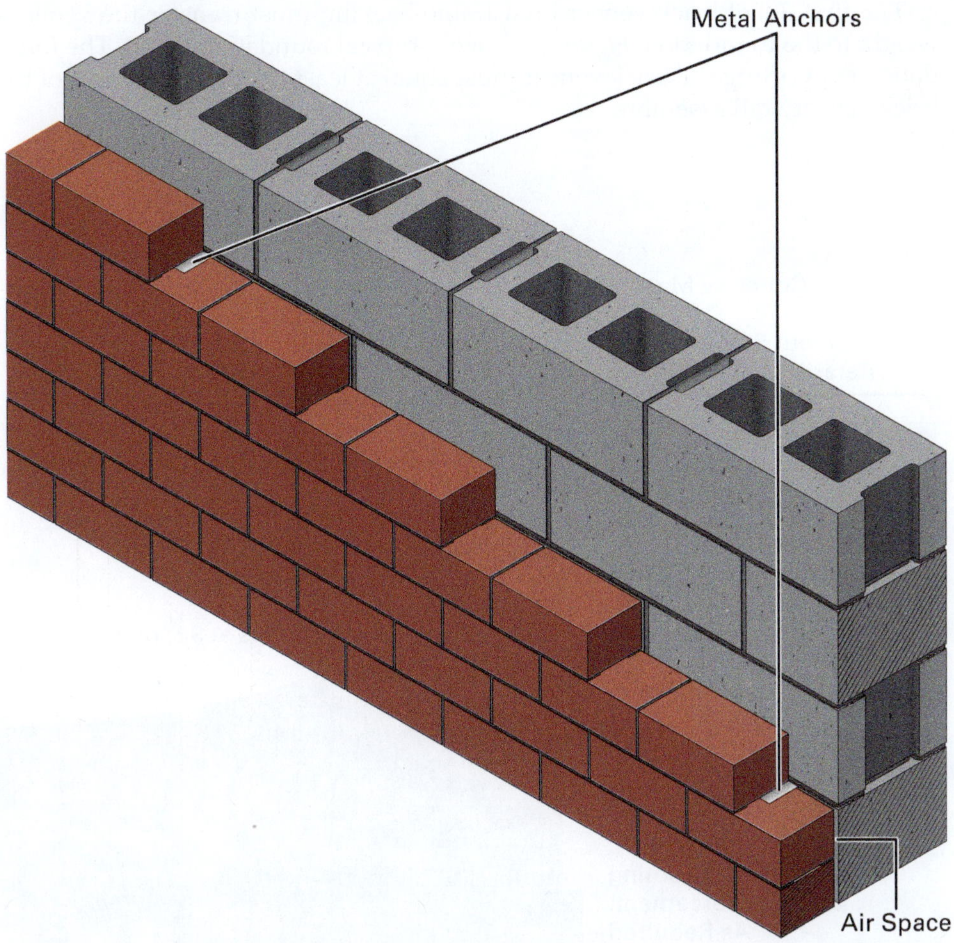

Figure 14 Brick veneer on a masonry surface.

Masonry veneer is attached to the structural frame by corrosion-resistant horizontal anchors. Building codes and project specifications dictate the spacing of metal anchors, which permit some slight horizontal and vertical movement of the veneer parallel to the face of the structural frame.

An air space is provided between the facing and backup. The backup is usually covered with a moisture-resistant material, and flashing and weep holes are included at the bottom of the air space to eliminate water that may pass through the masonry veneer.

1.5.1 Building Code Requirements

Building codes regulate the design of anchored veneers by setting requirements based on empirical data.

Although some codes require only a 1" clear cavity, the minimum recommended width of the open cavity between a veneer and the backing is 2". It is more difficult to keep a narrower cavity clear of mortar during construction. When rigid insulation is installed, the clear distance between the face of the insulation and the back of the facing wythe must be 2".

1.5.2 Brick Veneer Construction Details

For many years, brick veneer construction was principally limited to wood frame houses. Now it is used in low-rise commercial construction and is frequently used for high-rise buildings, especially with concrete masonry or steel stud backing systems.

Veneer construction looks like cavity wall construction in its final form. However, its construction and function are very different. For example, the supporting structure is completed before attaching the veneer.

The first 30' of brick veneer on a frame backing must transfer the veneer's weight to the foundation. *Figure 15* shows a typical foundation detail. The foundation supporting the brick veneer must equal at least the total thickness of the brick veneer wall assembly.

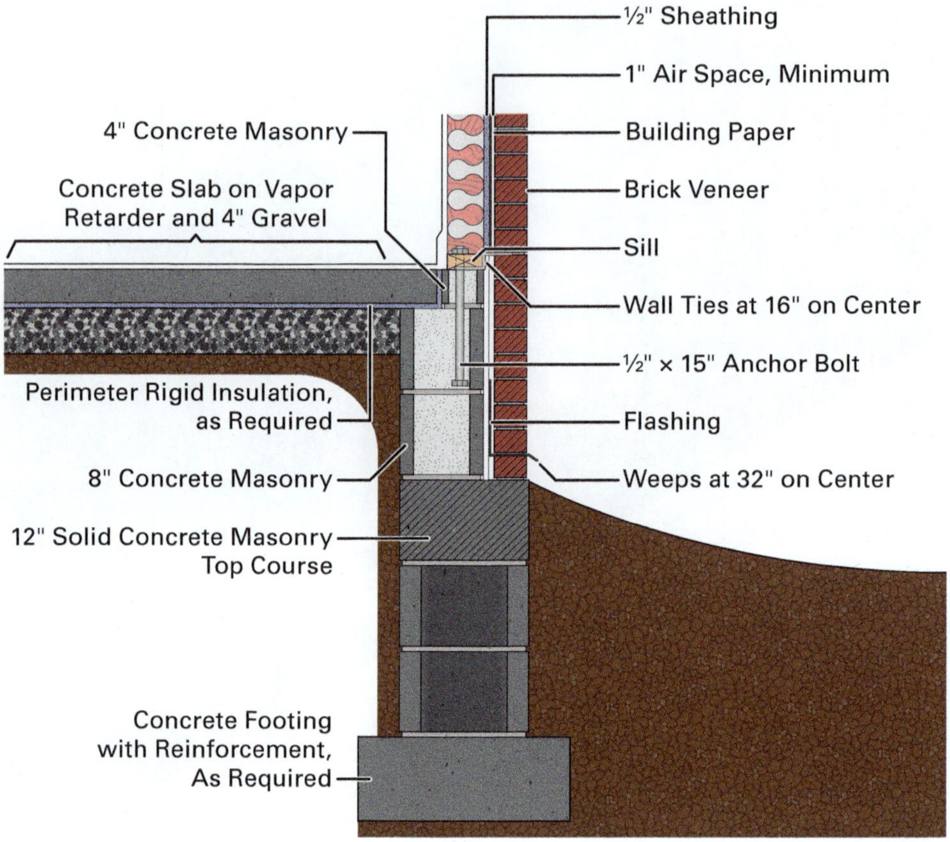

Figure 15 Brick veneer details.

1.6.0 Primary Function of Retaining Walls

A retaining wall is an unsupported wall that holds back soil and water. These materials constantly press against the wall, so they must be designed and constructed to withstand this force without collapsing or bowing.

There are five basic types of traditional masonry retaining walls:

- Gravity
- Cantilever
- Segmental
- Counterfort
- Buttress

Figure 16 shows these different types of retaining walls. Counterfort and buttress walls are not used very much today. However, they may be required where the lateral force on the wall is extremely high and there is limited space behind the wall to build a gravity-type structure.

A gravity-type wall depends primarily on its own weight to hold back soil pressure. A cantilever-type wall uses the weight of the soil, together with its own strength, to get the same results.

Reinforced cantilever walls offer the most economical design. The concrete footing anchors the stem and resists overturning and sliding due to both vertical and lateral forces.

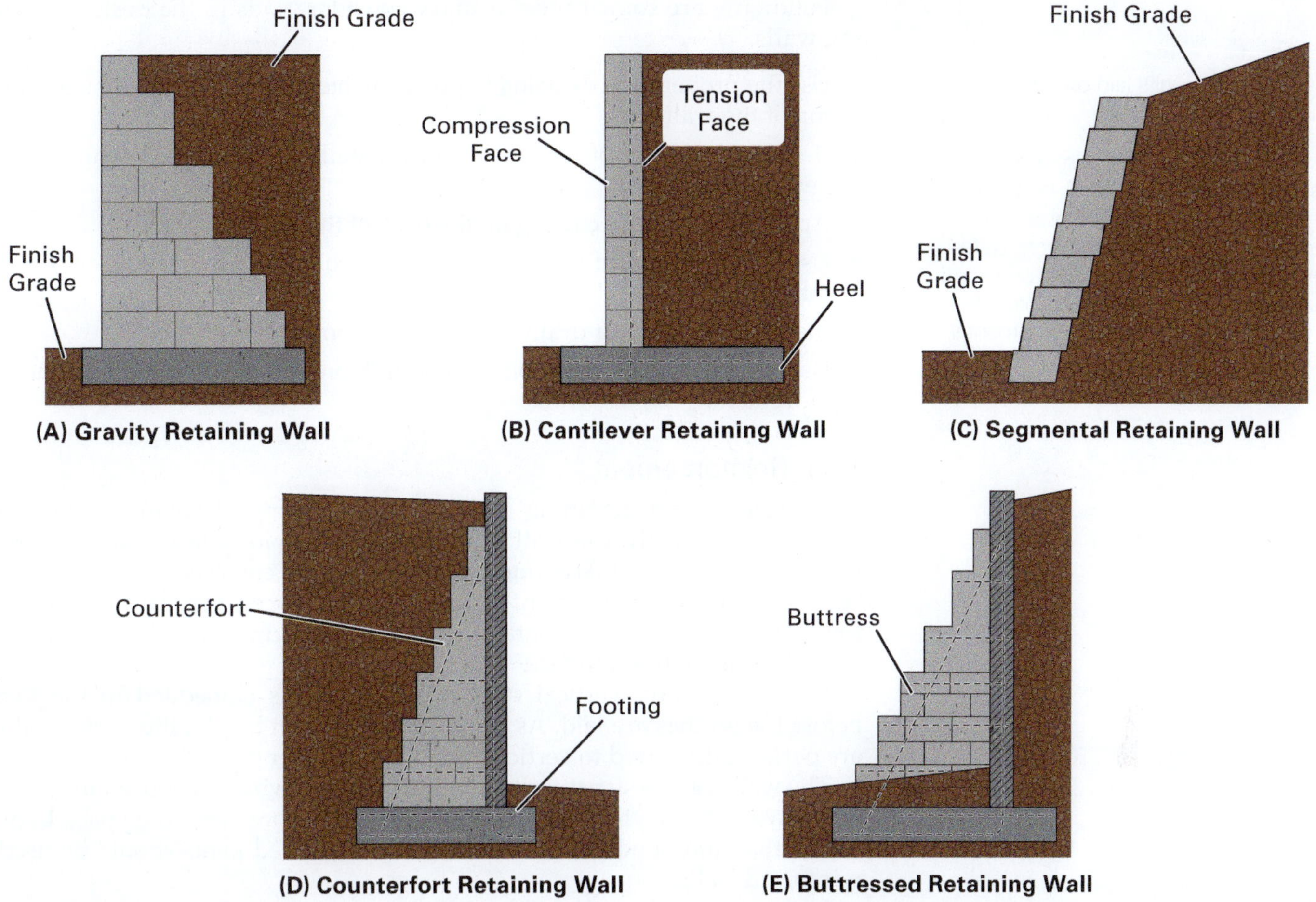

Figure 16 Types of retaining walls.

Figure 17 shows proprietary systems of interlocking CMUs called *segmental blocks*. These systems offer economical and attractive solutions for unreinforced retaining walls of moderate height.

Figure 17 Interlocking block retaining wall.
Source: Shutterstock

Cap: Masonry units laid on top of a finished wall.

Coping: The materials or masonry units used to form a cap or finish on top of a wall, pier, chimney, or pilaster to protect the masonry below from water penetration. Coping is usually projected from both sides of the wall to provide a protective covering and an ornamental design.

Following are some of the primary considerations in the design of retaining walls:

- Use the proper **cap** or **coping** to prevent water from collecting or standing on top of the wall.
- If coping consists of masonry units, install through-wall flashing directly beneath the coping.
- Apply a waterproof coating on the back of the wall.
- Use permeable backfill behind the wall to collect water and prevent soil saturation and increased hydrostatic pressure.
- Install weep holes or drain lines to drain moisture.
- Use expansion or control joints to permit longitudinal thermal and moisture movement.

1.6.1 Reinforcement

Reinforcement for retaining walls can be accomplished in several different ways. Externally, the wall can be reinforced with pilasters and buttresses placed along the wall face. Internal reinforcement consists of steel reinforcing bar, grout, or both. Steel reinforcement can be placed vertically and tied into the footing or placed horizontally with grout to form a bond beam at different elevations along the wall face.

For cavity walls, vertical steel reinforcement is embedded in the footing before the wythes are laid. As the wall increases in height, the horizontal rods are periodically wired to vertical rods for extra strength.

The walls can be structurally bonded by grout, which is poured into the cavity between the wythes of masonry. The grout core seals the space between the wythes and bonds the reinforcing steel. Full bed joints should be used for reinforced walls.

It should also be noted that grouting techniques for reinforced masonry walls vary in different parts of the country. Check the plans, specifications, and local building codes to determine the specific requirements for your project. The requirements for the size and placement of reinforcing steel may also vary according to the local codes.

Figure 18 shows a cross section of a typical cantilever retaining wall constructed of reinforced block.

1.6.2 Construction Techniques

The cantilever retaining wall uses the weight of the soil pushing down on the footing along with the strength of the vertical wall to resist the soil and water pressure pushing against the back face of the wall.

When building a cantilever retaining wall, be sure to provide standard moisture protection. Build weep holes and proper drainage to allow water to flow away from the wall.

Single-wythe walls are constructed using cored masonry units laid with face shell mortar bedding. The vertical cores are aligned to form continuous vertical spaces that receive grout.

Concrete footings for retaining walls should be placed on firm, undisturbed soil. In areas subject to freezing, they should also be placed below the frost line to avoid excess movement due to temperature changes.

Failure to drain the backfill area behind retaining walls causes a buildup of hydrostatic pressure. In areas where precipitation is heavy or poor drainage conditions exist, a continuous longitudinal drain of perforated pipe should be placed near the base, with discharge areas at the ends of the wall (*Figure 19*).

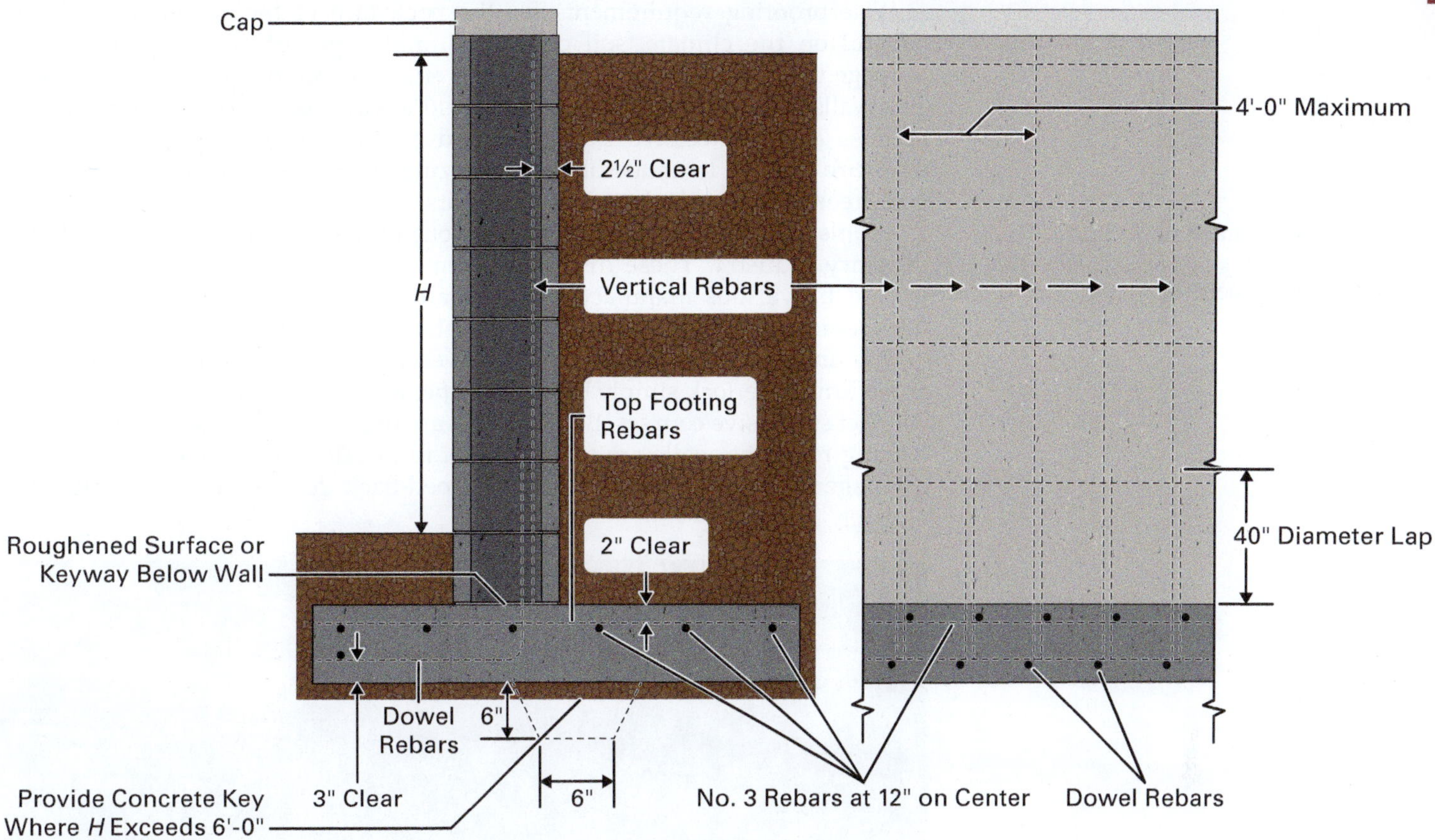

Figure 18 Cantilever retaining wall.

Figure 19 Drainage pipe installation for retaining wall.
Source: Shutterstock

Waterproofing requirements for the back face of the retaining wall will depend on the climate, soil conditions, and types of masonry units used. Seepage through a brick wall can cause efflorescence if soluble salts are present. Walls of porous concrete units should always receive waterproof backing because of the excessive expansion and contraction accompanying variable moisture content. In areas subject to freezing, a waterproof membrane can prevent freeze-thaw cycle damage to the units.

Segmental retaining wall (SRW): A wall made of segmental block stacked on top of each other without mortar bonding.

Segmental retaining walls (SRWs) are one of the newest developments in the masonry industry. These dry-stacked, interlocking-block retaining walls are offered by various manufacturers in many different designs and configurations. An example of this type of wall is shown in *Figure 20*.

The units are stepped back slightly in each course toward the embankment. Some units interlock simply by their shape, while others use pins or dowels to connect successive courses. Because they are dry stacked without mortar, interlocking retaining walls are easy and fast to install. The open joints allow free drainage of soil moisture, and the stepped-back designs reduce overturning stresses.

Figure 20 Segmental retaining wall.
Source: Shutterstock

1.7.0 Common Characteristics of Freestanding Walls

Freestanding walls include garden walls, fences, and screens of brick, tile, and CMUs. They can take many forms. Screen wall masonry units, such as those shown in *Figure 21*, are often used for privacy fences.

Fences and walls should be able to safely withstand wind loads, with many codes specifying a minimum resistance of 20 psf (pounds per square foot). In hurricane areas, higher wind pressure resistance is needed.

1.7.1 Reinforcement

Local building codes will determine how high garden walls and fences can be built and whether they must be reinforced or not. Building a garden wall or fence is the same as building other types of solid or composite masonry walls. Reinforcing requirements will depend on the specifications for wind load, earthquake resistance, and soil conditions.

For walls that require some type of reinforcement, the main concern is the embedding of the reinforcing steel. The reinforcing steel must be embedded into the footer or pier deep enough to anchor the wall firmly to the foundation and to resist any lateral pressure.

Figure 21 Screen wall.
Source: Alamy Stock Photo

1.7.2 Construction Techniques

Screen walls and garden walls are normally constructed the same way as interior partition walls. Screen walls are governed by the same height-to-thickness ($^h/_t$) ratio lateral support requirements as interior partitions, but walls with interrupted bed joints should be designed more conservatively because of reduced flexural strength and lateral load resistance.

Perforated walls (*Figure 22*) can also be built with any of the standard brick types. These walls are constructed by omitting the mortar from head joints and separating the units to form voids. The walls may be laid up in single- or double-wythe construction. In double-wythe walls, separate header or rowlock courses alternate with stretcher courses to form different patterns.

Figure 22 Perforated wall.
Source: Alamy Stock Photo

Double-wythe walls are more stable than single-wythe designs because of the increased weight, wider footprint, and through-wall bonding patterns. Piers may be either flush with the wall or projecting on one or both sides. The coursing of the screen panels must overlap the coursing of the piers to provide adequate structural connection.

Regardless of the exact design, the pattern of units in a perforated wall must provide continuous vertical paths for load transfer to the foundation, and the bearing width of the paths or columns should be at least 2".

All freestanding walls require some type of covering over their top. This is usually accomplished with a cap. A cap is usually made from masonry units that cover the top of the wall and are the same thickness. Coping applies to the covering over the top of a wall. These coverings can be of single or multiple units. The tops can also slope in one or both directions.

Figure 23 shows four types of coping on top of different types of garden walls. Copings normally do not serve any structural function. Typical materials include brick, precast or cast-in-place concrete, stone, terra-cotta, or metal.

Brick pier and panel walls are composed of a series of thin panels braced intermittently by reinforced masonry piers or pilasters. Details of this type of construction are shown in *Figure 24*. Reinforcing steel is added to each panel at the top and bottom and tied into the piers. Pier construction can be of several different designs.

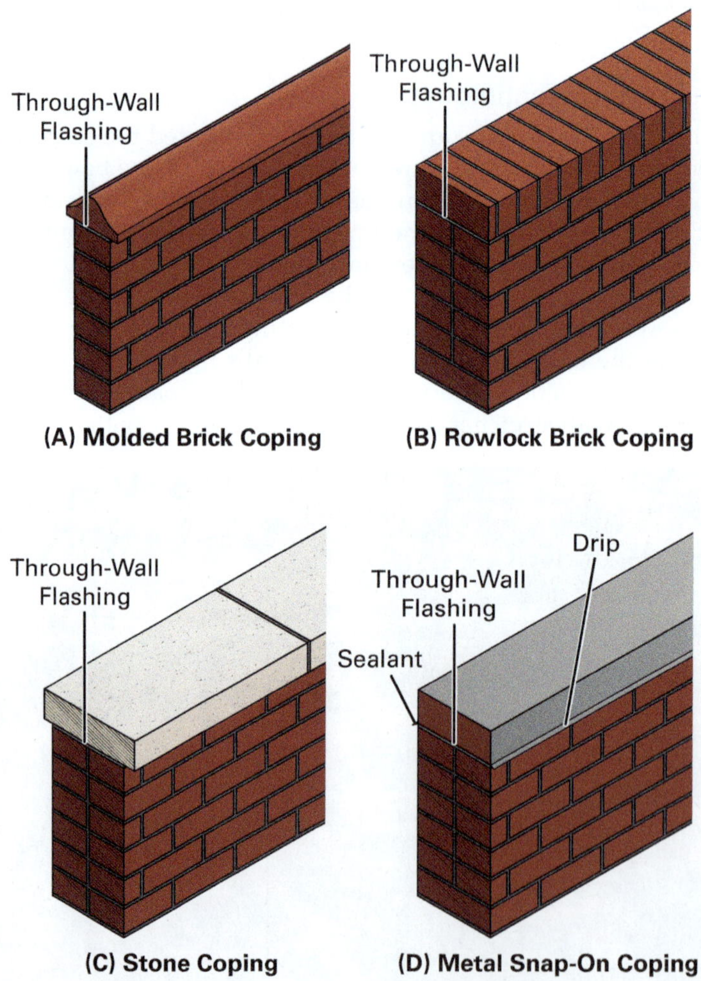

Figure 23 Four types of coping.

Figure 24 Pier and panel wall.
Source: Shutterstock

Because the panel section is not supported on a continuous footing, it is more like a beam spanning the clear distance between foundation supports. The piers and panels are constructed simultaneously, course by course. Initially, a 2" × 4" form is constructed under the first course of the panel and kept in place for a minimum of seven days until the mortar has sufficiently cured. Requirements for steel reinforcement will be determined by local codes. A serpentine garden wall does not require vertical support, because the shape of the wall gives it stability and resistance to wind pressures. However, each end of the wall needs to be anchored in a pilaster or other intersecting structure, or in a sharp curve (*Figure 25*).

Figure 25 Serpentine garden wall.
Source: Shutterstock

Although serpentine walls do not require any reinforcing because of the curves, they must be laid out properly. A rule of thumb is that a serpentine wall should not be built higher than 15 times the unit's width. For example, if a 6" wide block is used, the wall should not be more than $7\frac{1}{2}$' high. Because the wall depends on its shape for lateral strength, it is important that the degree of curvature be sufficient.

Once the design has been established, the wall must be laid out starting with the footing. This is not an easy task, because the footing must be curved in the same shape as the wall and placed so that the wall will always be in the center line. After the footing has been dug and poured, mark the center line of the wall and dry-bond the first course to determine unit placement and joint spacing on both faces.

Because of the curve, this spacing will always be different on opposite sides of the wall except for the two joints at each point where the curve changes direction. Sometimes it is helpful to make a plywood form cut to the shape of the curve to use as a guide for dry bonding and as a check after the course has been laid.

1.0.0 Section Review

1. In 8" walls, the type of course that can be used for the full width of the wall section is the _____.
   a. rowlock course
   b. soldier course
   c. sailor course
   d. header course

2. Horizontal anchors are usually placed in _____.
   a. alternate courses
   b. every course
   c. every third course
   d. a vertical line

3. In most cavity walls, the vertical load of the structure is carried by the _____.
   a. horizontal joint reinforcement
   b. middle wythe
   c. inner wythe
   d. outer wythe

4. Which of the following is *not* a common method for tying multi-wythe masonry walls together?
   a. Use of a full vertical collar joint
   b. Grouting the air space
   c. Use of masonry headers
   d. Use of horizontal anchors

5. What is the *minimum* recommended width of the open cavity between a veneer and the backing?
   a. 1"
   b. 2"
   c. 3"
   d. 4"

6. Which of the following is *not* an example of a traditional masonry retaining wall?
   a. Trellis wall
   b. Gravity wall
   c. Cantilever wall
   d. Buttress wall

7. What is the *minimum* safe wind load resistance for fences and walls, according to most codes?
   a. 5 psf
   b. 10 psf
   c. 15 psf
   d. 20 psf

2.0.0 Control Joints and Expansion Joints

Objective

Identify the requirement for and function of control joints and expansion joints.

a. Identify the effects of temperature and moisture on control joints and expansion joints.
b. Identify the uses of control joints.
c. Identify the uses of expansion joints.

Performance Task

1. Lay out and construct a composite wall with control joints and expansion joints.

Even though a building seems to stand still, it is in a constant state of motion. All building materials expand and contract due to changes in temperature, and many materials move with changes in moisture content. A structure's performance depends on construction techniques that allow for some degree of movement of the materials.

2.1.0 Effects of Temperature and Moisture on Control Joints and Expansion Joints

Cracking can be prevented in several ways. Either the temperature variation must be controlled, the structure must be strong enough to resist cracking, or the effective length of the structure must be limited.

In masonry walls, the proper placement and construction of joints limits the effective length of the structure and reduces its tendency to crack. As a rule of thumb, joints used on brick walls are called *expansion joints*, while those used on block walls are called *control joints*.

Another problem can result from the surface temperature of a material. If the interior wall temperature is controlled and limited, the wall may warp, resulting in joint cracks.

If building materials are used in combination, they must be compatible with the expected temperature changes. If one material expands more quickly than the other, problems may arise.

Many building materials tend to expand when moisture content increases and contract when water is lost. For example, brick expands slowly when exposed to water or very humid air. This expansion is very difficult to determine and is unpredictable. With this uncertainty, it is best to plan for some expansion or movement.

2.2.0 Control Joints

The proper positioning of control joints in block walls has evolved with changing local practices, construction materials, and local codes. The following factors should be considered when planning joint construction:

- The expected movement of the wall
- The resistance of the wall to horizontal tensile stress
- The number and locations of wall openings

Considering these factors will normally result in using a control joint where different materials join, where different wall heights or thicknesses join, or where walls intersect with other structural elements (*Figure 26*).

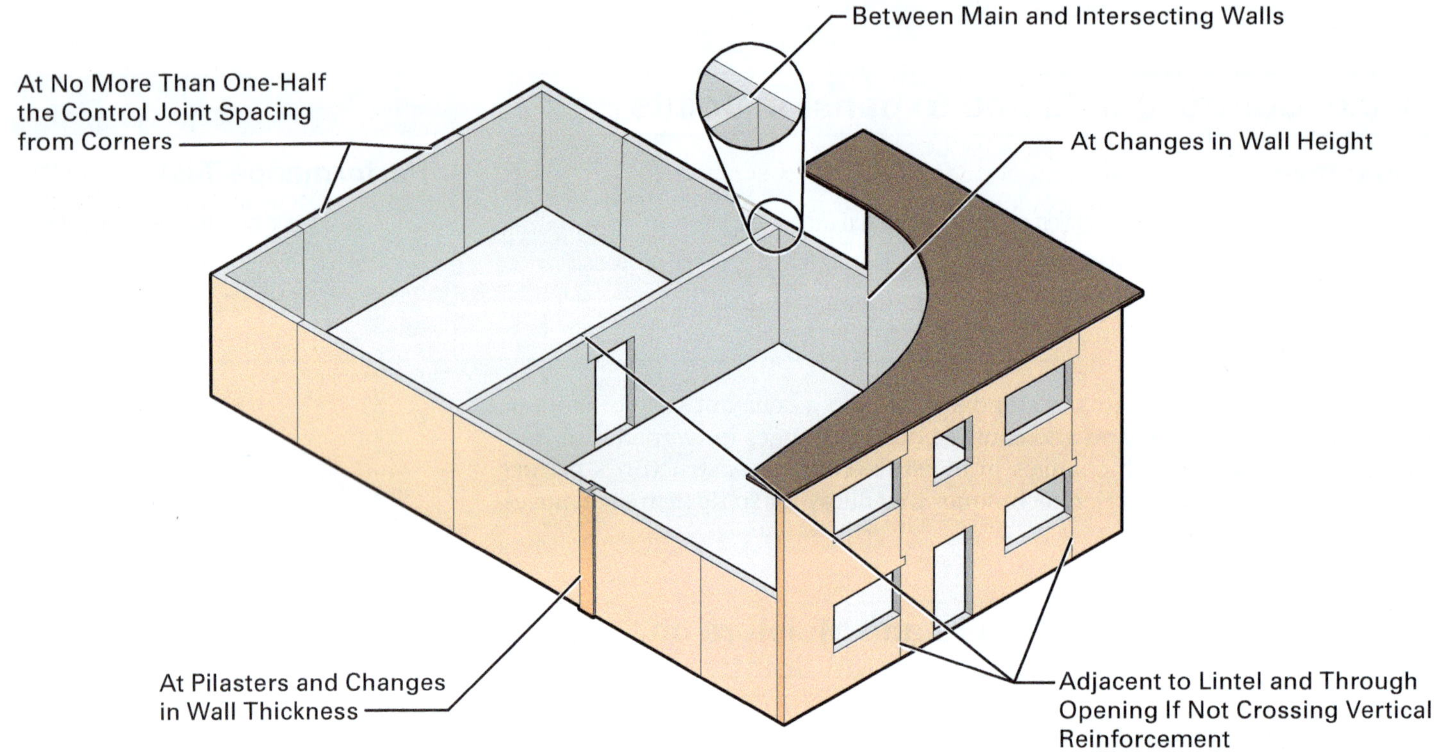

Figure 26 Locations for control joints.

Control joints are also placed at specified intervals in long, continuous block walls. They divide the wall into panels and limit stress. These control joints should be placed every 25' to 30' in a continuous block wall. The locations of control joints are also dependent on the length-to-height ratio of the wall and the use of joint reinforcement (*Table 1*).

TABLE 1 Spacing of Control Joints

Recommended Spacing of Control Joints	Vertical Spacing of Joint Reinforcement			
	None	24"	16"	8"
Expressed as a ratio of panel length to panel height (L/H)	2	$2\frac{1}{2}$	3	4
With panel length (L) not to exceed	40'	45'	50'	60'

For example, to determine the joint spacing for a wall 8' high and 100' long, the following analysis is necessary:

Step 1 Determine if horizontal joint reinforcement is specified, and if so, at what spacing. If specified at 16" on center (OC), a length-to-height ratio (L/H) of 3 should be used.

Step 2 The wall height is 8', so the maximum panel length is:

$$L/H = 3 \text{ and } H = 8'$$
$$L/8 = 3$$
$$L = 3 \times 8'$$
$$L = 24'$$

The maximum control joint spacing for this wall is 24' OC.

Not all block walls require control joints. For reinforced walls, the amount of horizontal reinforcing steel is generally more than adequate to control cracking. For concrete masonry used as a backing for other materials, joints may or may not be required. If required, the joint should extend through the facing wythe and the backup wythe if they are masonry-bonded or use full collar joints. If the two wythes are bonded only by horizontal anchors, joint construction does not have to be continuous through both wythes.

2.2.1 Types of Control Joints

There are at least five different types of control joints: a control joint using standard block, a control joint using a special tongue-and-groove control joint block, a premolded gasket control joint, a Z-type unit tie joint, and a **pencil rod** control joint. The first three types are commonly used in the masonry trade.

Pencil rod: A type of metallic tie that is similar to the shape of a straight wooden pencil; used for control joints in concrete masonry construction.

The tongue-and-groove control joint uses a specially designed full- or half-length control joint block, as shown in *Figure 27*. The tongue of one special unit fits into the groove of another, or into the open end of a regular flanged unit. Place the control joint block, focusing on head joint alignment. The head joint mortar will later be partially raked out and replaced with flexible sealant.

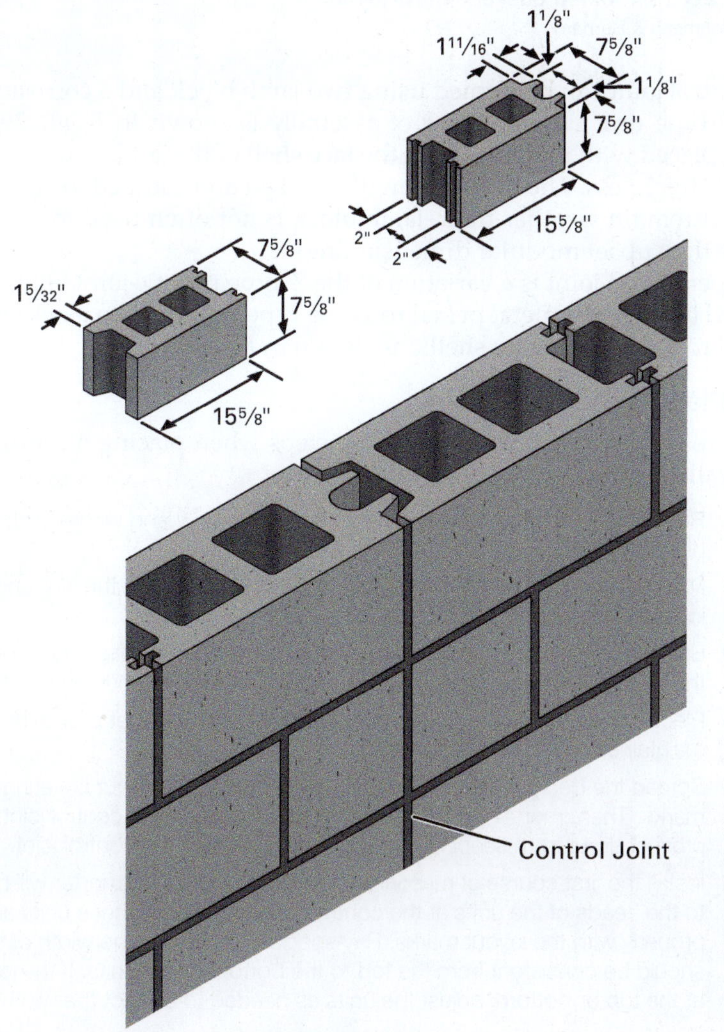

Figure 27 Special block for control joints.

A control joint can be formed by inserting a premolded rubber gasket between the sash blocks (*Figure 28*). The gasket is stiff but will allow some sideways movement. It also helps prevent moisture penetration.

Figure 28 Premolded gasket control joints.
Source: Hohmann & Barnard

A control joint can be formed using two jamb block and a corrosion-resistant metal Z-type unit tie. This type of assembly is shown in *Figure 29*. The jamb block is placed with mortar only at the face shell of the jamb and not throughout the block thickness. The head joint will be raked and caulked to allow for movement, yet remain weather-tight. Jamb block is not often used for control joints, because they upset modular dimensioning.

The pencil rod joint is a variation of the Z-type unit tie joint but is made with standard block and a metal pencil rod or Z-type unit tie. The block are tied with a pencil rod across the face shells, as shown in *Figure 30*.

2.2.2 Placing a Control Joint

To ensure proper function, follow these steps when placing a control joint in a block wall:

Step 1 Review the construction documents to identify how and where joints should be placed.

Step 2 Mark the location of the joint at the proper distance from the wall end, corner, intersection, or adjacent control joint.

Step 3 Establish the width of the joint by placing a mark for the first unit on either side of the joint. This is typically $\frac{3}{8}$" but can be less or more depending on the design.

Step 4 Place bond marks on either side of the joint, ensuring that proper bond is maintained around the structure.

Step 5 Spread the bed joint for the first course of masonry without covering the layout marks. Then, mortar the cross web on both sides of the control joint. This will prevent grout from seeping from adjacent cells into the control joint.

Step 6 Install the first course of masonry at the control joint. No mortar will be applied to the heads of the units at the control joint. Make sure these units are aligned properly with the layout marks. Pay special attention to the width of the joint; it should be consistent from the top to the bottom of the units. If the joint is flared at the top or bottom, adjust the units as needed to correct the width of the joint.

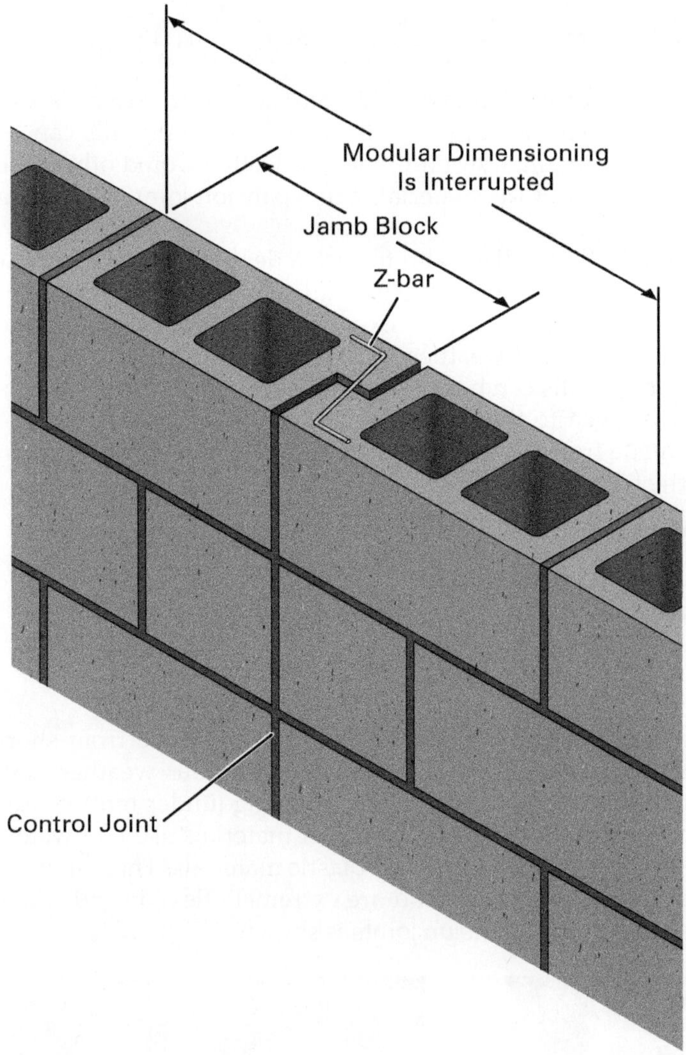

Figure 29 Z-type unit tie control joint.

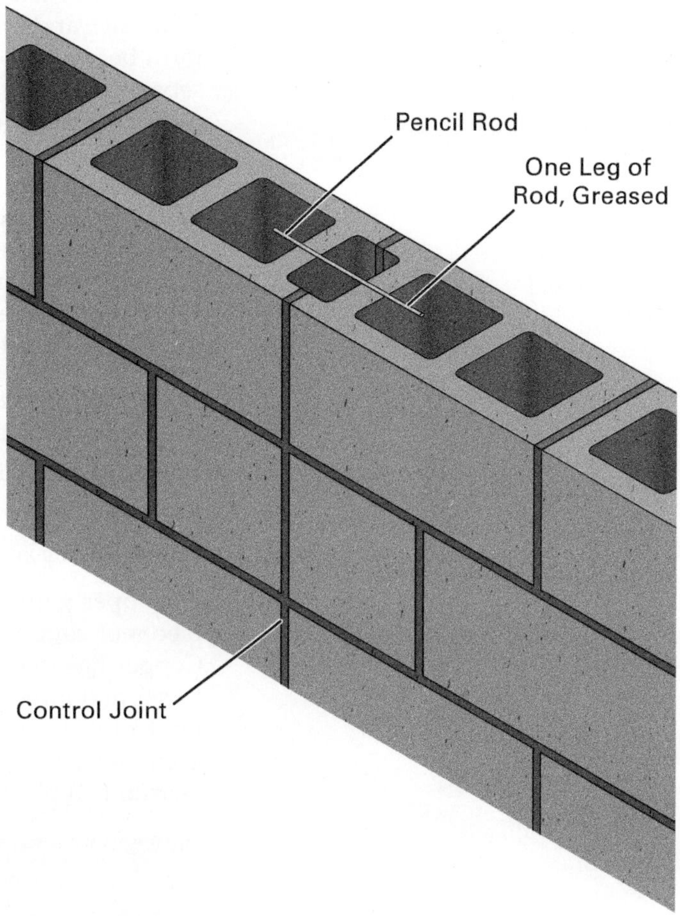

Figure 30 Pencil rod control joint.

Step 7 Install a second course of masonry at the control joint. Make sure to follow proper bond. Mortar the cross webs against the control joint before installing the units. Ensure that the joint is kept plumb in both directions and is of uniform width.

Step 8 Tool the adjacent masonry joints when they are thumbprint hard. The control joint should be kept free of all mortar and debris. Use a trowel or striking iron to clean out any excess mortar in the control joint.

Step 9 Install the horizontal joint reinforcement at the specified intervals. This is typically every 16' vertically, but the spacing can vary by project; refer to the project drawings and specifications for the correct spacing. The joint reinforcement is typically not continuous at the control joints. If this is the case, use small bolt cutters to cut reinforcement at the control joint, and make sure the ends of the reinforcement are kept about 1" away from the control joint. Occasionally, the design will require the joint reinforcement to be continuous through the control joint; again, be sure to check the drawings and specifications for this information.

Step 10 Install the remaining courses of masonry. As the work progresses, ensure that the joint is plumb, uniform in width, and clean.

2.3.0 Identifying the Uses of Expansion Joints

One of the most effective means of preventing damage due to expansion in a brick wall is by using an expansion joint. Although expansion joints can be expensive and difficult to maintain, they eliminate cracks that would otherwise form because of the movement of the material. An expansion joint is simply a separation of parts.

A brick wall, for example, that will expand to such a degree that cracks occur can be divided into multiple wall segments. The wall segments will have a gap or joint between them that allows room for expansion without cracking. The joint, of course, must be flexible and weather resistant.

Expansion joints in brick walls can be constructed in several ways so long as primary consideration is given to the flexibility and weather resistance of the joint. The joint can be made from a variety of materials. The five most common types of material are the following:

- Backer rod and sealant
- Copper
- Premolded foam pad
- Neoprene
- Extruded plastic

Copper joints have been used for many years. These are made from short pieces of copper sheets, which are lapped to form a continuous weather seal. Copper has the advantages of being flexible, noncorroding (under most conditions), and impervious to moisture. Some of the newer materials used for watertight joints include specially formed rubber and plastic materials. These may be premolded for the correct size and shape and are extremely flexible and water resistant. Typical construction of expansion joints is shown in *Figure 31.*

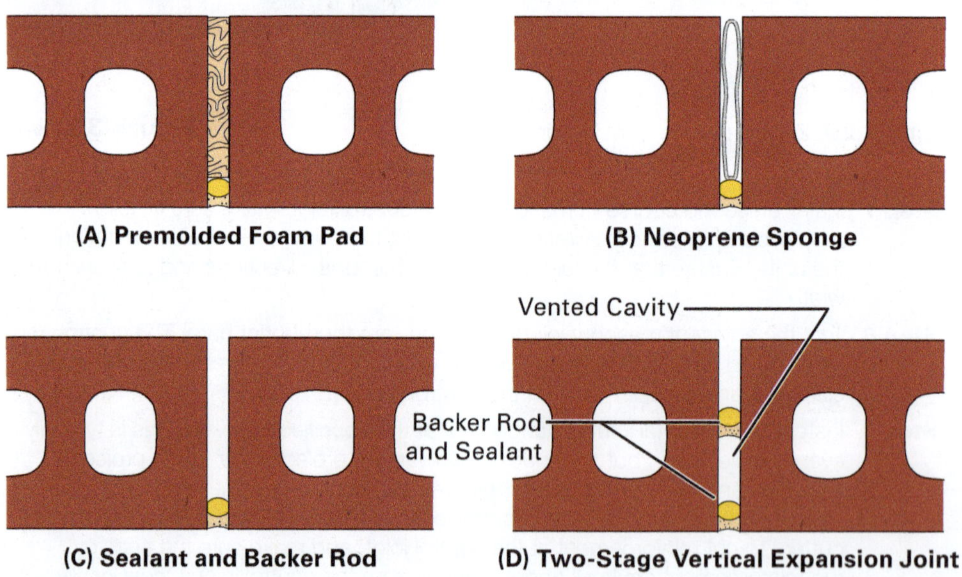

Figure 31 Expansion joints.

The use of any expansion joint requires the use of a sealing compound for the exposed side of the joint. This sealing compound must be an elastic material that will not become brittle with age and must blend with the finish treatment of the wall for an attractive appearance. This compound may be a standard caulking compound, such as silicon caulk, or a specially formulated sealing compound offering elasticity and durability (some of these are classified as polymer materials). Occasionally, a cover plate is used in place of, or in conjunction with, the sealing compound.

2.3.1 Placement of Joints

The placement of expansion joints in brick construction is determined by considering several factors. Among these are the loads carried by the wall; the strength of the wall; the maximum and minimum temperature exposures of the wall; the length, height, and thickness of the wall; and the locations of any intersecting walls. Due to these factors, there are no hard-and-fast rules that can be used for determining the position and spacing of expansion joints. *Figure 32* illustrates building locations where expansion joints are typically placed.

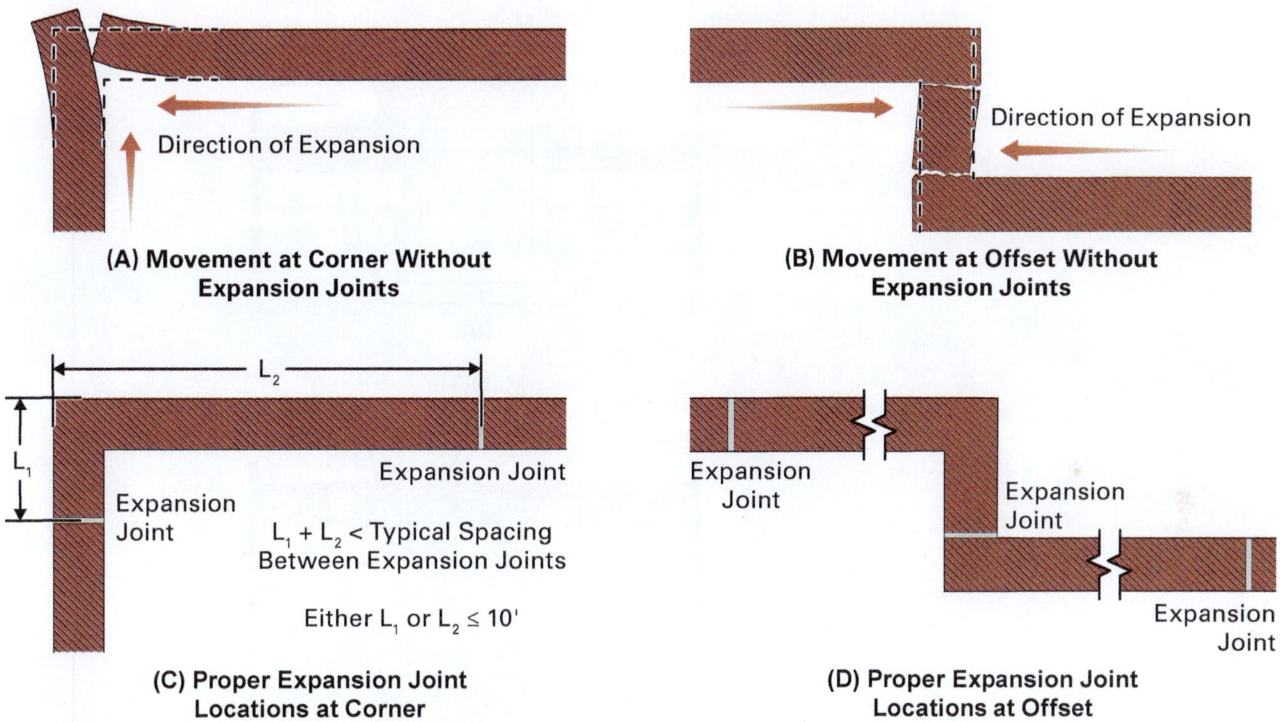

Figure 32 Typical locations of expansion joints.

2.3.2 Joint Details

Expansion joints should be designed for the specific situation. Each building should be analyzed and separate details developed to meet the requirements for each joint. Many details are very similar in construction, and the diagrams in *Figure 33* are provided as suggested construction details. These are typical of the many possible joints used in a variety of situations.

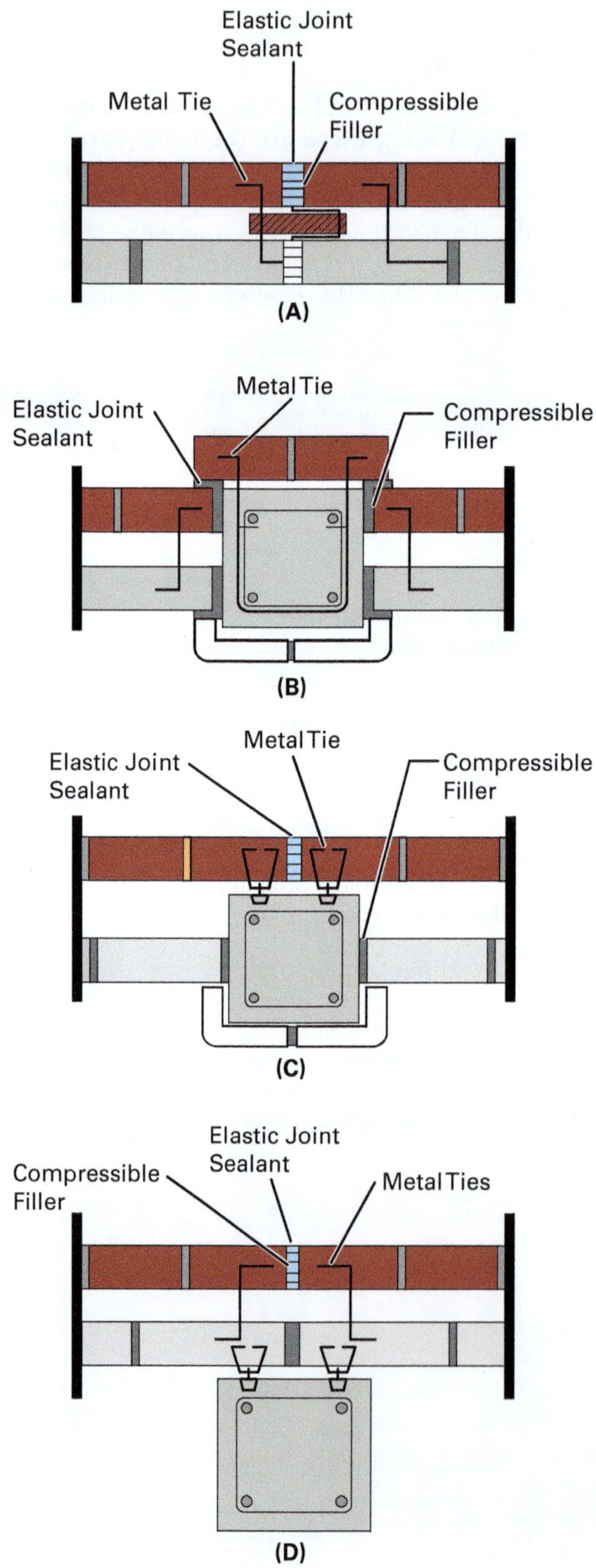

Figure 33 Expansion joints in straight walls.

In general, expansion joints, also sometimes referred to as *slip planes*, should be located at offsets and at junctions of walls in *L*-shaped, *T*-shaped, or *U*-shaped buildings (*Figure 34*). This joint prevents the expansion of one wall from exerting pressure on the other wall.

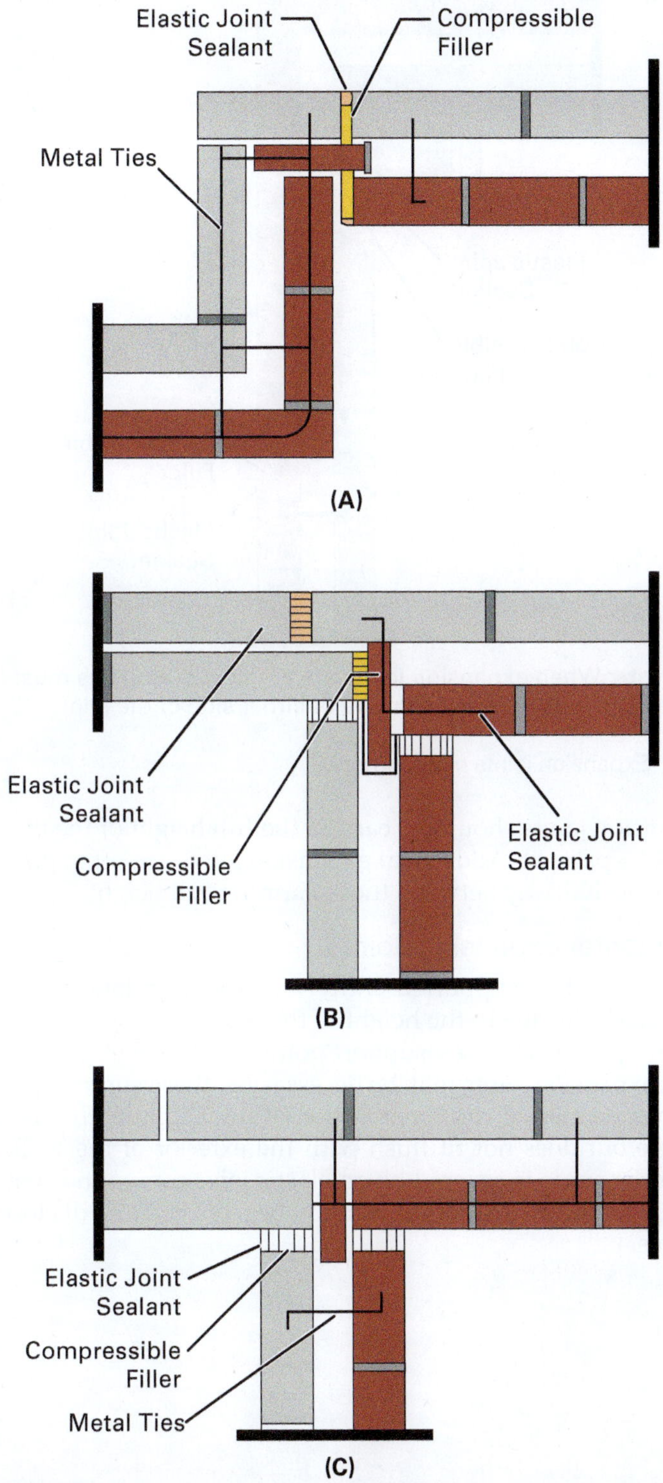

Figure 34 Expansion joints at offsets and junctures.

Expansion joints are frequently placed at or near external corners in both cavity and solid walls. This is particularly true with solid walls that rest on concrete foundations and extend above grade (*Figure 35*).

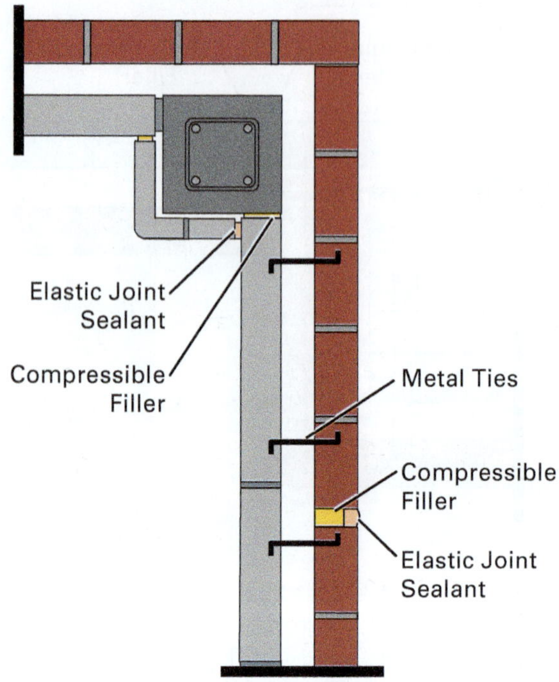

Note: When expansion joints are employed, wall ties must be placed within 12" (maximum) of either side of the joint.

Figure 35 Expansion joints near corners.

All expansion joints should be carried the full height of the masonry, even if this includes a parapet. Additional expansion joints are often placed in unreinforced parapets halfway between those running full height.

2.3.3 Horizontal Expansion Joints

Brick walls should be supported entirely on the foundation. However, this is not always possible, due to the height of the wall or because the wall may have many openings. In such cases, support points such as shelf angles are required. Horizontal expansion joints may be necessary at these support points. *Figure 36* shows a cross section of the construction of a wall panel that is supported by a shelf angle but does not fit flush with the exterior of the wall. In this case, the edge of the angle is covered by an elastic joint compound that also fills the whole horizontal joint. The sealant should be a color that will closely match the mortar joints.

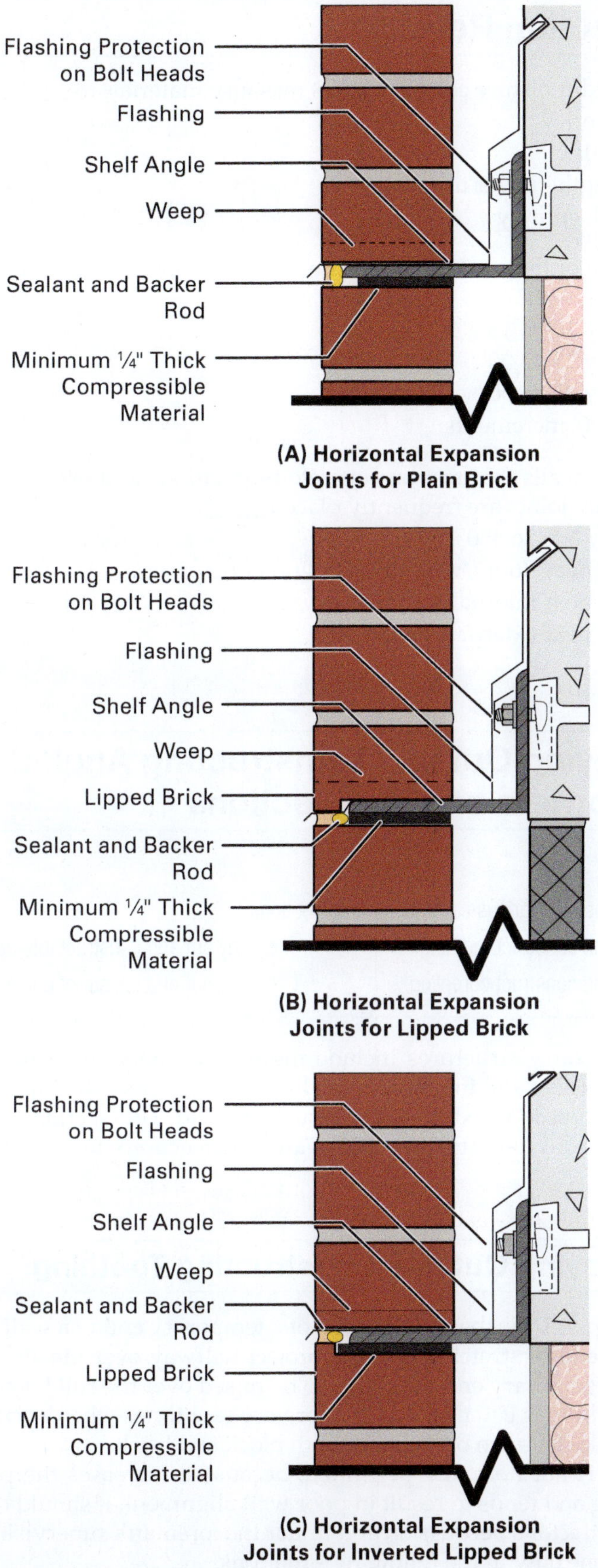

Figure 36 Horizontal expansion joints at shelf angle.

2.0.0 Section Review

1. Increased moisture content causes masonry materials to _____.
   a. expand
   b. shrink
   c. strengthen over time
   d. break entirely

2. As a rule of thumb, how often should control joints be placed in a continuous block wall?
   a. Every 10' to 15'
   b. Every 25' to 30'
   c. Every 45' to 50'
   d. At 100' increments

3. On solid walls that rest on concrete foundations and extend above grade, expansion joints are frequently placed _____.
   a. every 400' to 450'
   b. at or near doors, windows, and other openings
   c. at or near internal corners
   d. at or near external corners

3.0.0 Laying Out and Constructing Angled Corners and Intersections

Performance Tasks

2. Lay out and construct intersections.
3. Lay out and construct angled corners.

Objective

Lay out and construct various corners and intersections.

a. Lay out and construct toothing.
b. Lay out and construct corbeling.
c. Lay out and construct intersecting walls.
d. Lay out and construct angled corners.

Complex masonry structures include more than one wall. Walls can be tied together at corners and intersections. These connections distribute loads evenly even when exposed to external forces such as wind. This section discusses how to design and lay out angled corners and intersections to meet today's strict standards.

3.1.0 Laying Out and Constructing Toothing

Toothing: Construction of a temporary end of a wall with the end stretcher of every alternate course projecting. The projecting units are called *toothers*.

Toothing (*Figure 37*) is the construction of a temporary end of a wall by allowing every other end of stretcher brick to project halfway over the stretcher below. However, a temporary end may have to be raised over the full height of an area of wall. Toothing is usually reserved for very rare times when normal stepping is not possible. This can be done for both block and brick.

Toothing is not normally permitted, because it increases the possibility of poor jointing and tends to result in poor wall alignment. It should only be done with the contractor's permission and under the foreman's supervision. Toothing is usually done only in the following situations:

- When a halt must be made in construction for a period of time while waiting for delivery of units
- When construction must be halted for an extended time but is expected to continue later

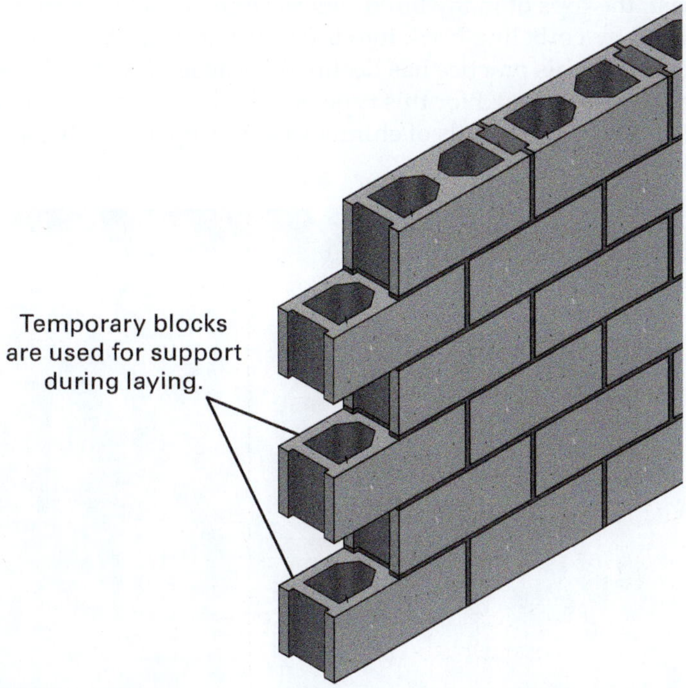

Figure 37 Toothing in block.

- When a masonry bond is needed between a wall section made of block and one made of brick
- When very high corner leads must be fully constructed before the line block is laid up
- When a walkway for the convenience of moving materials must be provided through the wall being built
- When window frames, doorframes, or jamb block have not yet arrived at the worksite

Toothing for block is done by laying up the block in alternate courses so that every other end block projects halfway over the block below. With this technique, the mason lays a block piece in the mortar at the end of the last course and then lays a full block directly over the piece. After the block is well set (but not completely set), the block pieces are carefully removed. Avoid disturbing the wall's alignment while completing this step.

Once construction begins again, check that the joints around the toothers are full of mortar and all surfaces are free of dirt and old mortar. Experience has shown that toothed areas are often the areas where the most water damage will occur, so pay extra attention to this.

Toothing for bricks is done by temporarily laying a bat in the mortar at the end of the course just laid, and then laying the next brick directly over it. After the mortar has set enough to be firm and rigid, the bats are carefully removed so as not to disturb the mortar bond of the permanent brick. Wooden wedges can be used for temporary support if needed.

3.2.0 Laying Out and Constructing Corbeling

Corbeling is the process of widening a wall by projecting masonry units outward to form a ledge or shelf. When corbeling, each brick course extends farther than the one below it. Generally, the unit should not project beyond the one below it more than one-third the unit width or one-half the unit height, whichever is less. Always follow good bonding practices and use well-fitted mortar joints to ensure that the projection will remain stable.

In the past, the tops of many buildings were constructed with masonry trim work designed by corbeling brick into intricate patterns. An example is shown in *Figure 38*. While this practice has declined because of cost and design considerations, there is still a need for this type of work in specialty areas and restoration projects. Also, inside walls of chimneys and fireplaces still use corbeling as a standard practice.

Figure 38 Corbeling on a masonry structure.
Source: Shutterstock

Brick should be dry and free of damage, because most edges are exposed. A dry brick also sets quickly, which eases the bricklaying process. The brick must be level, plumb, and ranged. Well-filled mortar joints give maximum strength and reduce the amount of tooling and jointing required. Building codes generally require the top course to be a full header course. Corbeling should not be done so quickly that the mortar and brick cannot take a firm initial set and remain in position.

To build corbeling for a brick wall, follow these steps:

Step 1 Snap a line on the foundation or base where the front edge of the wall is to be constructed.

Step 2 Lay out the first course as a dry course to check spacing. A running bond will be used with a header course every seventh course.

Step 3 Lay the first course with headers on the front wythe and stretchers on the back. Use a full mortar bed. Level, plumb, and range the brick. Check the spacing.

Step 4 Lay the second through sixth courses. Check the height of each course as it is laid and ensure that the bed joints are uniform and level.

Step 5 Lay the seventh course as a header course with full mortar joints.

Step 6 Continue to lay the wall until you reach the height where the corbel course is to begin. Check your progress often to be sure the wall is plumb and straight.

Step 7 Begin the corbel course by projecting headers out $^5/_8$" beyond the course below. Fill the wide head joint with mortar.

Step 8 Lay the second corbel course using stretchers along the front wythe and headers along the back wythe. Fill in the space between the bats. Be sure the course is level and straight.

Step 9 Lay the third corbel course using stretchers on the front and back wythes and $^3/_4$" headers between the two wythes. Each corbel should project $^3/_4$" beyond the course below.

Step 10 Lay the next course the same as the first course of the wall: headers on the front wythe and stretchers on the back wythe. Continue the wall to the desired height.

Step 11 Finish all joints when the mortar has set to thumbprint hard.

Step 12 Clean the wall with your trowel and bricklayer's brush.

3.3.0 Laying Out and Constructing Intersecting Walls

When intersecting walls are tied to a main wall, they provide additional support to the main structure. The two basic methods of anchoring intersecting walls are true masonry bonding and flexible connection (*Figure 39*).

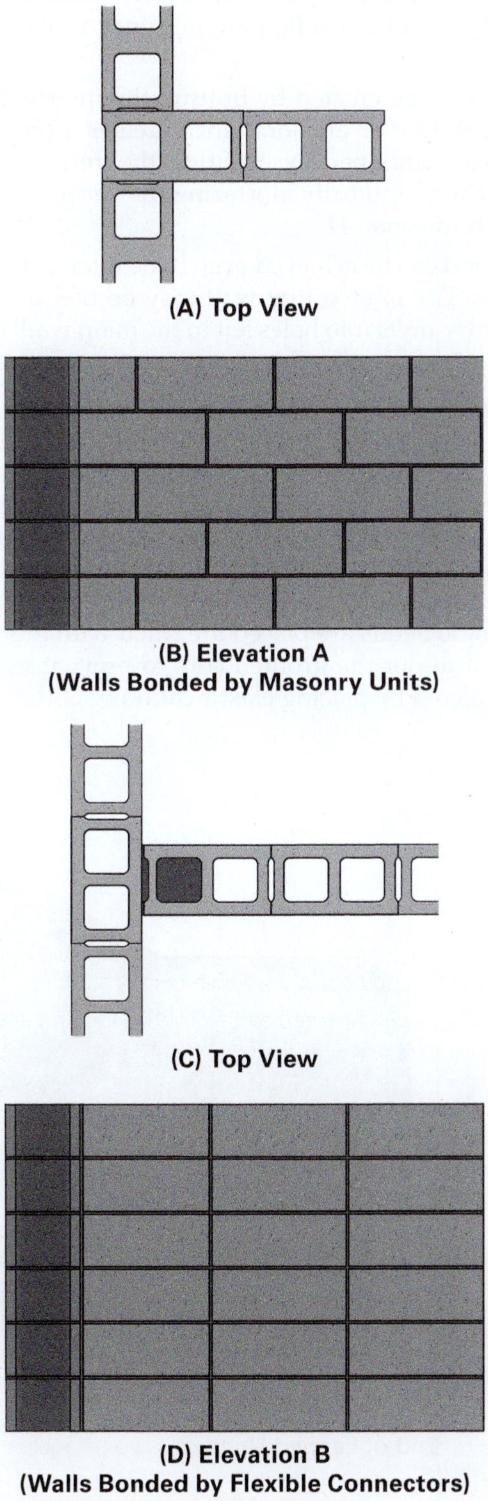

Figure 39 Masonry bonding and flexible connection.

An important requirement in masonry work is that load-bearing walls should be anchored where they meet at corners.

Various tie bar shapes are available, but all should extend at least 8" into the holding wall. If the walls are being carried up together, the anchors are installed when both walls reach bonding height. These anchors are generally provided at least every 36" OC.

When intersecting block walls are load-bearing and depend directly on each other for support, the bond between the two must resist all movement. For this reason, the bond at such an intersection is usually a rigid, true masonry bond.

This rigid bonding can be created in the following ways:

- If the courses of intersecting walls are laid up simultaneously, a true masonry bond can be established by overlapping masonry units at the wall intersection (*Figure 40*).
- A rigid bond can also be created by butting the intersecting walls together and applying a metal tie or anchor, which creates a physical bond. In such joints, the bond is maintained by grouting the joining cores, applying the various ties or anchors, and fully mortaring the vertical joints at the intersection. This is shown in *Figure 41*.
- A true masonry bond can be achieved even if the intersecting walls are not laid up simultaneously. The intersecting wall may be tied into the main wall by bonding the masonry units into holes left in the main wall. In such cases, metal bar anchors must be solidly placed in mortared cores to strengthen the bond.

Newer jointing techniques have been developed that use flexible anchoring and control joints that allow for slight movement but provide lateral support.

In one method of creating a flexible joint at intersecting walls, a steel tie bar is used for lateral support (*Figure 42*). These tie bars are made of heavy-gauge steel that is $\frac{1}{4}$" thick, $1\frac{1}{4}$" wide, and 16" to 28" long. The bars usually have 2" right-angle bends on each end that are locked into the grouted cores.

When tie bars are used, they are usually placed at 16" or 24" OC, and the cores in which the right-angle bends are placed are filled with mortar or grout. Metal lath should be placed under the grouted core to prevent mortar from seeping out. The joint is completed by placing elastic caulking at the wall intersection to create a control joint.

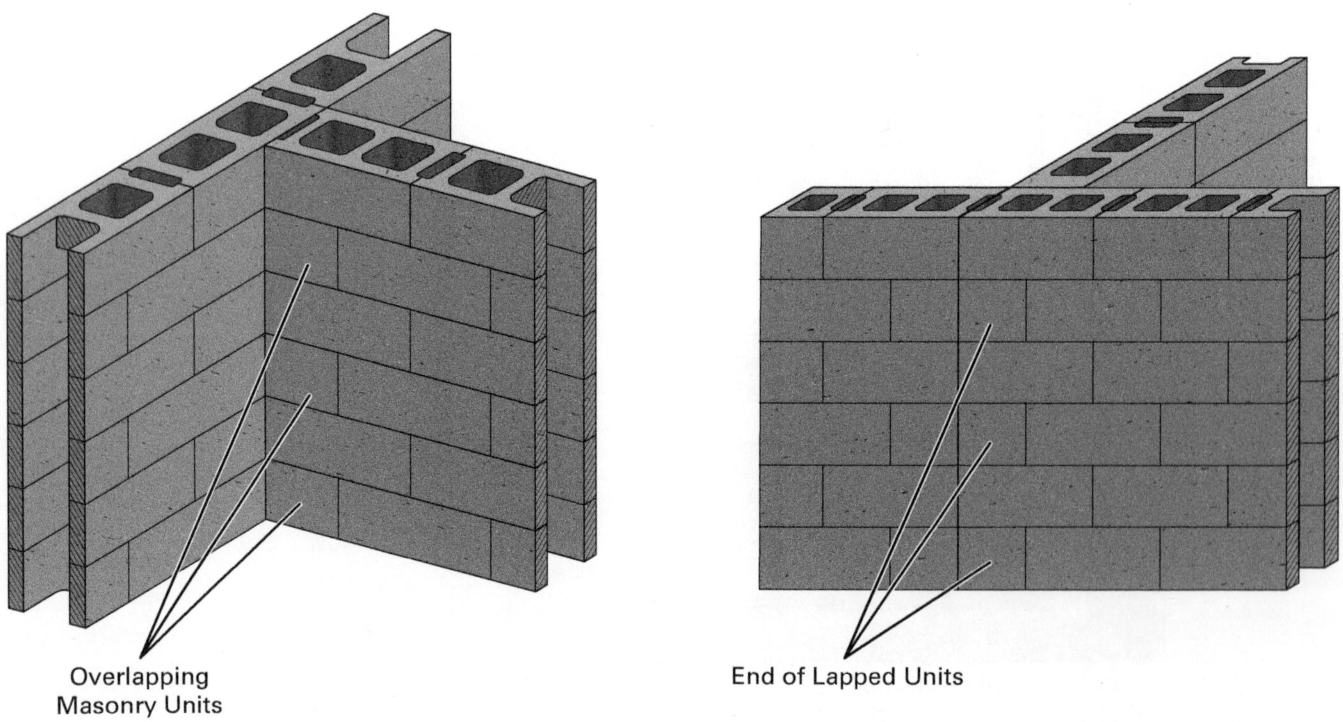

Figure 40 True masonry bond at intersecting walls.

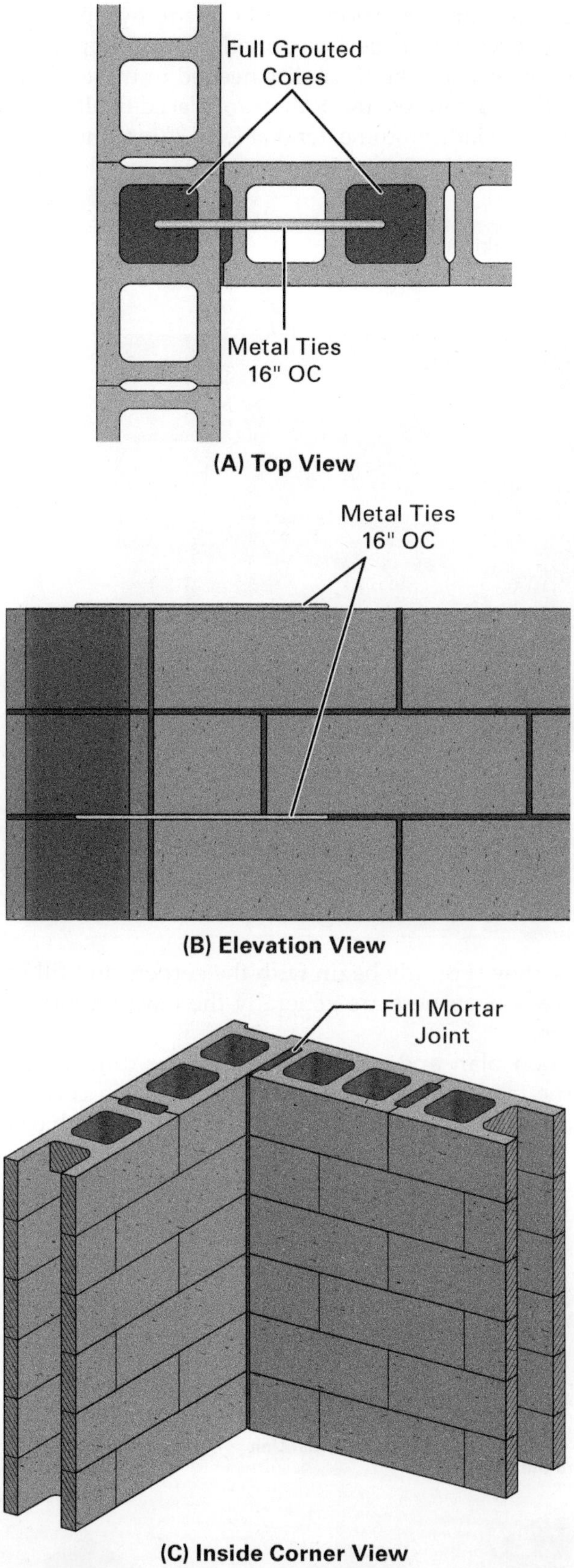

Figure 41 Rigid connections for intersecting walls.

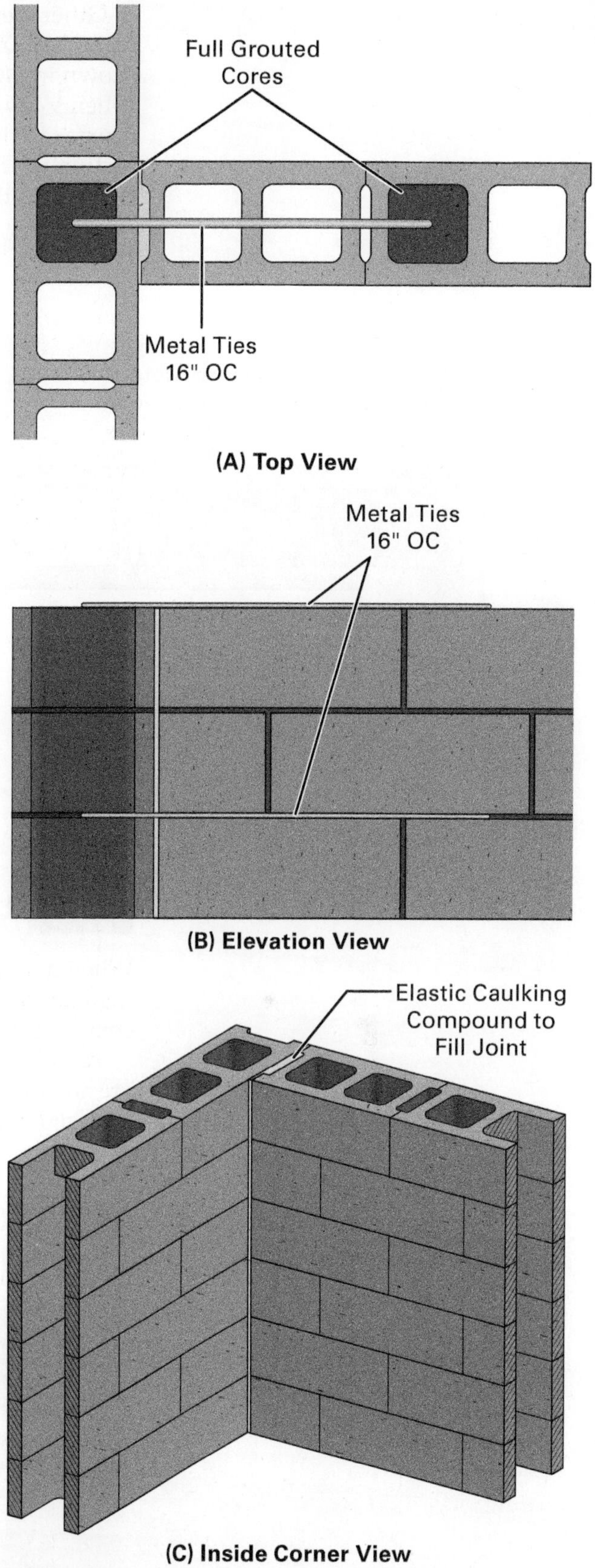

Figure 42 Flexible joints using tie bars.

Other flexible intersecting wall joint variations can be made by applying metal lath, Z-type unit ties, or *T*-shaped wire mesh to provide lateral support, as shown in *Figure 43*. Control joints provide the flexibility needed in these joints. When $\frac{1}{4}$" wire mesh or strips of metal lath are used, they are placed in alternate courses in the wall or at 16" OC. The lath should extend at least 8" into the intersecting wall. For safety, the lath should be folded down against the wall until it is tied into the intersecting wall.

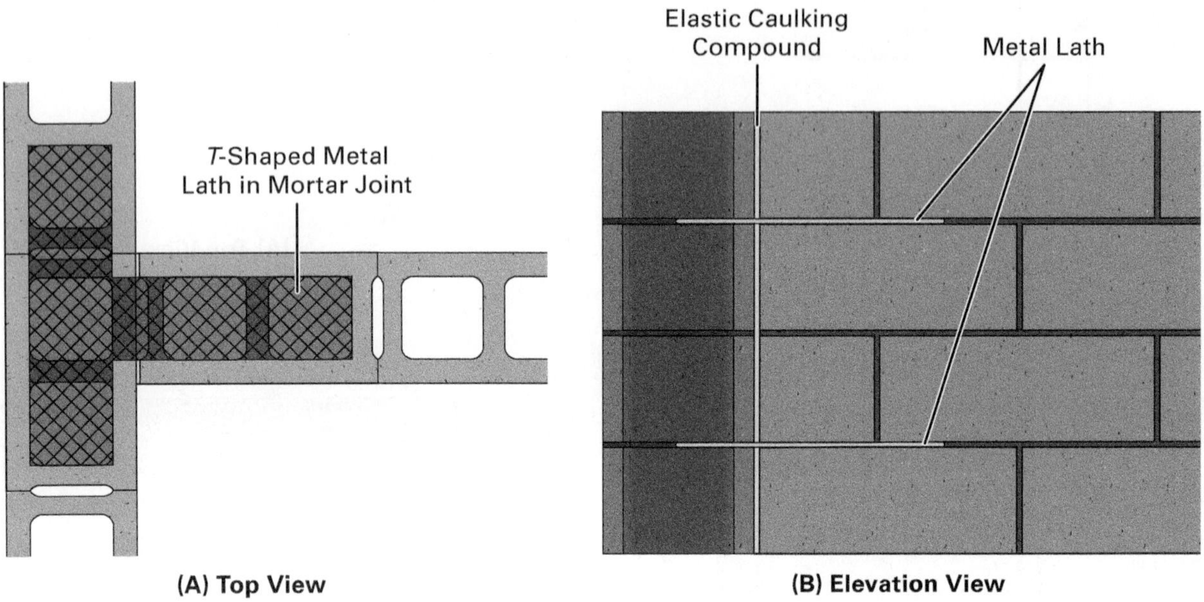

Figure 43 Flexible joints using metal lath.

3.4.0 Laying Out and Constructing Angled Corners

When masons construct walls, they typically begin with the corners and fill in the rest of the wall. For this reason, corners are generally the most important component of wall construction.

Always follow the foundation plan and ensure that modular dimensions are maintained while constructing corners and walls. Block used in corners is designed with square ends (*Figure 44*). However, specialized corner block may have ends that are beveled or molded with a 45° miter. These features are intended to reduce cutting while also maintaining modular dimensions. *Figure 45* shows the corner details for 4", 6", 8", and 12" block walls.

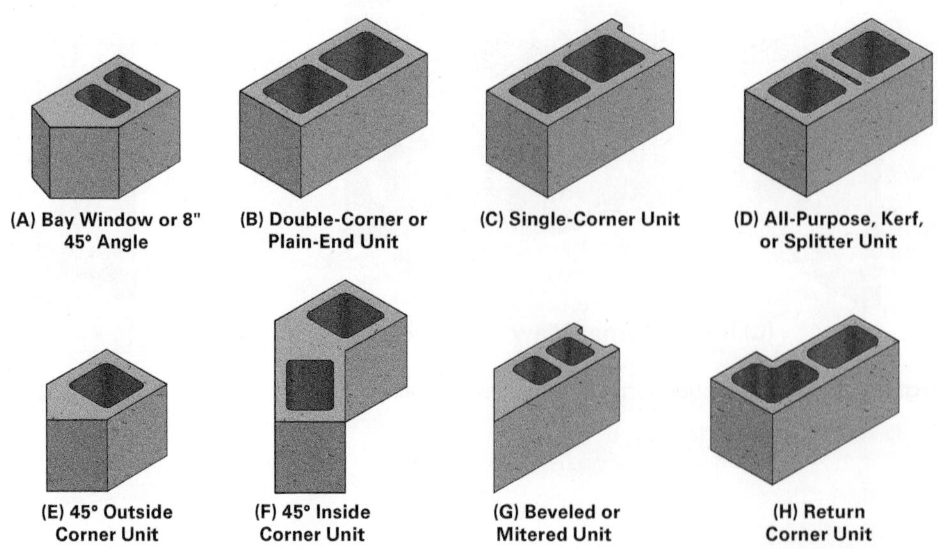

Figure 44 Block used in angled corners.

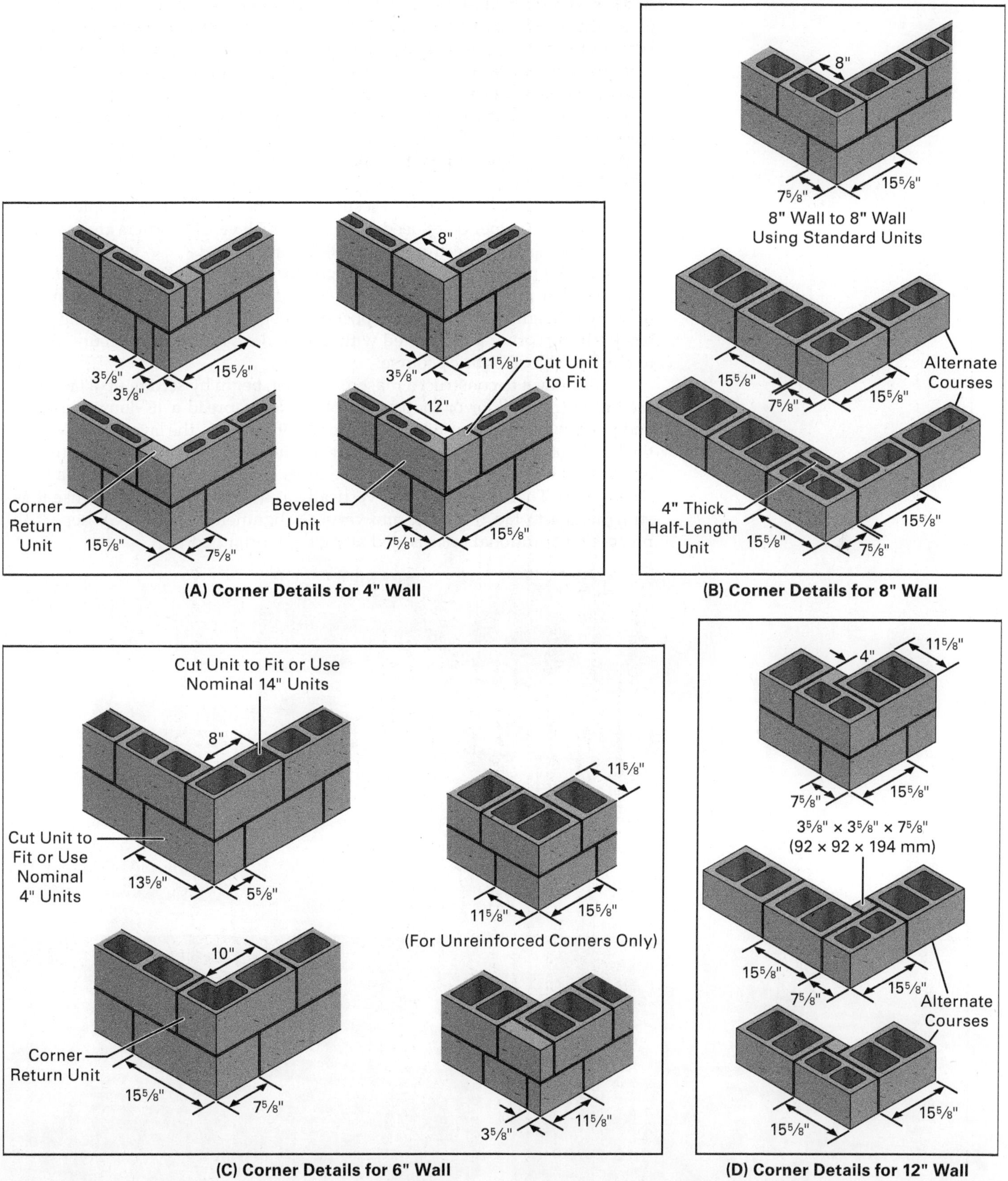

Figure 45 Corner details in 4", 6", 8", and 12" walls.

Corners must be designed to withstand stresses caused by structural loads. *ACI CODE-530/ASCE 5/TMS 402, Building Code Requirements for Masonry Structures,* lays out the stress requirements for corners in masonry construction. It specifies that stresses may be transferred from one wall to an intersecting wall using a running bond, steel connectors, or bond beams.

There are two ways to construct masonry corners: using a corner pole or using a level.

When using a corner pole to build a corner, begin by placing the corner pole at the location where the plans indicate the corner is to be constructed. If there is no existing structure to which to secure the corner pole, use telescopic braces to brace the post. Corner poles used for placing brick veneer corners are fitted with adjustable brackets to secure the pole to the structure.

Follow the manufacturer's instructions for securing the pole in position and ensuring that it is plumb. Then, use a modular spacing rule to begin marking off-course heights on the corner post and begin stringing lines using corner blocks. If the corner pole is fitted with line holders, use them to secure the lines to mark the block or brick course.

To use a level to construct a masonry corner, begin by setting up a layout line according to the floor or foundation plan. Next, spread a 16" long mortar bed and place two masonry units in a corner position along the layout line. Measure the height of the brick using a modular spacing rule (*Figure* 46A). Place the level across the brick center line and tap the units level using the handle of a trowel (*Figure* 46B). Then, turn the level vertically at the plumb corner and use it as a straightedge to tap the brick into vertical alignment (*Figure* 46C). Repeat this process for additional brick added along the mortar bed.

(A)

(B)

(C)

Figure 46 Using a level to place corner block.

Lay the second course using the same technique but reversing the direction to reflect the bond pattern (*Figure 47*). Tool the joints as you proceed, using the techniques you learned to lay block and brick in a wall. Continue this process until reaching the desired number of courses (*Figure 48*).

Figure 47 Laying the second course using a level.

Figure 48 A completed corner.

3.0.0 Section Review

1. A method of temporarily ending a wall, which is generally not permitted because it increases the possibility of poor jointing, is known as _____.
   a. stepping
   b. cornering
   c. corbeling
   d. toothing

2. When should you finish tooling joints?
   a. As you lay materials
   b. When the mortar is thumbprint hard
   c. When the mortar has completely cured
   d. Never

3. What is the typical spacing of tie bars?
   a. 16" to 24" OC
   b. 8" to 12" OC
   c. One tie bar every 1'
   d. One tie bar every 6"

4. What is a common tool for laying masonry corners plumb and square?
   a. Speed square
   b. Z-tie
   c. Corner clamps
   d. Corner poles

Module 28205 Review Questions

1. To completely bond two wythes of masonry in a multi-wythe wall, fill the collar joint with _____.
   a. foamed-in-place insulation
   b. reinforcing bars
   c. mortar
   d. rubble

2. The horizontal spacing between steel anchors used to bond wythes of masonry is typically _____.
   a. 36"
   b. 32"
   c. 28"
   d. 24"

3. Improved insulating value, reduced rain penetration, and prevention of condensation on interior walls are the main advantages of _____.
   a. a cavity wall
   b. solid masonry
   c. a hollow masonry wall
   d. a partition wall

4. Flashing is installed at the base of a wall cavity to _____.
   a. direct moisture to weep holes
   b. bond base courses together
   c. improve air circulation
   d. prevent entry by insects

5. Weep holes should be provided on center in head joints immediately above flashing sheets every _____.
   a. 5" to 7"
   b. 7" to 14"
   c. 16" to 32"
   d. 24" to 48"

6. A retaining wall that depends primarily upon its own weight to hold back the soil pressure is called a _____.
   a. gravity wall
   b. counterfort wall
   c. cantilever wall
   d. buttress wall

7. Seepage of water through a retaining wall can cause an aesthetic condition known as _____.
   a. blooming
   b. efflorescence
   c. inflorescence
   d. scale

8. The rule of thumb for building a serpentine wall is that the height should be no greater than the width of the structural unit times _____.
   a. 10
   b. 12
   c. 15
   d. 18

9. Limiting the size or locations of cracks in block masonry structures is the reason for using _____.
   a. continuous reinforcement
   b. collar joints
   c. adjustable wall ties
   d. control joints

10. Control joints are designed to allow longitudinal movement of up to _____.
   a. $\frac{1}{8}$"
   b. $\frac{3}{16}$"
   c. $\frac{1}{4}$"
   d. $\frac{5}{16}$"

11. The type of expansion joint material made from short pieces of material lapped to form a continuous weather seal is _____.
   a. copper
   b. extruded plastic
   c. premolded plastic
   d. neoprene

12. Toothing should only be used in special circumstances, because it can result in _____.
   a. additional expense
   b. wall collapse
   c. poor wall alignment
   d. wasted material

13. A common projection for each course of corbeling with standard brick is _____.
   a. $\frac{1}{2}$"
   b. $\frac{5}{8}$"
   b. $\frac{7}{8}$"
   d. $1\frac{1}{8}$"

14. A flexible joint must allow slight movement while providing _____.
   a. longitudinal support
   b. compressive strength
   c. torsional stability
   d. lateral support

15. When using a corner pole to position masonry units in a corner position, the length of the initial mortar bed should be _____.
    a. 8"
    b. 16"
    c. 24"
    d. 32"

Answers to odd-numbered questions are found in the Review Question Answer Key in the back of the book.

Answers to Section Review Questions

Answer	Section	Objective
Section One		
1. d.	1.1.0	1a
2. a.	1.2.0	1b
3. c.	1.3.0	1c
4. b.	1.4.0	1d
5. b.	1.5.1	1e
6. a.	1.6.0	1f
7. d.	1.7.0	1g
Section Two		
1. a.	2.1.0	2a
2. b.	2.2.0	2b
3. d.	2.3.2	2c
Section Three		
1. d.	3.1.0	3a
2. b.	3.2.0	3b
3. a.	3.3.0	3c
4. d.	3.4.0	3d

MODULE 28206

Effects of Climate on Masonry

Source: Shutterstock

Objectives

Successful completion of this module prepares you to do the following:

1. Identify the types of insulation used in conjunction with masonry construction and explain installation techniques.
   a. Explain the concept of heat transfer.
   b. Explain the purpose of and installation procedures for internal insulation.
   c. Explain the purpose of and installation procedures for external insulation.
2. Identify the need for moisture control in various types of masonry construction, and describe the techniques used to eliminate moisture problems.
   a. Explain the purpose of and installation procedures for flashing.
   b. Explain the purpose of and installation procedures for weep vents.
   c. Explain the purpose of and installation procedures for waterproofing.
3. Explain the various techniques used to provide adequate protection during hot- and cold-weather masonry construction.
   a. Explain the role played by weather data and information in masonry construction.
   b. Explain the various techniques used to provide adequate protection during hot-weather masonry construction.
   c. Explain the various techniques used to provide adequate protection during cold-weather masonry construction.

Performance Task

Under supervision, you should be able to do the following:

1. Install a 4' section of base flashing.

Overview

There is more to masonry than laying block and brick to create horizontal and vertical surfaces. Masons must understand how heat transfer and moisture absorption can affect masonry elements and the soundness of masonry structures and must also consider the effects of weather. This module describes techniques used to apply insulation, eliminate problems caused by moisture, and protect masonry construction, materials, and equipment during hot and cold weather. These skills are just as essential to good masonry as knowing how to lay block and brick.

Digital Resources for Masonry

Scan this code using the camera on your phone or mobile device to view the digital resources related to this craft.

SCAN ME

NOTE

Codes vary among jurisdictions. Therefore, it is essential to consult the applicable code whenever regulations are in question. Referring to an incorrect set of codes can cause as much trouble as failing to reference codes altogether. Obtain, review, and familiarize yourself with your local adopted code.

1.0.0 Insulation in Masonry Construction

Performance Tasks

There are no Performance Tasks in this section.

Objective

Identify the types of insulation used in masonry construction and explain installation techniques.

a. Explain the concept of heat transfer.
b. Explain the purpose of and installation procedures for internal insulation.
c. Explain the purpose of and installation procedures for external insulation.

Insulation is any material that reflects heat or simply provides a barrier that retards heat penetration. Insulation is also defined as any material that resists the flow of heat.

Generally, denser materials are less effective in resisting heat transfer. For instance, lightweight block are more resistant to heat movement than normal-weight block. The insulation materials that best resist heat transfer contain millions of tiny air cells or spaces that trap heat as it moves through the material.

Since the early 1970s, insulation has become a major design consideration for all building construction. Due to increased energy awareness, the general public now insists that the construction industry provide insulation in almost every type of construction in all areas of the country. Federal and state government agencies have also developed recommendations and regulations concerning insulation.

While most building plans provide information on the type of insulation to be used, masons should recognize the basic physical properties of heat transfer and be able to recommend an insulation type if asked.

1.1.0 Understanding Heat Transfer

The comfort of any structure depends on its physical features and the natural laws of heat transfer. Heat transfer is the movement of heat energy from a warmer area to a colder area by conduction, radiation, or convection.

Heat moves through different building materials at different rates. *Table 1* shows several types of materials and their heat resistance. Thermal resistance (or R-value) measures the heat transfer rate through a material. The higher the R-value, the better a material's insulating ability.

Indoor and outdoor temperatures fluctuate throughout the day, especially in deserts and mountains, which experience extreme changes.

TABLE 1 Heat Transfer Coefficients

Material or Product	Resistance per Inch of Thickness	Resistance per Thickness Shown
Concrete (per 1" thickness)	0.08	0.08
Face brick (per 1" thickness)	0.11	0.11
Hollow concrete block, 8" thickness	—	1.11
Stucco (per 1")	0.20	0.20
Metal lath and plaster, 3/4"	0.17	0.13
Gypsum board, 3/4"	0.90	0.45
Plywood, 1/2"	1.32	0.62

As the temperature changes, so does the direction of the heat flow. Depending on what materials are used in a wall, the heat flow from the outside of a building to the inside can vary. This is commonly referred to as **thermal mass**. As a result, the wall's thermal performance will differ from the steady-state R-value assigned to the materials used in the wall.

Thermal mass: The ability of a building wall to alternately absorb and release heat energy in response to temperature changes.

1.1.1 Conduction

Conduction is the passage of heat or electricity through a medium. Some materials transmit heat more quickly, meaning they have a higher conductivity.

Heat naturally flows from warm areas to colder areas to create balance. In the winter, when outside temperatures are lower than indoor temperatures, the heat from the room flows through the walls toward the colder outside areas (*Figure 1*). The wall temperature itself is lower than the room air temperature.

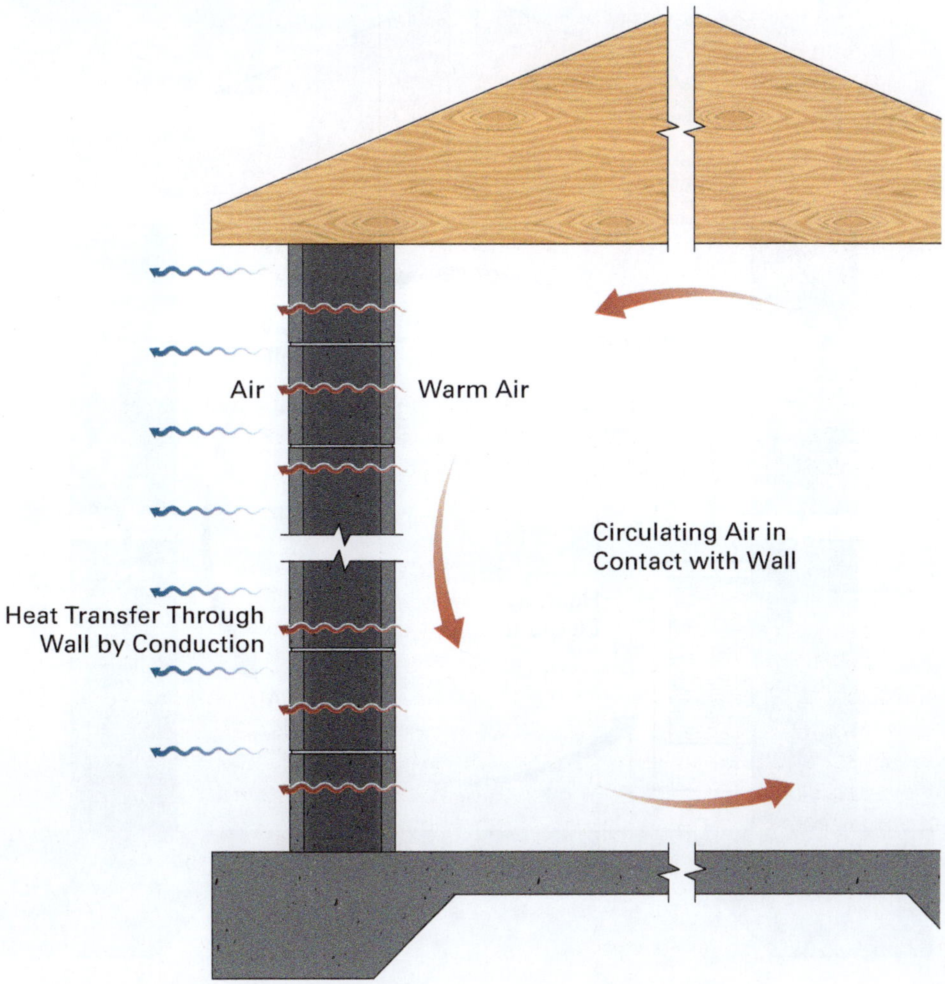

Figure 1 Heat transfer by conduction.

According to the law of conductivity, the heated air moves through the wall and mixes with the outside air. In the summer, the process and the direction of heat flow are reversed.

Generally, dense building materials conduct heat more quickly and are called *conductors*. For example, you can easily feel how quickly metals transfer heat. On the other hand, less dense materials, such as concrete masonry units (CMUs), wood, and plastics, are often called *insulators* because they prevent or slow heat flow. A wooden match, for example, does not become too hot to hold after it is lit.

1.1.2 Radiation

The surface of any object, building material, or human body transfers heat to nearby colder surfaces. Heat transfer by radiation is by invisible light waves and heat waves. For example, the sun heats Earth by radiation. A radiator heats a room with radiant heat instead of blowing warm air. A warmer wall loses heat to colder outside air by the same law of radiation.

1.1.3 Convection

Heat also moves from one area to another by a process called *convection*. Hot air expands and rises as cool air falls, creating convection loops that heat and cool an area, as shown in *Figure 2*. In fact, most heating and cooling systems directly depend on convection.

Heat transfer is typically a combination of conduction, radiation, and/or convection. For example, the air that encounters a heat source is heated by conduction and rises along a wall. As the air loses its heat, it becomes cooler and heavier until it falls to the original heat source in the convection loop.

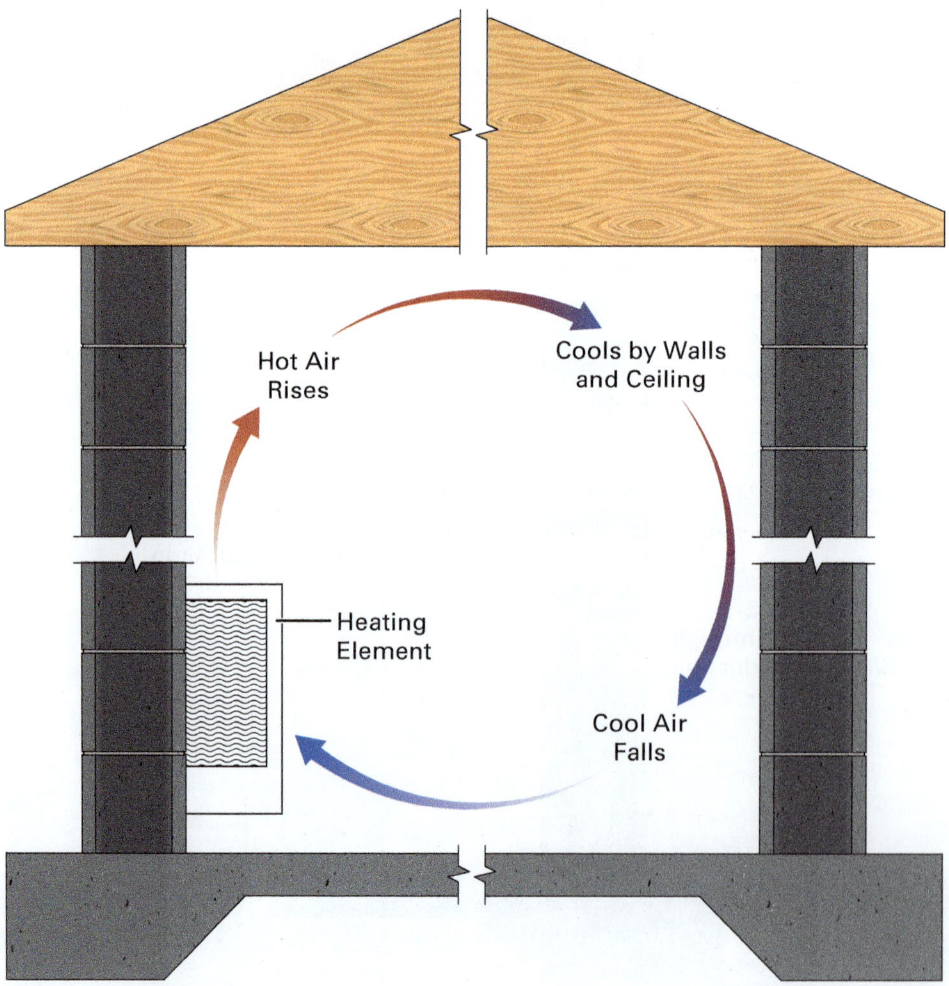

Figure 2 Heat transfer by convection.

Masons must be aware of which construction materials will be most effective at resisting heat flow through walls and which additional materials, such as insulation, they can use to create an adequate building envelope.

Insulation materials should be chosen based on the following factors:

- The resistance or R-value of various insulation types
- Material cost, application, and maintenance
- Insulation location and installation method
- Changing climates and temperatures
- Code specifications

There are several basic categories of insulation and several types of insulation within each category. Following are the basic types of insulation:

- Loose-fill
- Rigid or slab
- Reflective
- Foamed-in-place
- Insulation insert
- Flexible blanket or batt
- Spray foam/injected

Most building material suppliers can advise you on the proper R-value for your specific application. The R-value differs according to the type and thickness of the insulation being used. *Table 2* gives values for several different types of insulation.

TABLE 2 R-values for Different Insulation Types

Product	R-value per 1" Thickness
Loose Fill	
Mineral fiber (rock, slag, or glass)	2.20–3.00
Cellulose	3.70
Perlite	2.70
Vermiculite	2.13
Flexible	
Mineral wool batts	3.10–3.70
Rigid	
Cellular glass	2.63
Expanded polystyrene (extruded)	5.00
Expanded polystyrene (molded)	3.57
Expanded polyurethane	6.25
Mineral fiberboard	3.45
Polyisocyanurate	7.20
Reflective	
Aluminum foil	3.48
Foamed-in-Place	
Urea formaldehyde	4.20
Polyurethane	6.25

For masonry work, these general types of insulation are classified by their position in a structure. When insulation is placed in individual cores or in cavities between wythes, it is often referred to as *internal insulation*. When insulation material is applied to a wall surface, it is often called *external insulation*.

1.2.0 Installing Internal Insulation

Insulation placed between the interior and exterior faces of masonry work is known as *internal insulation*. This may be loose-fill insulation placed in block cores or cavity walls, insulation inserts in the cores of CMUs, or rigid insulation placed between the masonry wythes in a cavity wall.

1.2.1 Loose-Fill Insulation

Loose-fill insulation (*Figure 3*) is manufactured as fibers, granules, or chips and is delivered by bag or in bulk. These loose fibers can be poured, placed by hand, or blown into cavities or cores. When choosing loose-fill insulation, ensure that the desired properties of the material are met.

Figure 3 Loose-fill insulation.
Source: Shutterstock

In most situations, loose-fill insulation must meet at least the following criteria:

- It must not transmit moisture from the exterior wythe to the interior wythe.
- It must not absorb water or prevent water from escaping the cavity.
- It must be strong enough to support its own weight. In other words, it should not settle or pack down over a period of time. Settling eliminates the dead air spaces between particles, reducing the insulation value.
- It must be composed of inorganic or treated organic material. Organic material should be treated to resist fire, decay, insects, and rodents.
- It must be easy enough to handle so that it can be poured into wall cavities in lifts of about 4′ high.

Two of the more popular loose-fill insulation materials are water-repellent vermiculite and silicone-treated perlite. Vermiculite is an expanded form of mica that is resistant to fire, decay, moisture, and vermin. Perlite is a volcanic glass that is also expanded by a heating process to create lightweight insulation. Both of these loose-fill insulation types resist packing when they are applied in 4′ lifts.

Loose-fill insulation can readily be placed in block cores and in the cavity between masonry wythes. The insulating value depends on the amount of insulating material used.

For example, a normal-weight 8" block has an R-value of about 1.11; a lightweight 8" block has an R-value of about 2.18; and an 8" lightweight block with loose-fill insulation in the cores has an R-value of about 5.03. The trade-off is between the block's initial cost and the savings that can be realized due to lower heating/cooling costs.

WARNING!

Always wear appropriate personal protective equipment (PPE), including respiratory protection, when installing loose-fill insulation, as the insulation materials can cause injury if inhaled. When installing loose-fill insulation in windy conditions, position workers upwind of the insulation outlet.

1.2.2 Rigid or Slab Insulation

Most masons prefer to use preformed, rigid, nonstructural insulation board rather than loose-fill insulation when building cavity walls. This type of insulation allows installation of the necessary flashing and weep vents at the base of cavity walls.

Rigid board insulation is manufactured from synthetic materials and is made in various sheet sizes ranging from $\frac{1}{2}$" to 2" thick (*Figure 4*).

The most common types of rigid, nonstructural insulation include the following:

- Rigid urethane foam
- Expanded, molded polystyrene foam
- Polyisocyanurate foam
- Expanded perlite
- Expanded polyurethane
- Cellular glass fiberboard
- Preformed fiberglass

When these rigid boards are used as internal insulation, they are fastened to the outside face of the inner wythe by applying mastic to the block or by using metal ties to clip the board in place. This method maintains an air space between the outer wythe and the insulation to ventilate the cavity. *Figure 5* shows a rigid board installed on an interior block wall.

Figure 4 Rigid foam board.
Source: Shutterstock

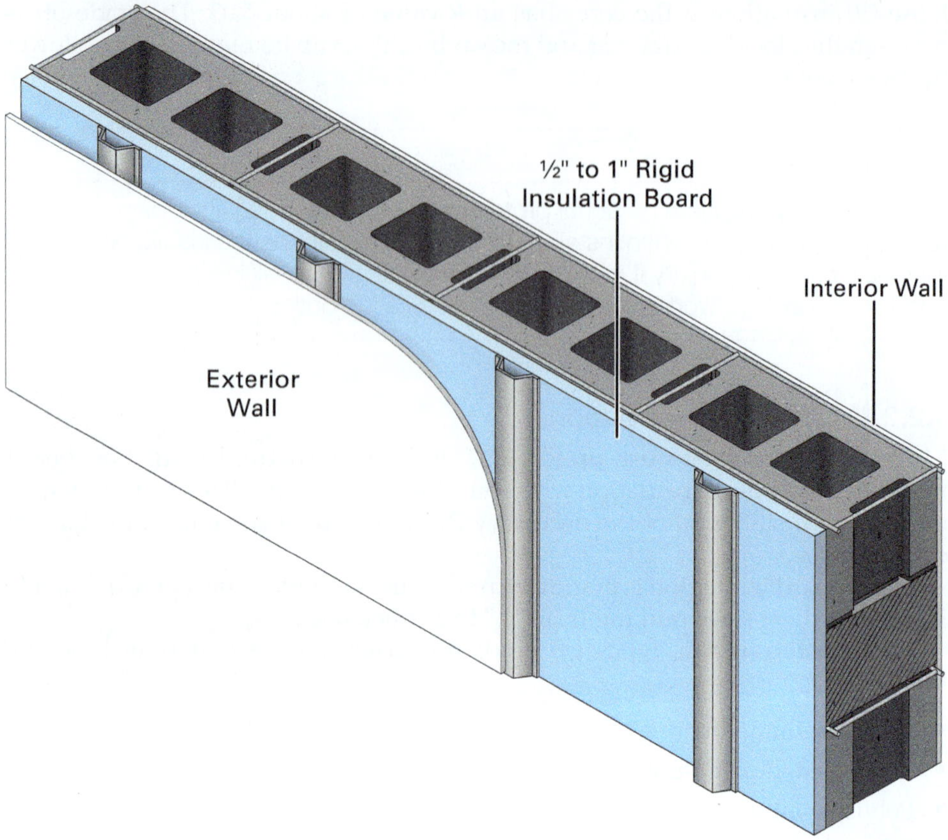

Figure 5 Installing rigid insulation board.

Urethane foam is a plastic noted for its resistance to heat transfer and durability. However, urethane expands when damp and burns when exposed to flames. Polyisocyanurate will also burn, but it has the highest R-value of any rigid board insulation.

Perlite board has one of the lowest R-values and must be kept dry because it absorbs water. By far, the most popular rigid board materials used in cavity wall applications are polystyrene foam boards, cellular glass fiberboard, and preformed fiberglass boards.

R-value

The actual R-value for insulation is calculated from laboratory experiments. It measures the material's overall thermal resistance.

Polystyrene boards are formed by molding or by extrusion. The molded type is often called *beadboard* because it is made from small plastic beads forced or molded together into a rigid board. Extruded polystyrene is more rigid than molded boards and has a higher R-value. Both types are moisture resistant but give off carbon monoxide gas when they burn.

Cellular glass fiberboard and fiberglass board insulation are even more resistant to water vapor and moisture absorption.

1.2.3 Reflective Insulation

Reflective insulation is often used inside heated spaces to reduce the amount of heat moving through masonry walls. This type of insulation is made from one or more layers of metal foil. The degree of insulation that reflective foil provides depends on how long it remains shiny and how it is used with an air space. To be effective, the foil must face an air space where heat can be reflected.

1.2.4 Foamed-in-Place Insulation

Foamed-in-place insulation can be used in the same situations where batts are used. These insulating products are made in a liquid foam or as expanding pellets. When they are poured or foamed in place, they expand to fill a cavity and then cure to form a solid mass.

In masonry, these foams are usually placed inside block cores on the interior wall face (*Figure 6*). Urethane foam is particularly efficient where a high R-value is required, but it will expand after it has been injected. Once the extruded foam has set, clean off any excess and fill in the remaining holes.

Figure 6 Injected insulation.
Source: Shutterstock

1.2.5 Insulated Block

Insulated block uses foam insulation inserts to provide better energy efficiency and soundproofing than standard block (*Figure 7*). The block is designed to provide a continuous and uninterrupted thermal barrier that absorbs and stores energy, allowing buildings to stay cooler in warm weather and warmer in cool weather. This allows designers to install smaller HVAC (heating, ventilating, and air conditioning) systems, which cost less to operate.

Because additional wall insulation is not required, the use of insulated block can also reduce construction time. The foam inserts are nontoxic and moisture resistant. Insulated block offers the advantage of allowing the insulation to be placed simultaneously as the masonry unit along with flashing or joint reinforcement. Care must be taken to avoid getting mortar droppings in the block cells.

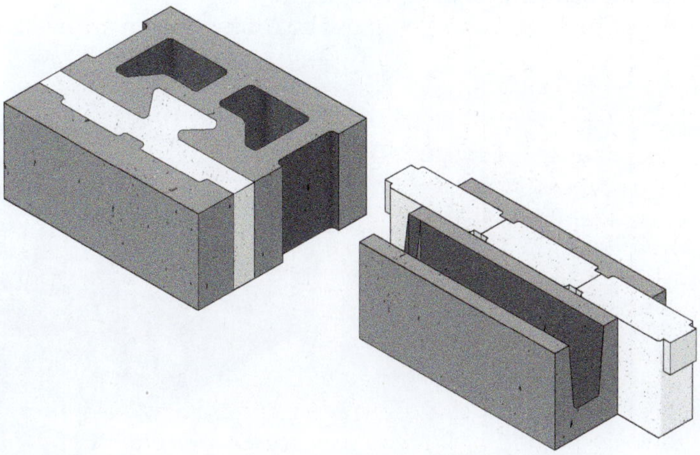

Figure 7 Common types of insulated block.

1.3.0 Installing External Insulation

External insulation is any type of insulation that is placed on the external faces of masonry units. The application of external insulation can be the task of the mason. However, masons must understand the basic application process and the value of these insulation types. The basic types of external insulation include the following:

- The addition of a coating on the exterior face of the masonry
- The addition of rigid board insulation on the exterior surface of the masonry
- The use of blanket or batt insulation between furring strips

1.3.1 Finishes as Insulating Agents

The insulation created by surface finishes is quite low, but many of these coatings protect insulation materials. These surface finishes are primarily used to change the color or texture of the project, to enhance the sound-insulating ability of a wall, or to prevent the passage of moisture.

Surface coatings include Portland cement paints, acrylic latex paints, oil or oil alkyd paints, and rubber-based paints; stains; Portland cement plaster finishes; clear coatings; and waterproof bituminous coatings.

The value of paint as an insulation material directly relates to the amount of moisture the paint will allow to pass through a wall. A type of paint that does not allow the passage of water vapor should be applied to the side of a wall where moisture enters.

Rubber-based paints are the most moisture resistant and are often used as primers. Clear coatings and stains are primarily used to increase the coloration of the wall and overall durability. Portland cement mixed as stucco plaster is the best coating application for its insulating value.

Plaster application can be done in two or three coats, by hand or by machine. Two coats are usually enough when applied to masonry units. These two coats usually have a rough-floated base and finish coats, as shown in *Figure 8*. Three coats are required if the masonry work is not very smooth or if metal lath is used as a base for the plaster. In both cases, the finish should be damp cured.

Figure 8 Block coatings.

1.3.2 Rigid Board Insulation

Rigid board insulation may be applied directly to the exterior face of a masonry wall using mastic or over furring strips attached to the interior face. Insulation boards are normally used as a base for plaster, bituminous coatings, or gypsum board, but manufacturers also make decorative rigid insulation boards that can be used as a finished surface. The most popular options are rigid glass fiberboard, polystyrene panels, and polyurethane boards or sheets.

1.3.3 Blanket and Batt Insulation

Blanket and batt insulation (*Figure 9*) are considered flexible insulation. They are usually applied between furring strips anchored directly to the interior of a masonry wall or between framing members on the interior of the wall.

These flexible insulation types are usually made of mineral wool or cellulose fibers that resist fire, moisture, rodents, and insects. Batts are manufactured to fit most standard framing patterns between structural studs. Blanket insulation is usually made in longer rolls that can be cut to size.

Flexible insulations come in three specific varieties:

- *Wrapped*—A tape binding on one or both edges and a vapor barrier on one or both sides
- *Faced*—A vapor barrier on only one side
- *Friction fit*—Without a covering, but designed to fit tightly in the frame or furring space

Faced and wrapped insulations usually have a stapling or tacking flange for installation. Builders should ensure that the vapor barrier is placed toward the living space and that the taped edge of wrapped insulation overlaps on the face of the furring strips. If the insulation does not have a vapor barrier, the builder will often place continuous polyethylene sheeting over the inside stud faces.

When placing the furring strips, follow the safety precautions necessary for attaching materials to masonry structures. The furring strips should be placed at 16" or 24" on center (OC). They can be attached by anchor bolts using a masonry drill or by specially hardened nails using a nail gun.

Furring strips on the interior face of a masonry wall (*Figure 10*) have several advantages for improving insulating efficiency. First, the furring strip adds an air space between the masonry and the wall finish, which provides improved heat and sound transfer resistance. Rigid insulating boards, batt insulation, or

Figure 9 Flexible insulation.
Source: Shutterstock

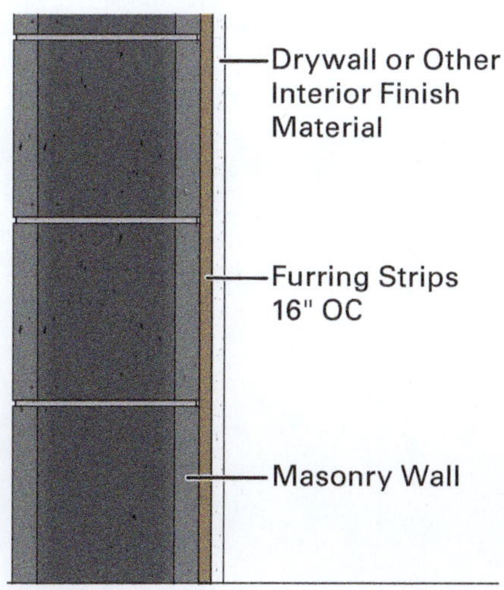

Figure 10 Using furring strips.

reflective films placed between furring strips further increase the wall's insulating qualities.

Proper insulating techniques reduce heat loss, control sound transmission from area to area, and maintain energy efficiency. However, the mason and builder must consider cost savings as well. Energy savings is not directly proportional to insulation thickness or type. For instance, using two types of insulation or two layers reduces heat transfer, but the savings may not be doubled. Normally, using lightweight block and popular insulation types will be the best insulating solution. Masons need to recognize the most efficient means of creating a well-insulated structure and be ready to recommend the appropriate applications.

Test and Prepare Masonry Before Plastering

The insulating value of plaster is due largely to the bond achieved between the plaster coating and the masonry. The surface preparation of the block is very important in achieving the proper bond. For new construction, the wall face must be free of oil and dirt and must be dampened before the plaster is applied. Older concrete masonry surfaces should be tested before the application process begins.

Fiberglass Insulation

Flexible fiberglass is probably the first thing that comes to mind when the average person thinks of insulation. Most people don't realize that this common material must be handled carefully. The tiny strands of glass in fiberglass insulation can irritate the skin, injure eyes, and cause various respiratory problems. Always wear proper PPE when handling fiberglass insulation.

However, fiberglass safety is more than just personal protection during installation. If debris from the installation is not properly removed or if existing insulation is disturbed, fiberglass particles could spread through the building. Fiberglass that enters the HVAC system will be carried to all parts of the building. Always use care when handling fiberglass insulation. This protects you and the building's current and future occupants.

1.0.0 Section Review

1. The passage of heat or electricity through a medium is called _____.
   a. convection
   b. conduction
   c. radiation
   d. resonance

2. When insulating cavity walls, most masons prefer to use _____.
   a. reflective insulation
   b. loose-fill insulation
   c. rigid board insulation
   d. foamed-in-place insulation

3. What is the maximum lift to reduce the risk of loose-fill insulation packing?
   a. 4′
   b. 5′
   c. 6′
   d. 7′

2.0.0 Moisture Control in Masonry Construction

Objective

Identify the need for moisture control in various types of masonry construction, and describe the techniques used to eliminate moisture problems.

a. Explain the purpose of and installation procedures for flashing.
b. Explain the purpose of and installation procedures for weep vents.
c. Explain the purpose of and installation procedures for waterproofing.

Performance Task

1. Install a 4' section of base flashing.

Moisture can have adverse effects on masonry construction. One of the most common issues is cracks in mortar and masonry units due to repeated expansion and contraction as water is absorbed and then evaporates. Moisture can also corrode exposed metal elements such as rebar, bolts, ties, and anchors. Lastly, moisture-rich minerals can cause chemical staining of mortar and masonry units as they evaporate.

Weather is perhaps the most common root cause. Throughout their life cycles, masonry structures are exposed to years of rain and snow, mist, fog, running water, and freeze-thaw cycles (*Figure 11*). Masons design structures to withstand these damaging effects by using the proper techniques and materials. This section explains how to use flashing, weep vents, and **waterproofing** to protect masonry structures against moisture damage.

Waterproofing: The process of treating masonry with a material that will retard penetration of moisture.

Figure 11 Masonry building in cold weather.
Source: Shutterstock

2.1.0 Installing Flashing

Flashing is the application of any material that prevents water penetration or provides water drainage. It is difficult to completely prevent rainwater from entering a wall at places such as roof intersections, parapets, sills, or other projections, but flashing will control moisture movement and penetration within masonry walls. It is installed at these vulnerable locations to collect water and divert it through weep vents to the exterior.

2.1.1 Flashing Materials

Suitable flashing materials must possess at least the following four important properties:

- Resistance to moisture penetration
- Resistance to corrosion
- Durability to resist puncture, abrasion, or other damage during installation
- Flexibility so that the flashing can be worked to the correct shape and retain this shape after installation

Flashings are usually formed from sheet metal, rubberized asphalt, plastics, or a combination of these materials. Material selection is determined by cost and suitability, but suitability should never be compromised because of cost. Replacement costs always exceed original costs, so it is wise to select a permanent flashing material for the original installation rather than must replace it.

The following are the most commonly used materials for flashing applications:

- *Copper*—Though costlier than most other flashing materials, copper is durable, is available in special preformed shapes, and has an excellent moisture barrier. It is not affected by the caustic alkalis in masonry mortars, so it can be safely embedded in fresh mortar and will not deteriorate unless excessive chlorides are present. When using copper flashing, the use of chloride-based additives in mortar (such as calcium chloride to speed the set of mortar in cold weather) should be prohibited.
- *Galvanized steel coatings*—These materials can corrode in fresh mortar. When placed in contact with mortar, it is usually necessary to protect them with a rubberized asphalt coating. Galvanized steel is widely used in residential buildings and generally performs adequately. However, exposed galvanized steel flashing requires painting maintenance.
- *Stainless steel*—Stainless steel flashings are available in several gauges and finishes. They are durable, corrosion resistant, excellent moisture barriers, and surprisingly workable.
- *Thermal plastics*—Plastics are becoming one of the most widely used flashing materials. These plastic flashings are tough, resilient, and highly resistant to corrosion. However, because the chemical compositions of plastics differ widely, it is impossible to lump all plastic flashings into one general group.
- *Rubberized asphalt*—Rubberized asphalt is commonly used in residential and commercial construction for various flashing applications. It is constructed using modified asphalt adhesives that nearly eliminate the leakage of bituminous fabrics used in the past. Primers, drip edges, and termination bars are typically recommended for use with rubberized asphalt.
- *Combination flashings*—Combination flashings consist of materials combined to gain the best properties of each, such as sheet metal coated with plastic or rubberized asphalt materials. Combination materials can provide a lower-cost flashing by reducing the thickness of the metal required, or they may permit the use of corrodible metals that might be inadequate unless coated.

2.1.2 Flashing Placement

When the construction calls for cavity walls or veneering over an existing framework, the application of flashing and weep vents is vital. Because it is impossible to keep all moisture out of an open cavity, and all moisture that does enter a cavity eventually travels downward, masons must provide a way for the moisture to be channeled to the exterior. Cavity walls are far more weather resistant than solid masonry walls if proper techniques are applied.

When properly installed, flashing provides a channel at the bottom of the cavity where moisture can collect and flow to a weep vent. To properly bed the flashing, most masons apply a thin coat of mortar at the bed joint, place the flashing, and finish the bed joint, making sure that the joint size remains the same for all joints. When the mortar sets, the flashing becomes a permanent part of the structure, and no open joint is left where water or moisture may drip past the flashing. The flashing is placed along the full length of the cavity.

Where flashing extends to the interior, place its end between the furring and the interior finish and turn it up at least 1" to collect moisture that may penetrate the wall.

As noted earlier, any moisture that enters a wall gradually travels downward. To divert this water to the exterior, place flashing above grade at the wall base (*Figure 12*). In areas where first-floor wood joists require protection from termites, use metal flashing that projects past the inside face of the wall and is bent down at an angle.

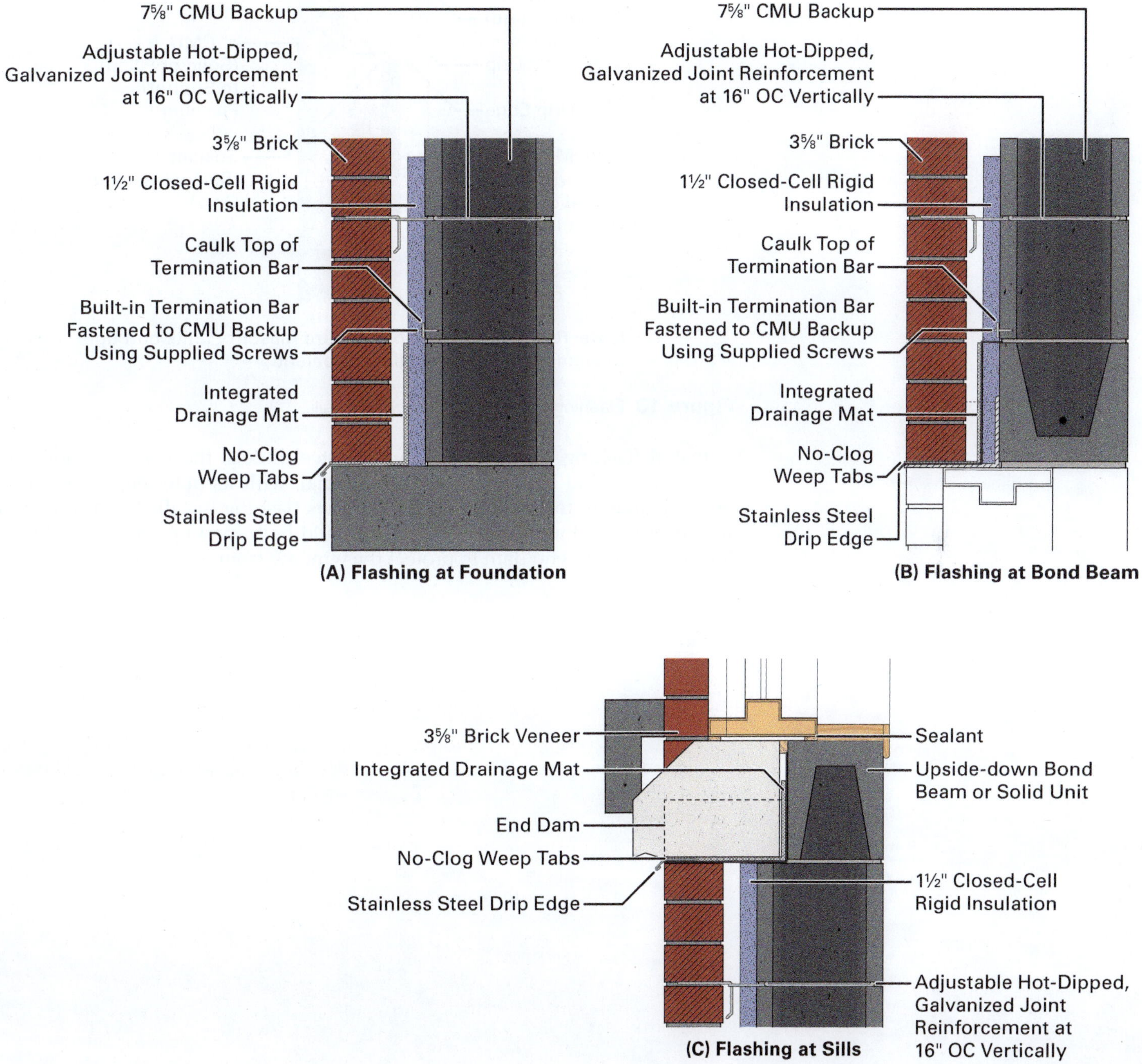

Figure 12 Flashing in a cavity wall.
Source: Mortar Net Solutions – Masonry Division of Keene Building Products

Place flashing under and behind all sills. Sill flashing ends should extend beyond both sides of the opening and turn up at least 1" into the wall. In addition, sills should slope and project away from the wall to drain water away. When the underside of a sill is not sloped, it should have a drip notch, or the flashing should be extended and bent down to form a drip (*Figure 13*).

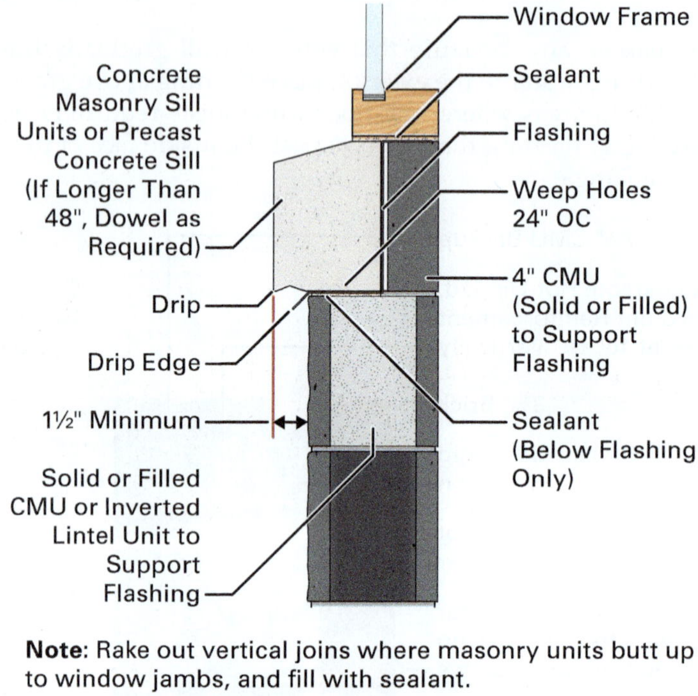

Figure 13 Flashing in a single-wythe windowsill.

Install flashing over all openings. At steel lintels, the flashing should be placed under and behind the facing material, with its outer edge bent down over the lintel to form a drip. *Figure 14* shows flashing installed in a composite wall over a window. Stepped flashing should be turned up on the inside and folded to form an end dam to protect the window from moisture.

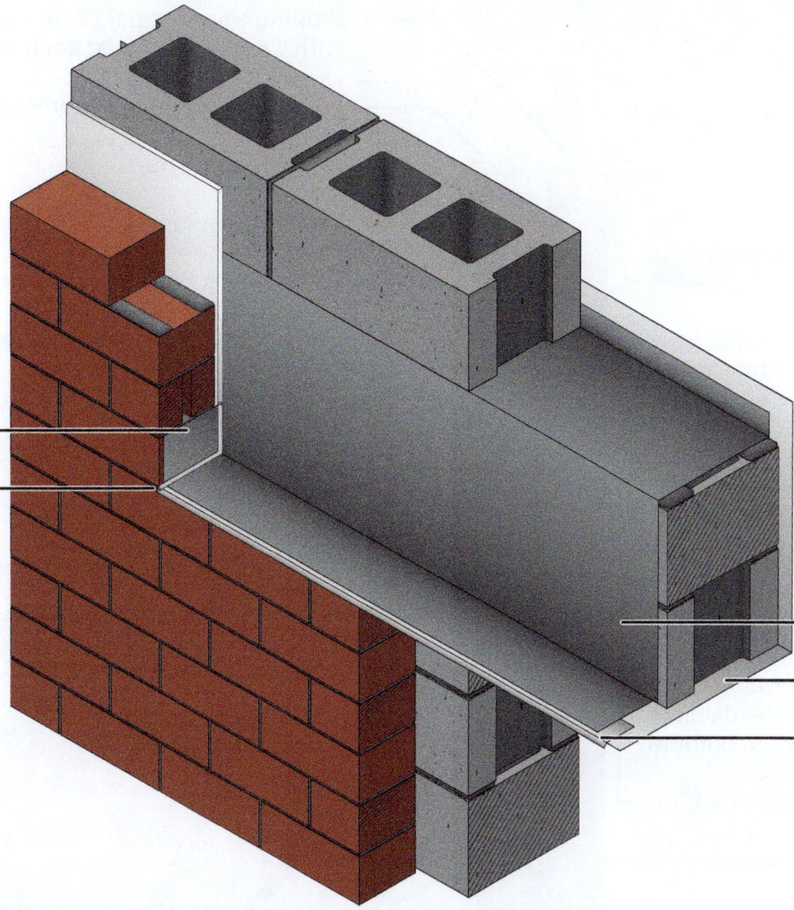

Figure 14 Flashing at a window.

Projections and recesses tend to hold rainwater and snow. They should have a top slope for drainage, and if possible, a drip to keep the water away from the wall surface below.

Face both sides of parapet walls with the same durable masonry materials. Using an inferior material on the side not exposed to view is a frequent cause of trouble. Do not paint or coat the backs of parapet walls; they must be left free to dry rapidly.

If required, place flashing in the mortar bed beneath the coping. Flashing may terminate at the wall surface wherever the coping provides a natural drip. However, where the copings are flush with the wall surface, extend the flashing at least $\frac{1}{4}$" on both sides and turn it down to provide a drip (*Figure 15*).

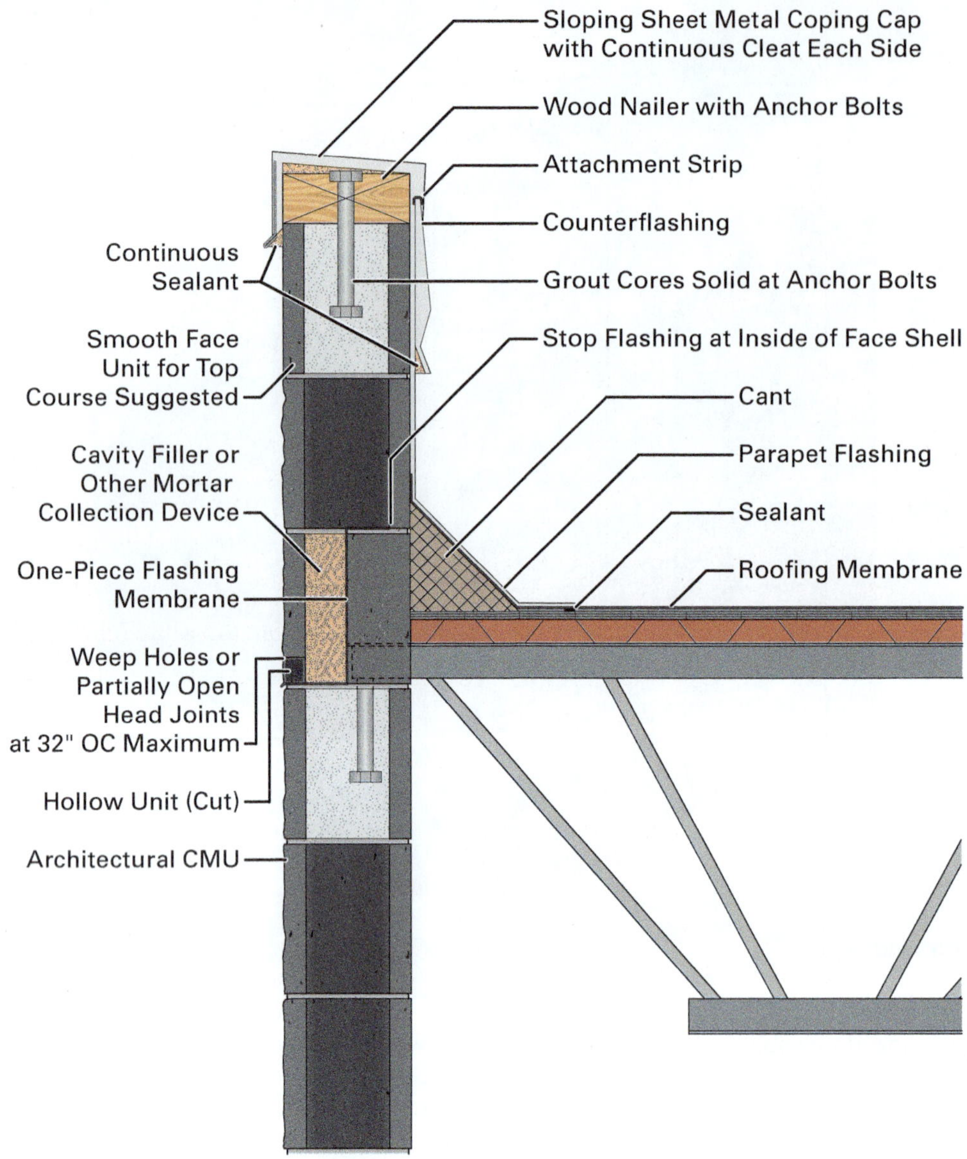

Figure 15 Flashing of coping and parapet wall.

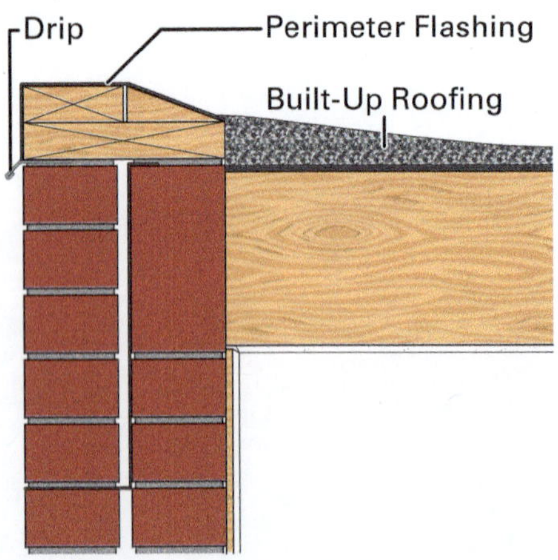

Figure 16 Perimeter flashing on a parapet wall.

Because perimeter flashing occurs at vulnerable points, it must be designed and installed carefully. The design of the base flashing depends upon the type of roofing material being used. Where the base flashing is metal, the counterflashing should also be metal, extending into the wall and overlapping the base flashing (*Figure 16*).

Flashing at chimney walls consists of base flashing placed under the roofing and turned up against the masonry. Counterflashing overlaps the base flashing and extends into the mortar joints, where it is securely embedded (*Figure 17*).

Flashing built on the back of a chimney will prevent the buildup of melting snow and ice. This flashing is held in place by a triangular wood structure called a *cricket* or a *saddle*. The triangular shape of the flashing allows water to flow around the chimney without pooling. The height of the flashing is typically half the width of the chimney, although local codes vary. The ridge is level and extends back to the roof slope.

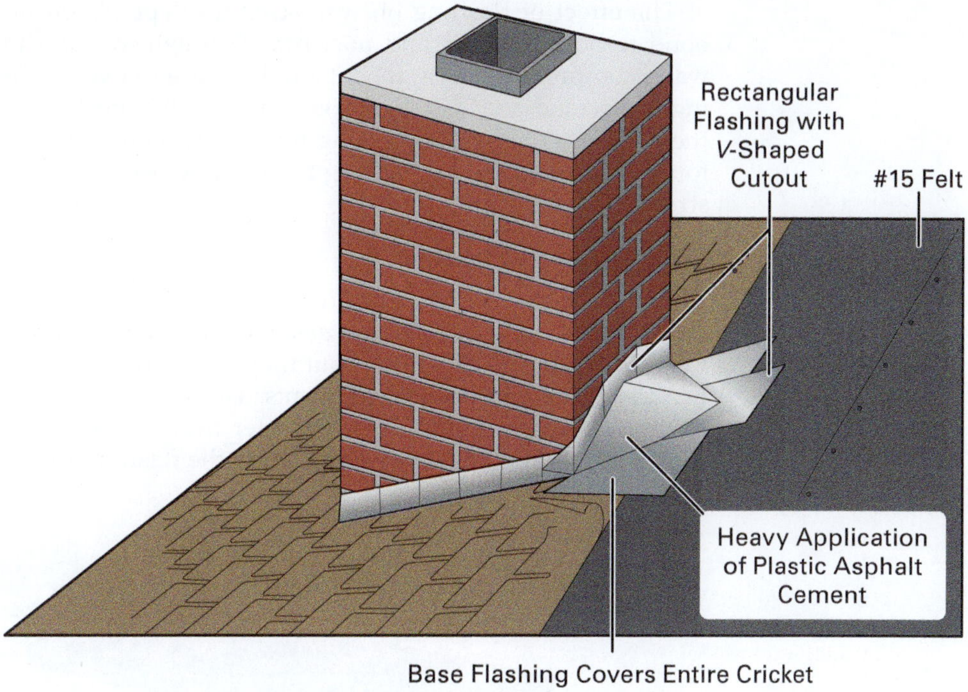

Figure 17 Chimney flashing.

The cricket consists of a horizontal ridge piece and a vertical piece at the back of the chimney (*Figure 18*). The support can be covered with heavy-gauge metal, which is nailed to the roof deck and sealed (refer to Figure 17). The support can also be covered with plywood and shingled.

Figure 18 Chimney cricket.

2.1.3 Installation of Flashing

Masonry must be relatively smooth and free of projections that might puncture the flashing and reduce its effectiveness. Through-wall flashing should be placed on a thin mortar bed with another thin bed laid on top to bond the next masonry course. Seams in the flashing must be completely bonded to prevent water penetration.

Although most sheet metal flashings can be soldered, lock slip joints are required at intervals to permit thermal expansion and contraction. Many plastics can be permanently and effectively joined by heat or an appropriate adhesive. The elasticity of plastic flashings eliminates the need for expansion joints.

The effect of flashing on wall strength depends on mortar placement and bond to both flashing and masonry. Through-wall flashing does not affect a wall's compressive strength, but it will reduce its overall shear strength. If flashing is placed directly on a masonry unit with a mortar bed on only one side of the flashing, the bending strength of the wall at that point is zero. It is therefore important that the flashing be properly bedded in mortar to preserve the strength of the wall.

2.1.4 Unitized Flashing

Unitized flashings (*Figure 19*) are factory fabricated and combine all the important components of a flashing system. Unitized flashings are typically ready to use directly out of the box, but they can also be produced in the field using manufacturer-approved components. Unitized flashings are typically precut for masonry openings to ensure better overall performance and are finished with a termination bar that tightly seals the flashing to masonry to block moisture penetration.

Figure 19 Unitized flashing.
Source: Mortar Net Solutions – Masonry Division of Keene Building Products

2.1.5 Single-Wythe CMU Through-Wall Flashing

Ensuring proper drainage in a single-wythe block wall using traditional construction methods usually requires additional steps and specially shaped materials to create a functional wall-flashing system. Traditional through-wall flashing for single-wythe walls can present a bond break at the flashing course.

An alternative option to the traditional single-wythe flashing method is to use a BlockFlash® pan, set on the bed of the single-wythe wall, that vents every cell of the wall with a pitched pan and attached weep vent (*Figure 20*). The BlockFlash pan does not require additional materials or special shapes and will not create a bond break at the flashing course.

Figure 20 BlockFlash pan set in a single-wythe wall.
Source: Mortar Net Solutions – Masonry Division of Keene Building Products

2.2.0 Installing Weep Vents

Flashing concealed in tooled mortar joints is not self-draining. Weep vents, also called *weeps, cell vents,* or *weep holes* (*Figure 21*), allow for the escape of collected moisture through the outer wythe of a cavity wall. Placed directly on the flashing at the base of a wall, shelf angle, sill, or masonry opening, weep vents allow air and water to pass from the cavity to the exterior. Some of the performance characteristics of a properly functioning weep vent include the following:

- Prevents insects from entering the cavity
- Reduces mold and mildew
- Helps prevent moisture condensation on structural components
- Helps prevent efflorescence
- Reduces masonry unit saturation and reinforcement corrosion

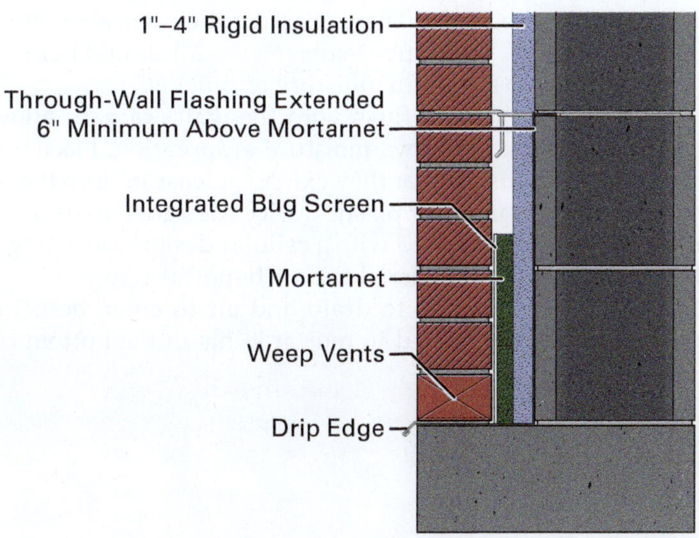

Figure 21 Weep vents with MortarNet™ in a multi-wythe wall foundation.
Source: Mortar Net Solutions – Masonry Division of Keene Building Products

Weep vents are typically spaced every 24" immediately above flashing sheets to prevent water buildup (*Figure 22*). They are also placed wherever there is a risk of water penetration. For instance, weep vents may be placed over the heads of windows, under windowsills, over steel lintels, or at every floor height on a concrete structure veneered with masonry work. Remember that weep vents should never be placed where they will be backfilled after the construction is complete.

Figure 22 Weep vents at a wall foundation.
Source: Mortar Net Solutions – Masonry Division of Keene Building Products

Weep vents can be made in several different ways. The recommended way to form a weep vent is to simply leave the head joints open. Install screens or mesh made of noncorrosive metal or plastic in open-head weep joints if called for in the project specifications. Open-head weep joints should be spaced a maximum of 24" apart.

Wick: A bundle of fibers that are loosely twisted or braided and woven together to form a cord that can carry water away from an area by capillary action, which occurs as long as the drip end is lower than the absorption end.

Several manufacturers make plastic tubes that can be inserted into the mortar to serve as small weep vents. These plastic tubes can also be fitted with **wicks** to improve drainage. Weep tube vents (*Figure 23*) should be spaced closer than vents because they do not drain moisture as quickly.

Position wicks so that they are flush with the wall's face, or allow them to extend from the wall face to improve moisture evaporation. Place wicks along the back of the brick and ensure that they extend at least 16" into the air space.

Cell vents (*Figure 24*) are metal or plastic devices designed to be the width of a mortar joint. They are manufactured with a cellular design consisting of a series of parallel channels and can be ordered to match mortar colors.

These channels allow moisture to drain and air to enter, boosting natural evaporation. Cell vents are installed in mortar joints in the bottom course of a brick wall.

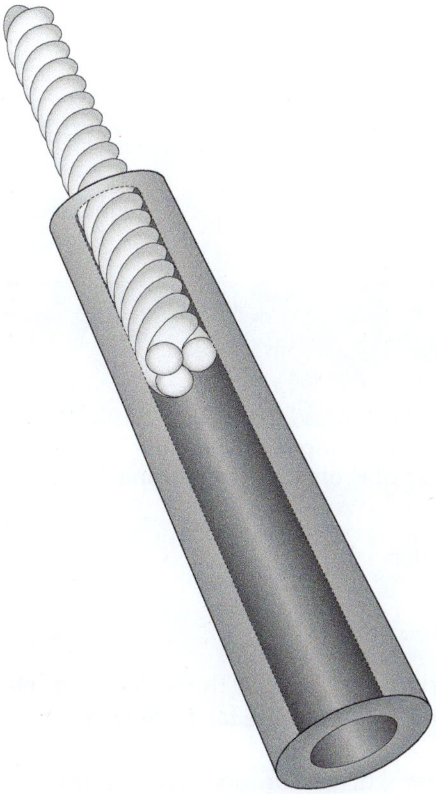

Figure 23 Weep tube vent.

(A) Different Varieties of Cell Vents

(B) Cell Vents Installed in the Bottom Vertical Mortar Joint of a Brick Veneer Wall

Figure 24 Cell vents.
Source: Mortar Net Solutions – Masonry Division of Keene Building Products

2.3.0 Installing Waterproofing

In masonry construction, waterproofing refers to treating a wall with material that will reduce moisture penetration and protect against seepage from hydrostatic pressure.

Below-grade foundation walls are the most critical areas for waterproofing applications. They require protection against rising water tables, hydrostatic heads, structural movement, and groundwater.

A liquid waterproofing system guarantees the high film thickness needed to cope with these problems. For all below-grade waterproofing applications, be sure to fill any cracks, crevices, and grooves. Ensure that the coating is continuous and free from breaks. Asphalt coatings and **membrane** systems, using thin, pliable waterproofing materials, are two widely used methods for waterproofing foundation walls.

Waterproof coatings should be carried over the exposed tops and outside edges of the footing to form a cove at the junction of the wall and footing (*Figure 25*). Spread the coating around all joints, grooves, and slots and into all **chases**, corners, **reveals**, and soffits. Bring the coating up to the finished grade. Apply waterproofing material on the outside face of backup block when plastering is eliminated. Apply waterproofing material to the inside block face once the wall has been plastered. Commercial admixtures are also available. These admixtures improve the waterproofing properties of mortar, stucco, and concrete.

Membrane: A layer of thin, pliable material used to waterproof masonry.

Chase: A continuous recess built into a wall to receive pipes, wires, or heating ducts.

Reveal: The visible portion of a masonry jamb between the face of the wall and the frame.

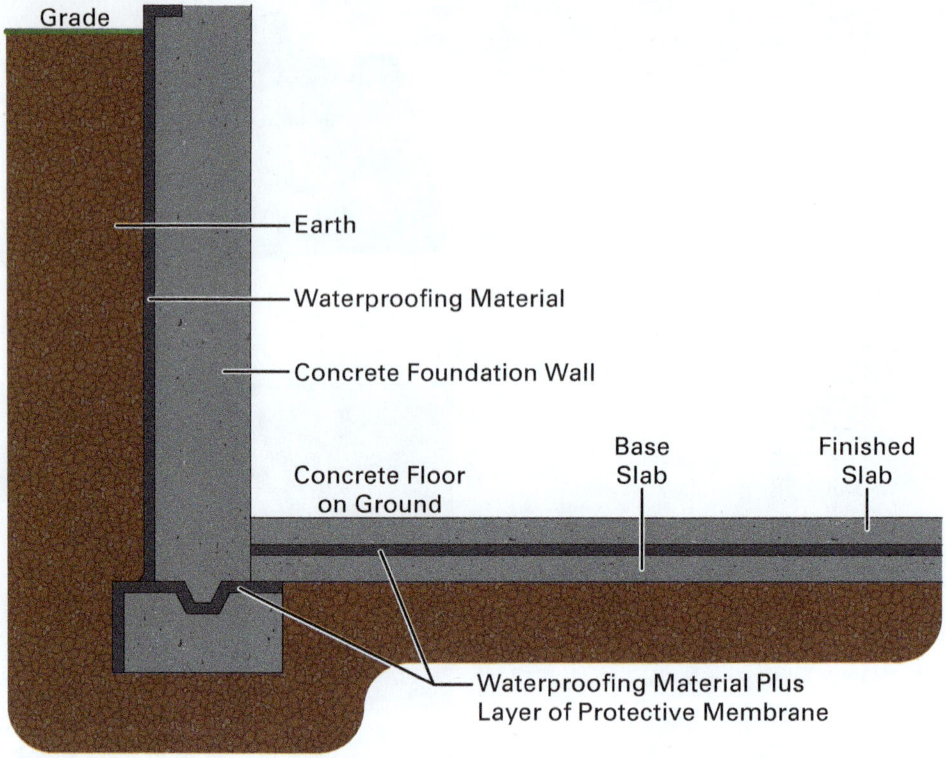

Figure 25 Below-grade waterproofing application.

2.3.1 Asphalt Coating

When treating a foundation with an asphalt coating, prepare the wall by removing all dust, dirt, or mortar drippings that may prevent the tar from sticking to block surfaces.

Apply the waterproof coating with a broom, brush, or mop. Wear appropriate PPE when working with hot asphalt. For mild climates that have minimal ground moisture, a single coating will likely be sufficient. However, you should always refer to your local applicable code for standards that apply in your area.

Hot tar preparations will dry to a fairly hard surface. However, many types of cold-coating materials are designed to remain flexible. This allows them to resist shrinkage and cracking, while filling smaller mortar cracks on the surface.

2.3.2 Membrane Systems

To install a membrane system, make sure the masonry has been cleared of dirt and mortar droppings, and then use a mop to apply a coat of coal tar, wood tar, or pitch primer. While the primer is still sticky, apply a layer of geotextile fabric. The project specifications may call for this process to be repeated for two to six layers. The final layer is a coat of primer that seals everything.

2.3.3 Spray-On Membrane Systems

Spray-on polymer/asphalt emulsions can be applied to below-grade foundation walls and allowed to dry to form a membrane. Spray-on membranes are applied with specialized applicators, and they require adequate ventilation and the use of appropriate PPE.

Spray-on membranes are designed to span cracks of up to $^1/_{16}$" and remain flexible at low temperatures. They are also designed to withstand hydrostatic pressure more effectively than other types of waterproofing.

Ensure that the foundation wall has been cleaned according to the manufacturer's instructions and standards. Cover brick and other elements to prevent them from contacting the membrane material. The membrane temperature,

Don't Break the Seal

Do not place backfill for 24 to 48 hours after applying waterproofing. Where possible, place backfill within seven days to avoid unnecessary damage to the waterproofing. Place the backfill carefully so that it will not rupture or damage the finish and break the seal.

when applied, should be between 110°F and 130°F (43°C and 54°C). The membrane should be applied to the required thickness, considering that the thickness will decrease as it dries. Ensure that the membrane is applied evenly to provide a smooth coat.

2.0.0 Section Review

1. A flashing material that can corrode in fresh mortar and that requires painting maintenance if installed exposed is _____.
   a. PVC
   b. galvanized steel
   c. copper
   d. stainless steel

2. What is the maximum distance between open-joint weep vents?
   a. 8"
   b. 12"
   c. 16"
   d. 24"

3. Apply waterproofing material to the inside face of block once the wall has been _____.
   a. grouted
   b. plastered
   c. painted
   d. tied

3.0.0 Extreme Weather in Masonry Construction

Objective

Explain the various techniques used to provide adequate protection during hot- and cold-weather masonry construction.

a. Explain the role played by weather data and information in masonry construction.
b. Explain the various techniques used to provide adequate protection during hot-weather masonry construction.
c. Explain the various techniques used to provide adequate protection during cold-weather masonry construction.

Performance Tasks

There are no Performance Tasks in this section.

Weather conditions can seriously affect masonry work. Always take precautions to avoid damage from hot, cold, wet, or extremely dry weather.

Weather affects mortar in several ways. For example, hot weather can cause mortar joints to set too rapidly, making tooling joints more challenging.

High wind velocities, high humidity, and too much direct sunlight lower mortar water retention, reducing its workability. In addition, its bonding ability decreases, making it more difficult to create weathertight joints.

Cold weather, ice, sleet, and snow can reduce mortar strength and durability. For instance, if the temperature falls below 40°F (4°C), hydration is impaired, and the mortar will become weak. If the masonry units are cold when they are laid up, they will drain the heat from the mortar, impairing strength development, initial set, and bonding ability.

Rainy weather will directly affect joint strength and the appearance of the completed structure. Rain can dilute the mortar, and if the chemical mixture is affected, its strength may be reduced. Partially completed work must be protected from rain so that it doesn't become saturated. If this occurs, the mortar can be washed from the joints, leading to efflorescence when the area dries.

Rapid drying of the mortar must be avoided during extremely dry weather. In addition to evaporation, dry masonry units will absorb water from the mortar. If mortar dries too rapidly, poor workability will result.

Recognition of these problems improves both the economy of the structure and the quality of the workmanship. Cold-weather construction techniques have enabled the construction industry to extend its season into the winter months.

Take the following additional steps to produce effective masonry structures under all weather conditions:

- Understand and review material performance at varying temperatures and weather conditions.
- Gather weekly and daily weather data.
- Plan material selection and storage methods according to the weather data.
- Modify the construction procedures and post-construction protection features as needed.

Admixtures can chemically change the properties of the mortar. They are often used in difficult construction situations, such as hot- or cold-weather construction, because they can speed up or slow down the setting time. This will partially protect mortar against the effects of extreme temperatures.

Different admixtures produce different results, so it is important to use the correct one. Choose an admixture that is compatible with the mortar, construction practices, and job specifications.

3.1.0 Using Weather Data and Information

Accurate weather forecasting is an important part of planning all-weather construction. Important weather factors include temperature, precipitation, and humidity. Although *ideal* is a relative term, for masonry construction, ideal temperature is generally considered to be between 40°F and 90°F (4°C and 32°C). Many codes specify a minimum working temperature and direct how to protect the work when the temperature falls below that mark, and some codes address procedures such as fog spraying for high temperatures.

Construction schedules are based on long-term weather expectations, local weather information, and the contractor's experience. Long-term weather expectations are called *climatology data,* while local weather information is called *meteorological data*.

Meteorological information can be obtained from the National Weather Service. Climatological information can be obtained from the National Climatic Data Center or apps that can be downloaded onto a cell phone. They usually provide climatic information through maps (*Figure 26*).

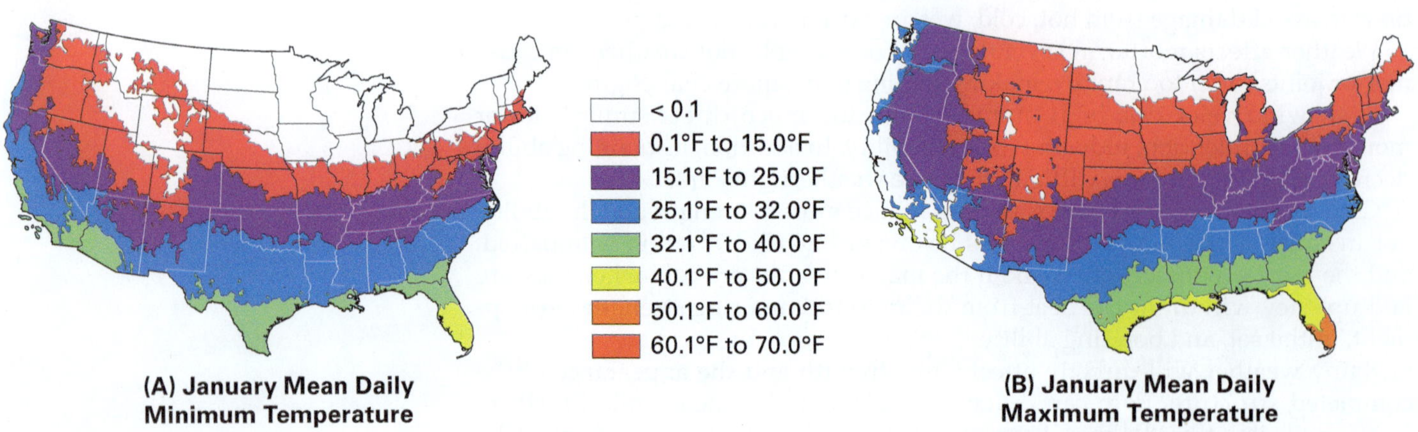

Figure 26 Examples of climatology data (Fahrenheit).

3.2.0 Protectin[illegible]g Hot-Weather Construction

Hot temperatures, particularly when combined with high winds and direct sunlight, have a greater effect on mortar than on masonry units. However, several effects on physical properties can occur for both.

3.2.1 Mortar and Masonry Unit Performance

Hot weather is generally defined as temperatures above 90°F (32°C). However, temperature, wind speed, relative humidity, and solar radiation all influence masonry materials. These factors affect the absorption of masonry units, and the set and drying rates of mortar. The primary concern in hot weather is evaporation of water from the mortar, reducing the bond between the mortar and the masonry unit.

Low temperatures and low humidity affect masonry more than high temperatures and high humidity. For example, hot, humid weather can increase the cement's hydration rate.

The masonry units themselves are the least affected by hot weather. However, the interaction between the masonry units and the mortar or grout is critical. Warmer units absorb more water from the mortar. If there is not enough water present in the mortar when the units are laid, the mortar will have a lower bond strength.

Mortar tends to lose its **plasticity** rapidly in hot weather. This is due to the evaporation of the water from the mix and the increased rate of hydration of the cement. The use of admixtures to increase plasticity is not recommended unless their full effect on the mortar is known. Instead of an additive, use mortar with high lime content and above-average water retention.

Plasticity: A material's ability to be easily molded into various shapes.

Tempering (*Figure 27*) keeps mortar at the proper consistency. During hot weather, this may be required at more frequent intervals. To avoid over-tempering, mix smaller batches so that the mortar does not lose water before it is applied.

Figure 27 Tempering mortar.
Source: McGee Brothers

Grout reacts to hot wea… idly in hot weather and reduces t… grout requires a high slump for placement, so you can maintain a high water-to-cement ratio by reducing evaporation and initially mixing grout with sufficient water.

3.2.2 Materials Selection and Storage

The primary objective in hot-weather masonry construction is to avoid moisture loss. Masons must maintain cool temperatures in all equipment, mortar ingredients, and masonry units. If possible, when preparing the jobsite, arrange for materials to be delivered to and stored in the shade, as well as on pallets for quick transport (*Figure 28*).

Sandpiles should be covered and shaded with tarps. This will lessen evaporation of the natural moisture of the sand. If the sand's free moisture content drops, it can be cooled by lightly sprinkling water on the sandpile. As a rule of thumb, 1 gal of water will cool 1 yd^3 of sand by 20°F (–7°C). Stockpiled masonry units may also be sprinkled with water. However, do not soak them, and refrain from ever laying up wet units.

Consider the comfort of workers and the protection of materials throughout the construction process. If necessary, set up sunscreens to protect materials, the mixing area, and the construction area.

Figure 28 Palletized materials.
Source: Shutterstock

3.3.0 Protecting Masonry During Cold-Weather Construction

Cold weather during masonry construction can also affect material and craftsmanship. Consider the effects of cold weather in planning, scheduling, and setup of masonry work. In addition to anticipating specific weather conditions, masons must determine the possible effects weather will have on the materials and schedule. Plan how to protect materials and workers, and strategize creative solutions for storing materials.

In the United States, all model building codes have requirements relating to the construction of masonry during cold weather. While not identical, each code has similar general guidelines. These address materials protection, materials heating, use of frozen materials, and protection of completed work.

3.3.1 Mortar and Masonry Unit Performance

The major concern during cold-weather construction is to keep all equipment, materials, and supplies at a temperature well above the freezing point. As is true with hot-weather construction, the temperature will affect mortar hydration and bond strength development in several ways:

- Mortar mixed at lower temperatures has lower water content, longer setting and hardening times, higher air content, and lower early strength than mortar mixed at normal temperatures.
- As the temperature falls below 40°F (4°C), the heat of hydration decreases or stops altogether. This results in decreased mortar strength.
- Although cold CMUs may appear normal, they will be slightly smaller in volume. When the temperature returns to normal, the block expand, leading to cracks along the joints.
- Wet or frozen masonry units have a reduced water absorption rate. This can increase the possibility of early freezing, mortar expansion, joint cracking, and potential bonding problems.

All these problems can be prevented by selecting mortar ingredients to maintain hydration, reducing the mortar water content to below 8%, heating mortar materials as needed, selecting appropriate masonry units, and protecting materials during storage and construction.

3.3.2 Materials Selection and Storage

Although specified by contract and code requirements, a mason may have several options for dealing with cold-weather problems. For example, high-early-strength (Type III) Portland cement substituted for normal (Type I) cement will increase the mortar's protection because it increases the rate of strength development without changing the other physical properties.

Understanding cold-weather effects makes it easier to make informed decisions about using particular masonry units and mortar admixtures. An absorptive masonry unit will suck water from the mortar and lessen the problem of initial freezing and mortar joint expansion. On the other hand, a masonry unit with a very low absorption rate will not collect enough water to prevent mortar joint expansion as ice forms. Considerations should include the unit's absorptive rate, volume, density, and temperature when used in cold-weather construction.

Carefully consider using admixtures in cold-weather construction, because most admixtures reduce mortar strength. Only two types of admixtures—accelerators and air-entraining agents—are recommended during cold weather. Accelerators offer protection against cold weather by increasing the high-early-strength development of the cement in the mortar.

The most frequently used accelerator is calcium chloride. Soluble carbonates, silicates, and fluorosilicates, calcium aluminate, and some aluminous cements have also been used as accelerators. Using calcium chloride may increase shrinkage, efflorescence, and corrosion of metal ties, joint reinforcement, flashing, and anchors.

Calcium chloride is not recommended in reinforced masonry construction. If it is used, it should be strictly regulated to 2% by weight of Portland cement or 1% by weight of masonry cement.

Air-entraining agents are sometimes used in cold climates to increase workability and freeze–thaw durability. Air-entraining admixtures cause millions of tiny air bubbles to form in the mortar. These air bubbles provide room for expansion during freezing cycles. This results in a more durable mortar. Avoid corrosion inhibitors and color pigment with dispersing agents. Experience has shown that these admixtures often slow hydration.

Regardless of the materials selected, they must be protected to prevent freezing. Build a temporary shelter to store mortar materials. If a shelter cannot be built, cover all materials when the temperature is below 40°F (4°C). Completely cover bagged materials with canvas or plastic sheeting (*Figure 29*). Store the bags on raised platforms so that they cannot absorb cold and moisture from the ground. Also, cover the mortar sand and masonry units to protect them from water, snow, and ice.

Figure 29 Covered work area.
Source: Lang Masonry Group

3.3.3 Construction Practices

There are several important factors in cold-weather masonry construction. First, maintain safe and comfortable working conditions. Also, maintain the physical properties of the mortar and masonry units. And finally, protect the finished portions of the construction. Heat is the most important ingredient in solving these problems.

Consider using windbreaks or temporary heated enclosures when low temperatures are expected, to protect materials, mortar-mixing areas, and work areas. In fact, US federal regulations now require heated enclosures on all government projects.

Figure 30 shows typical scaffold enclosures. They are erected over an exterior wall using the tubular scaffold as part of the system. Several factors must be considered when creating an enclosure. Think about material strength, durability, transparency, fire resistance, flexibility, ease of installation, cost, weather severity, type of masonry structure being built, and Occupational Safety and Health Administration (OSHA) safety requirements.

Often, an unheated structure or covering will be sufficient. It will reduce the windchill factor and prevent the masonry from freezing. Enclosures typically use lumber, steel, canvas, building paper, fiberglass, and clear plastic sheets.

A widely used technique is to make polyethylene panels. These can be fastened and braced to the scaffolding around the work area. The panels are made of lumber framing with plastic film and a reinforcing fiber or wire mesh. They are particularly effective because they resist the wind but allow the light and heat of the sun to pass through.

Figure 30 Scaffold enclosures.
Source: Shutterstock

If the enclosure is heated, ensure that the temperature is not too high to cause mortar to dry out too quickly. When planning a structure, consider wind force, loading, counterweights, and bracing and their effects on the scaffolding. Use frames or ties to keep the plastic from blowing or flapping.

OSHA regulations require that all temporary enclosures heated with combustible fuel have separate venting systems for the heater, to avoid carbon monoxide buildup. Properly vented oil or gas heaters with blowers are recommended for projects with complete enclosures where heat is unavailable from a stationary heating plant. Ideally, locate the heater outside the enclosure and blow hot air inside. This way, the heater can blow warm air across both sides of the constructed wall. Workers can be provided with fresh air without pollution created by burning fuel.

WARNING!

Carbon monoxide is a gas produced by combustion. Breathing carbon monoxide fumes can cause serious illness and death.

Jobsite preparation also includes setting up heat sources to heat mortar materials and masonry units, specialized tools for cold-weather masonry, and post-construction protection for the structure. The final steps in preparing the work area should be to provide walkways over icy or hazardous areas and to thaw the base where the masonry structure will be constructed.

It's Cooler at Night

On days with extremely high temperatures and high winds, avoid working during the midday and afternoon hours. Work schedules can be altered to early morning or late evening hours. This may create difficulties with the other construction trades, but it benefits worker comfort, safety, and efficiency.

3.3.4 Heating Materials

When the site has been prepared and construction begins, the immediate objective is to ensure proper hydration of the mortar. This may require heating various materials immediately before application.

Mixing water is the easiest mortar material to heat, and it can also store the most heat. Do not heat the water to a temperature above 160°F (71°C). Heat the water in a 55 gal drum. An immersion heater, flame guns, or steam probes can also be used to heat the water. Always be careful not to pollute the water.

Heat the mixing water to produce mortar temperatures between 70°F and 120°F (21°C and 49°C). Maintain that temperature in later batches. If the mixing water is too hot, the mortar may flash-set when the Portland cement is added. Prevent this by combining sand and water in the mixer before adding the cement. This will lower the temperature and prevent flash-setting.

Masonry sand contains moisture, which will turn to ice at temperatures below 32°F (0°C), even when covered. When the temperature of the sand is below freezing, it should be heated slowly and evenly. Take care to avoid scorching the sand through excessive heating. A temperature of 45°F to 50°F (7°C to 10°C) is acceptable, but temperatures can reach as high as 180°F (82°C) without scorching. A practical and easy method of checking the sand temperature is simply to touch it. If the sand can be held in your hand, the temperature is not too hot.

Whatever method is used, avoid contaminating the sand with foreign substances. Mix the sand periodically to produce uniform heating and to avoid scorching.

Antifreeze

Do not use an antifreeze agent as an admixture. Antifreeze agents, such as alcohol, are sometimes used to lower the freezing point of mortar. Experience has shown, however, that so much of the antifreeze admixture is needed to lower the freezing point that the mortar's strength is reduced.

When dry masonry units reach 20°F (–7°C), they should be heated to 40°F (4°C). This increased temperature prevents the sudden cooling of mortar when it is laid up with the units. When the weather is damp and cold, heating the units will increase efficiency. Thaw wet-frozen masonry units, but do not overheat them.

The most effective method of heating masonry units is to direct the blower from an oil, gas, or electric heater directly against the pile. If heated enclosures are used for construction, a supply of masonry units can be thawed on the scaffold.

3.3.5 Mortar Mixing and Tooling

During cold-weather construction, mix mortar in smaller quantities than usual. Limit the mortar batch size to an amount that can be used in a one-hour period. Once the mortar is mixed, carry it immediately to the mason. Do not leave it to cool in a mortar box or wheelbarrow. If the mortar must be placed in a mortar box, use a metal box that has been raised off the ground and leveled.

Keep the following points in mind when preparing cold-weather mortar:

- Maintain the mortar temperature between 70°F and 90°F (21°C to 32C).
- When mixing the mortar, do not add heated water that will raise the temperature of the heated mix above 105°F (41°C). This prevents the danger of flash, or false, set when the water and cement are mixed.
- Never use frozen or scorched sand in the mix.
- Never use antifreeze admixtures to lower the freezing point of mortar.
- Follow the *International Building Code* requirements for protection against freezing. Masonry must be protected against freezing for at least 24 hours after being laid. Grouted masonry must be protected for up to 48 hours, unless the only cement in the grout is Type III Portland cement.

Tooling joints in cold-weather masonry construction creates few challenges. In fact, you may be able to lay more units before tooling. However, allow the same set time for each section of completed masonry, to ensure uniform coloring of joints.

3.3.6 Protecting Completed Structures

Protect each masonry wall section from freezing as it is completed. This allows it to cure properly without being damaged. When the temperature is above 25°F (–4°C), cover the wall with a tarp. This will allow the mortar to set properly if it has been previously heated. When the temperature falls below 25°F (–4°C), use an insulated blanket or heater to warm the mortar.

Use the following checklists to review the jobsite conditions and specifications for cold-weather masonry construction.

During the preliminary planning phase, do the following:

- Check provisions for winter construction in contract specifications.
- Lay out the site to accommodate material deliveries during poor weather conditions.
- Review code requirements for cold-weather masonry protection.
- Review available weather data for the project area:
  - Expected daytime and nighttime temperatures
  - Expected windchill factor and wind speeds
- Select the type of required protection:
  - Enclosure for mortar preparation
  - Enclosure for masons on the wall if heat is required
  - Protection of completed walls
- Arrange for short-term, daily weather forecast service during winter months.

During the execution phase, do the following:

- Anticipate temperature, wind, and rain.
- Do not lay masonry on frozen surfaces.
- Provide an enclosed area for storage of mortar materials and mortar preparation.
- Set the work area for the masons:
  - Employ windbreaks when the windchill factor approaches zero.
  - Employ a complete enclosure when the daytime temperature is between 10°F and 15°F (–12°C and –9°C).
  - Heat the enclosed work area when required for the protection of completed work.
- Review the procedures for heating materials:
  - Heat the mortarboard and mortar to a temperature between 70°F and 90°F (21°C and 32°C).
  - Use heated mix water (70°F to 105°F or 21°C to 40°C) when the air temperature is below 40°F (4°C).
  - Do not use frozen sand.
  - Heat and completely thaw sand when the temperature is below 32°F (0°C).
  - Heat units as required so that the temperature of units when laid is not less than 20°F (–7°C).
- Protect the wall from early freezing, if not enclosed.
- Avoid rapid dry-out of mortar due to excessive heating.
- During the protection phase, do the following:
  - Provide minimum protection by covering the top 2′ of completed wall with plastic or canvas, to prevent water entry.
  - When the night temperature is expected to be below freezing, cover the newly built wall completely.
  - If the expected temperature is between 10°F and 15°F (–12°C and –9°C), or if the expected windchill is zero or less, cover the completed masonry and maintain it above freezing for 48 hours or more with insulated blankets or auxiliary heat.

Cold-Weather Clothing Tips

Use the following tips to prevent injury due to cold weather:

- Dress in layers.
- Wear thermal underwear.
- Wear outer clothing that will resist wind and moisture.
- Wear a face mask and head and ear coverings.
- Carry an extra pair of dry socks when working in snowy or wet conditions.

3.0.0 Section Review

1. When developing a construction schedule, obtain meteorological information from _____.
   a. the US Coast Guard
   b. the National Weather Service
   c. the Weather Channel
   d. the National Climatic Data Center

2. In hot-weather masonry construction, the *primary* objective is to _____.
   a. avoid moisture loss
   b. facilitate moisture loss
   c. increase the use of admixtures in mortar
   d. decrease the use of admixtures in mortar

3. A set accelerator that may increase shrinkage and efflorescence is _____.
   a. calcium carbonate
   b. calcium silicate
   c. calcium chloride
   d. calcium aluminate

Module 28206 Review Questions

1. The length of time it takes a given amount of heat to move through a material is a measure of its thermal _____.
   a. density
   b. resistance
   c. conductivity
   d. inductance

2. When rigid boards are used as internal insulation, they are applied to _____.
   a. the outside face of the inner wythe
   b. the inside face of the outer wythe
   c. the inside face of the inner wythe
   d. both sides of the cavity

3. The type of material that is *most often* used when reinsulating existing buildings is _____.
   a. slab insulation
   b. expanded polystyrene foam
   c. reflective insulation
   d. foamed-in-place insulation

4. Flexible insulation with a vapor barrier on only one side is described as _____.
   a. reflective
   b. friction-fit
   c. faced
   d. wrapped

5. Unless the structure has been designed to withstand the damaging effects of humidity, repeated absorption and evaporation of moisture in masonry structures can cause _____.
   a. cracking
   b. erosion

c. fungus growth
d. warping

6. Flashing and weep vents are vital to controlling moisture in _____.
a. foundation walls
b. cavity walls
c. garden walls
d. retaining walls

7. Weep vents are typically placed immediately above the flashing, spaced every _____.
a. 12"
b. 16"
c. 24"
d. 33"

8. A membrane waterproofing system involves applying one or more layers of _____.
a. geotextile fabric
b. heavy-gauge plastic sheeting
c. roofing paper
d. fiberglass membrane

9. Fresh mortar washed from joints by rain may dry on masonry surfaces and show _____.
a. fluorescence
b. yellow staining
c. efflorescence
d. mildew deposits

10. Ideal weather for masonry construction is generally considered to be a temperature range between 40°F and _____.
a. 60°F
b. 70°F
c. 80°F
d. 90°F

11. During hot weather, the *primary* concern is _____.
a. overheating of exposed surfaces
b. evaporation of water from the mortar
c. excessive mixing of mortar
d. expansion of masonry materials

12. OSHA requires that all temporary enclosures heated with combustible fuel _____.
a. be inspected daily
b. include at least two exits
c. be supplied with fire extinguishers
d. have a separate venting system

Answers to odd-numbered questions are found in the Review Question Answer Key in the back of the book.

Answers to Section Review Questions

Answer	Section	Objective
Section One		
1. b.	1.1.1	1a
2. c.	1.2.2	1b
3. a.	1.3.3	1c
Section Two		
1. b.	2.1.1	2a
2. d.	2.2.0	2b
3. b.	2.3.0	2c
Section Three		
1. b.	3.1.0	3a
2. a.	3.2.2	3b
3. c.	3.3.2	3c

MODULE 28207

Construction Inspection and Quality Control

Source: Shutterstock

Objectives

Successful completion of this module prepares you to do the following:

1. Describe how standards and specifications are used to ensure quality control throughout the masonry industry.
   a. Describe the standards and specifications that apply to masonry units, mortar, grout, and accessories.
   b. Describe the standards that apply to laboratory and field testing of masonry construction.
2. Describe how masonry sample panels and prisms are built and tested to ensure quality control on a project.
   a. Describe how to build sample panels.
   b. Describe how to build hollow masonry prisms.
   c. Describe how to build grouted masonry prisms.
   d. Describe how to prepare and test mortar and grout prisms.
   e. Describe how to conduct masonry tests.
3. Describe how mortar is tested to ensure quality control on a project.
   a. Describe how to perform sand tests.
   b. Describe how to perform mortar consistency tests.
   c. Describe how to perform brick absorption tests.
   d. Describe how to perform laboratory tests.
4. Describe how field inspections and observations are used to ensure quality control on a project.
   a. Describe why and how inspections of standards and codes are performed.
   b. Describe why and how inspections of materials are performed.
   c. Describe the types of observations that are undertaken during construction.
   d. Describe why and how construction tolerances are monitored.

Performance Tasks

Under supervision, you should be able to do the following:

1. Build a prism for mortar testing.
2. Perform a slump test.

Digital Resources for Masonry

SCAN ME

Scan this code using the camera on your phone or mobile device to view the digital resources related to this craft.

NOTE

Codes vary among jurisdictions. Therefore, it is essential to consult the applicable code whenever regulations are in question. Referring to an incorrect set of codes can cause as much trouble as failing to reference codes altogether. Obtain, review, and familiarize yourself with your local adopted code.

Overview

Quality inspection and testing are important for several reasons. Building owners want to be sure they get what they have paid for in terms of building quality. Architects and engineers want to be sure they get what they have specified in terms of load-bearing, performance, and appearance. Contractors want to be sure they get what they have paid for regarding materials and craftworker skills. Masons want to be sure the materials they are using are the right materials and the techniques they are using will give the right results for the job. This module introduces the concepts and tasks related to quality inspection and testing.

NCCER Industry-Recognized Credentials

If you are training through an NCCER-accredited sponsor, you may be eligible for credentials from NCCER's Registry. The ID number for this module is 28207. Note that this module may have been used in other NCCER curricula and may apply to other level completions. Contact NCCER's Registry at 888.622.3720 or go to **www.nccer.org** for more information.

1.0.0 Standards and Specifications in Masonry Construction

Performance Tasks

There are no Performance Tasks in this section.

Objective

Describe how standards and specifications are used to ensure quality control throughout the masonry industry.

a. Describe the standards and specifications that apply to masonry units, mortar, grout, and accessories.

b. Describe the standards that apply to laboratory and field testing of masonry construction.

Quality: The totality of features and characteristics of a product or service that bear on its ability to satisfy stated or implied needs.

Quality assurance: All planned and systematic actions necessary to provide adequate confidence that an item or a facility will perform satisfactorily in service.

Quality is defined by the International Organization for Standardization as the totality of features and characteristics of a product or service that bear on its ability to satisfy stated or implied needs. On most construction jobs, the quality standards are set in the contract documents and specifications. These usually incorporate specifications set in national and local building codes and in ASTM International (ASTM) material specifications. ASTM also publishes procedures for testing materials to make sure they meet the specifications.

Architects and project engineers set up procedures for checking materials and work in progress against project specifications. This is known as **quality assurance**, which is defined as actions planned to provide adequate confidence that an item or facility will perform satisfactorily in service. The architect or engineer can try to ensure quality, but assurance is not the same as control.

One way architects and engineers ensure quality is by requesting a list of submittals. These are documents, samples, test results, and reports listed in the contract to be delivered over the course of the work. They can include shop drawings, product data, and samples. A typical submittal would include the following:

- *Product data*—Submit manufacturer's catalogs, test reports, and data for all proposed products, including installation, cleaning, and maintenance recommendations.
- *Samples*—Submit three units showing colors and textures.
- *Accessories*—Submit small samples of anchors, insulation, adhesive, vapor barrier, and other accessories for approval by the architect or engineer.

Quality control is usually defined in specifications as verification that eliminates the possibility of error. Quality control tasks are the procedures for testing, inspecting, checking, and verifying to ensure that the work meets the required quality standard specified in the contract. Because the contractor has control over when and how the work is done, the contractor is responsible for quality control. Quality control and quality assurance both include field inspections and testing.

Quality control: Procedures for testing, inspecting, checking, and verifying to ensure that the work meets the required standard of quality established by the contract documents.

Testing requirements can be specified in the contract submittals. Typically, submittals include samples, test reports, and certificates. Contractors must provide samples of masonry units showing extreme variations in color and texture. The contractor must certify that the materials comply with the applicable grade, type, or class specifications.

Test reports must be completed for each type of building and face brick. The testing and reporting must be done by an independent laboratory. Typical tests include the following:

- Compressive strength
- Twenty-four-hour cold-water absorption
- Five-hour boil absorption
- Saturation coefficient
- Initial rate of absorption

1.1.0 Identifying the Standards and Specifications for Masonry Units, Mortar, Grout, and Accessories

Most of the masonry specifications in construction contracts reference building codes or standards. A standard usually has a single subject. The standard is based on studies, research, and advances in materials and construction techniques. Building codes are enforceable standards.

Model codes are developed by national or international organizations. These model codes are then adopted by states and municipalities. Once adopted, they become enforceable laws. The *International Building Code®* (*IBC®*) and the National Fire Protection Association's *NFPA 5000®*, *Building Construction and Safety Code®*, are the primary codes used in the United States.

NOTE

Always check what building codes are used in your area. Codes are updated periodically; make sure you are using the latest edition.

Standards are technical specifications that establish minimum requirements for all aspects of masonry units and masonry work. ASTM standards are the most commonly used. The following organizations publish consensus standards:

- ASTM International (ASTM)
- The American Society of Civil Engineers (ASCE)
- The American Concrete Institute (ACI)
- The American Institute of Steel Construction (AISC)

Trade associations also publish technical notes and advisories. While these are not standards, they are considered best practices. Some contracts require the use of best practices. These technical notes are available online at the organizations' websites. The following trade associations publish technical notes:

- The Mason Contractors Association of America (MCAA), *www.masoncontractors.org*
- The Brick Industry Association (BIA), *www.gobrick.com*
- The Concrete Masonry & Hardscapes Association (CMHA), *www.masonryandhardscapes.org*

General contract specifications usually include clauses such as "materials and equipment must be of good quality and new, and work must be free of defects and conform to specifications." A standard of quality can be set in one of several different ways, depending on the method used to describe it:

- Descriptive specifications name exact properties of materials and methods of installation without using brand names.
- Proprietary specifications list specific products by brand name, model number, and so on.
- Performance specifications describe the results of the work, the measures or criteria by which the results will be judged, and the methods by which the evaluation will be done.
- Reference standard specifications name and incorporate established industry quality and performance standards for products and processes.

There are more than 100 ASTM standards for masonry and masonry-related products, and others are currently in development. They are listed in the *Appendix 28207A*. Some of these standards apply to specialty products such as sewer block and firebrick, while others are designed for research rather than construction work. Nevertheless, many standards do apply to masonry work on the typical job. The most intensive use of reference standards is on large commercial projects.

The following sections list the most commonly referenced detailed standards for clay and concrete masonry units (CMUs), mortar and grout, accessories, and product testing.

1.1.1 Clay Masonry Units

Most clay masonry units fall under *ASTM C216, Standard Specification for Facing Brick (Solid Masonry Units Made from Clay or Shale)*, and *ASTM C652, Standard Specification for Hollow Brick (Hollow Masonry Units Made from Clay or Shale). ASTM C216* lists the following three types of face brick:

- Face brick standard (FBS), the industry standard
- Extra select brick (FBX), with tighter size tolerances
- Architectural brick (FBA), with greater size irregularities

If no type of brick is specified in a contract, the default is type FBS.

Brick is also graded according to its resistance to damage from freezing. The grade is tested in accordance with *ASTM C67, Standard Test Methods for Sampling and Testing Brick and Structural Clay Tile*. The grades are as follows:

- *Severe weathering (SW or SX)*—Grade SW is specified when brick is likely to freeze when it is saturated with water. It can withstand freeze-thaw cycles in colder climates.
- *Moderate weathering (MW or MX)*—Grade MW is specified when brick will not be exposed to freeze-thaw cycles when it is wet.
- *Negligible weathering (NW or NX)*—Grade NW is specified when brick will not be exposed to the weather. It is for indoor use.

ASTM C216 also sets compressive strength standards for face brick and requires efflorescence testing according to procedures in *ASTM C67*. If the contractor does not specify the grade, the default is SW.

Common or standard brick shapes, tolerances, and performance are set under *ASTM C62, Standard Specification for Building Brick (Solid Masonry Units Made from Clay or Shale)*. Brick grade or weathering regions are shown in *Figure 1*. The compressive-strength requirements for face brick listed under *ASTM C216* also apply to common brick.

ASTM C652 covers hollow brick in several classes and grades. *ASTM C126, Standard Specification for Ceramic Glazed Structural Clay Facing Tile, Facing Brick, and Solid Masonry Units*, covers glazed brick and structural clay tile.

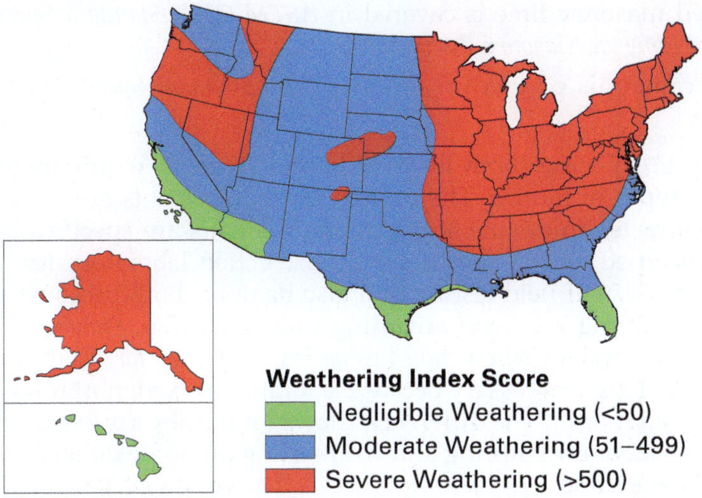

Figure 1 Brick weathering regions.

1.1.2 Concrete Masonry Units

Most CMUs fall under *ASTM C90, Standard Specification for Loadbearing Concrete Masonry Units*. There are two types of units: Type I and Type II. Type I units have limits on shrinkage and moisture content and must be kept dry at the plant, in transit, and at the jobsite. Type II units have limits on shrinkage but not on moisture content.

Weight classifications include light, medium, and normal. These weight classifications affect fire resistance and water and sound absorption. Compressive strength requirements are the same for all types and weights: 1,900 pounds per square inch (psi). Compliance with *ASTM C90* is verified by testing in accordance with *ASTM C140, Standard Test Methods for Sampling and Testing Concrete Masonry Units and Related Units*.

ASTM C129, Standard Specification for Nonloadbearing Concrete Masonry Units, specifies unit types and weight categories for non-load-bearing CMUs. These have the same Type I and II classes and three weight categories as for load-bearing block. The major difference is in the load-bearing requirement, which is only 600 psi for these units.

ASTM C55, Standard Specification for Concrete Building Brick, specifies standards for load-bearing and non-load-bearing concrete brick. Grading is based on strength and weather resistance, with Grade N giving high strength and freeze resistance and Grade S giving moderate strength and freeze resistance. Each grade can be either Type I, moisture controlled, or Type II, not moisture controlled. Minimum compressive strength is 3,500 psi for Grade N and 2,500 psi for Grade S. Sampling and testing are done according to *ASTM C140*.

1.1.3 Masonry Mortar and Grout

The standards for masonry mortar are specified in *ASTM C270, Standard Specification for Mortar for Unit Masonry*. These standards cover four types of mortar: M, S, N, and O. Reference specifications cover the following cementitious agents:

- Portland cement is covered in *ASTM C150, Standard Specification for Portland Cement*.
- Masonry cement is covered in *ASTM C91, Standard Specification for Masonry Cement*.
- Hydraulic and slag cement is covered in *ASTM C595, Standard Specification for Blended Hydraulic Cements*.
- Quicklime is covered in *ASTM C5, Standard Specification for Quicklime for Structural Purposes*.

Masonry Building Code

The ACI and the ASCE collaborated with other organizations and issued the *Building Code Requirements and Specification for Masonry Structures*, through the Masonry Standards Joint Committee (MSJC). They are commonly referred to as the *MSJC Code* and the *MSJC Specification*, respectively. Both the IBC and NFPA 5000 reference the *MSJC Code*.

- Hydrated masonry lime is covered in *ASTM C207, Standard Specification for Hydrated Lime for Masonry Purposes.*
- Mortar cement is covered in *ASTM C1329, Standard Specification for Mortar Cement.*

ASTM C270 specifies both proportion and strength requirements for each of the four types of mortar. The proportion requirements are used by default in most contracts, unless the strength requirements are specifically included. Strength-specified mortar requires preconstruction laboratory testing according to *ASTM C270.* If field testing will also be done, both preconstruction and construction testing are done according to *ASTM C780, Standard Test Method for Preconstruction and Construction Evaluation of Mortars for Plain and Reinforced Unit Masonry.* This is necessary because laboratory-mixed mortar has less water than job-mixed mortar. The difference allows for the absorption of the masonry units. If both laboratory testing and field testing are to be done, the water content must be the same.

Grout mixtures are specified by proportion or compressive strength in *ASTM C476, Standard Specification for Grout for Masonry.* Two types of grout, fine and coarse, are specified by proportions of ingredients. *ASTM C476* allows pumping admixtures for high-lift grouting. Strength testing for grout is done in accordance with *ASTM C1019, Standard Test Method for Sampling and Testing Grout.*

1.1.4 Masonry Accessories

Several different specifications cover masonry accessories. In most cases, the manufacturer certifies that the materials conform to the standards. The following are some of the most common standards for masonry accessories:

- *ASTM A36, Standard Specification for Carbon Structural Steel,* covers size and strength requirements for the steel used for angle lintels and shelf angles in masonry construction. It also applies to bent bar or strap anchors. Coatings must be specified separately.
- *ASTM A82, Standard Specification for Steel Wire, Plain, for Concrete Reinforcement,* covers steel wire used in joint reinforcement, some anchors, and ties. It includes strength and size requirements, but not coatings.
- *ASTM A951, Standard Specification for Steel Wire for Masonry Joint Reinforcement,* covers materials, fabrication, tests and tolerances, and some coatings for prefabricated wire joint reinforcement.
- *ASTM A153, Standard Specification for Zinc Coating (Hot-Dip) on Iron and Steel Hardware,* covers galvanized coatings for steel joint reinforcement, anchors, and ties. It also includes various masonry accessories.
- *ASTM A167, Standard Specification for Stainless and Heat-Resisting Chromium-Nickel Steel Plate, Sheet, and Strip,* covers the steel most commonly used in anchors, ties, and flashing. It lists more than 20 types of stainless steel according to chemical composition.

1.2.0 Identifying the Standards for Laboratory and Field Testing of Masonry Construction

Laboratory and field testing are mostly concerned with the compressive strength of mortar, grout, and masonry performing as a unit and mortar or grout performing alone. On large commercial projects, these materials are tested before and during construction. Test samples are built before and after the actual laying has started.

ASTM C780 is a complex standard that covers methods of sampling and testing mortar for plastic and hardened properties. This standard calls for a high water-to-cement mix design in laboratory testing to simulate actual construction conditions.

The tests performed under *ASTM C270* use a low water-to-cement ratio for the test mortars. These tests cannot be used to compare results with tests on mortars mixed on the job. Job-mixed mortars use a much higher water-to-cement ratio to overcome the absorption of the masonry units. The formulas in *ASTM C780* calculate a wetter mix design for laboratory testing to simulate job-mix conditions. These tests can then be used to accept or reject job-mixed mortars during construction.

ASTM C1019 covers both field and laboratory sampling for grout. This standard is referenced when compressive strength is to be tested by unit strength or by the **prism** method.

Prism: A sample prepared under controlled conditions specifically for testing that uses the same materials, techniques, and craftsmanship as those used at the jobsite.

ASTM C1314, Standard Test Method for Compressive Strength of Masonry Prisms, is another complex standard. Project engineers specify the compressive strength of masonry for a particular job. The contractor verifies this in one of two ways: by unit strength or by prism test.

Verifying by unit strength calls for the contractor to submit the manufacturer's certification on the masonry units and specify that the mortar is mixed according to *ASTM C270* proportions. These documents allow the engineer to check compressive strength in lookup tables in *ASTM C1314*, without any further testing.

Verifying by prism test calls for a mason to build a test panel, or prism, using the selected units and mortar. The prism is then tested according to *ASTM C1314* procedures. This test can be used for preconstruction and construction evaluation of masonry strength.

ASTM C67 sets procedures for sampling and testing individual clay or shale brick and structural clay tile. This standard covers procedures for sampling and testing brick for modulus of rupture, compressive strength, absorption, saturation coefficient, effect of freezing and thawing, initial rate of absorption (suction), and efflorescence. *ASTM C67* also includes procedures for measuring size and warpage. *ASTM C140* defines the procedures for testing CMUs. These and other test specifications are designed to measure the quality of the ingredients used in masonry construction.

1.0.0 Section Review

1. Which organization develops the most commonly used material standards?
   a. OSHA
   b. BIA
   c. ACI
   d. ASTM

2. Compared to the water-to-cement ratio used in mortar on the job, the water-to-cement ratio used for test mortar in tests performed under *ASTM C270* is _____.
   a. high
   b. low
   c. identical

2.0.0 Building and Testing Masonry Sample Panels and Prisms

Performance Task

1. Build a prism for mortar testing.

Objective

Describe how masonry sample panels and prisms are built and tested to ensure quality control on a project.

a. Describe how to build sample panels.
b. Describe how to build hollow masonry prisms.
c. Describe how to build grouted masonry prisms.
d. Describe how to prepare and test mortar and grout prisms.
e. Describe how to conduct masonry tests.

The testing of masonry requires masons to perform several tasks. They build sample panels, which may be submittals under the contract documents; they make masonry, grout, and mortar prisms for laboratory testing; and they perform field tests and field observations. The following sections describe these tasks.

2.1.0 Building Sample Panels

On many jobs, the masonry contractor must build a sample panel, or mock-up, at the construction site before the masonry work begins. The sample panel serves two purposes:

- It can be observed throughout the construction period for any change or damage due to changing weather conditions or poor craftwork.
- It demonstrates the appearance and quality of the materials and the masonry skills that will be used on the project.

The size of the sample panel can vary with the complexity of the project. A typical sample panel is shown in *Figure 2*. The sample panel is made from the same materials that will be used in the structure; therefore, it will include all structural elements that are visible, as well as those that are behind the wall.

Because the panel is an example of what the finished structure will look like, the contractor typically sets the requirements for constructing the panel. In general, the panel should do the following:

- Demonstrate the average skill of the masons working on the project
- Use the mortar type, ingredients, and admixtures specified for the project
- Use masonry units specified for the project and selected from the production run intended for shipment to the job
- Have the same coursing, bonding, and wall thickness as those specified for the project
- Have the same joint tooling as that specified for the project
- Have the same finishing operations or coatings as those specified for the project

The masonry units used in the sample should be randomly selected from those delivered to the construction site. Some units should be selected from each cube delivered so that any variations in color, texture, or strength can be determined before construction begins.

(A) Front Sample Panel

(B) Sample Panel Detail Showing Pintle

(C) Sample Panel Detail Showing Flashing and MortarNet™

Figure 2 Sample panel.
Source: Florida Masonry Apprentice & Educational Foundation

Because the sample panel will show the effects of weathering on the finished structure, it should be placed where it will be subject to the same weather conditions as the finished construction.

No special curing, protective steps, or skills should be applied that will not be used in the finished structure. Approval of the sample panel is the responsibility of the quality assurance representative, the architect, or the owner.

Masons can also use the sample panel to check their work and to foresee any problems that may arise. For instance, if efflorescence appears in the sample, the masonry supervisor may need to ask the contractor about changes in the mortar mixture.

2.2.0 Building Hollow Masonry Prisms

Flexural: Capable of turning, folding, or bending.

Tensile: Capable of being stretched or extended, ductile.

Masonry prisms are samples used for testing in a laboratory. The tests will determine the **flexural**, **tensile**, and compressive strength of the combined masonry and mortar. Just as with sample panels, prisms should be built using the same materials, techniques, and craftsmanship as those to be used in the final structure. A mason may be asked to make as many as 10 test prisms for laboratory testing before construction begins. Prisms can be made in different sizes, as shown in *Figure 3*.

Test prisms are also made during construction for quality control of laying procedures and field operations. Typically, a minimum of three test prisms are made for every 5,000 ft^2 of wall area, or every story built.

Two sets of three or five prisms each are sometimes required. These will be tested at three, seven, and 28 days after building. The seven-day test is often used to approximate the 28-day strength; tests have shown that the seven-day strength usually equals 90% of the 28-day strength.

$$S_7 = 0.9 \times S_{28}$$

or

$$S_7 \div 0.9 = S_{28}$$

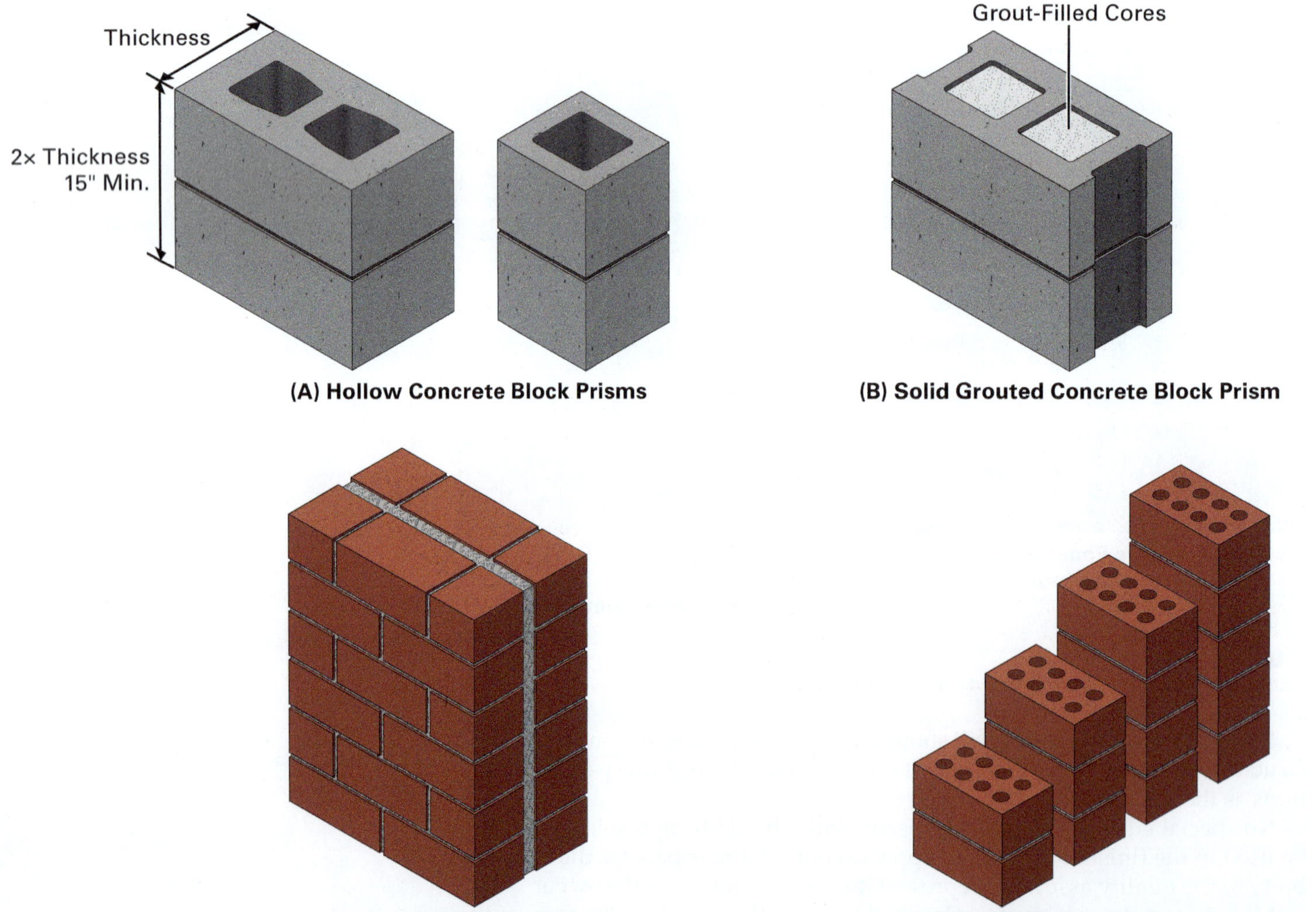

(A) Hollow Concrete Block Prisms

(B) Solid Grouted Concrete Block Prism

(C) Concrete Brick Prism

(D) Prisms with Different Slenderness Ratios (Height-to-Thickness Ratio)

Figure 3 Block and concrete brick test prisms.

For example, if the seven-day test results in a strength of 180 psi, the 28-day test can be calculated at 200 psi:

$$(180 \div 0.9 = 200)$$

The size may be specified for any prism. The thickness of the prism should be the same as the thickness of the masonry wall in the structure. The length of the prism should be equal to or greater than the thickness.

The tests accurately reflect expected wall performance if the prism is five times as high as it is thick, as shown in *Figure 4*. If there are no specifications, select the prism height using the parameters for the seven-day test; however, check with the laboratory to make sure the testing equipment is large enough to take a prism with the 5:1 dimension.

The mortar bedding, thickness and tooling of joints, moisture content of the masonry units at the time of laying, and bonding should be as close as possible to the completed structure. Unless specified otherwise, the prism should also be laid up in a stack bond.

All masonry materials in the prism should be randomly selected from the construction stockpile. Block should be laid with the thicker end of the face shell up. Full mortar bedding should be used in the prism construction if uncored masonry units are used. Hollow units should have only face shell bedding.

Note that the walls may have collar joints mortared or grouted, as in the project specifications. Hollow units may also be grouted as specified. However, no structural reinforcing rods should be used in the prisms. Metal ties or joint reinforcement can be used for the walls, but vertical or horizontal structural reinforcement is normally not included in the prism. A block prism should be grouted according to the job specifications, but no reinforcement should be used. Most prisms are tested by crushing, and steel bars could break the tester.

The prisms must be built on a flat surface where they remain undisturbed for at least 48 hours. A prism can be built on a foundation of an unattached $^3/_4$" plywood board. After the prisms are complete, mark each one for identification.

After 48 hours, the prism can be bundled to protect the bond. The prism is topped with another $^3/_4$" plywood board. The boards are bound tightly with two wires to keep the joints from being disturbed.

Preconstruction prisms are then sent to the laboratory, where they are cured at a temperature between 65°F and 85°F. Quality control prisms are cured at the jobsite. They must be left in a place where they will be undisturbed but subject to the same weather conditions as the structure on the job itself. Normally, the prisms are cured for 28 days, but some tests are made after only three days or seven days of curing.

2.3.0 Building Grouted Masonry Prisms

Test prisms are also made with grout if the project calls for grouting. The prisms should be grouted only when the actual walls are to be grouted. Do not include any reinforcement in the prism except for the metal ties necessary for bonding the walls. When prisms are to be grouted, follow these steps:

Step 1 Make the block prism using the materials and techniques that will be used on the job. Place a nonabsorbent board or piece of $^3/_4$" plywood under the masonry.

Step 2 Allow the mortar joints to cure for 24 hours before pouring the grout.

Step 3 Pour the grout into the cores or collar joints in one layer up to $1^1/_2$" from the top of the prism.

Step 4 Rod the grout 25 times with a 1' × 2" wooden stick, or vibrate it to ensure that the void is filled. (If a vibrator is to be used on the job, the prism should be vibrated, not rodded.)

Step 5 Let the prism stand for about 30 minutes to allow the units to absorb some water from the grout.

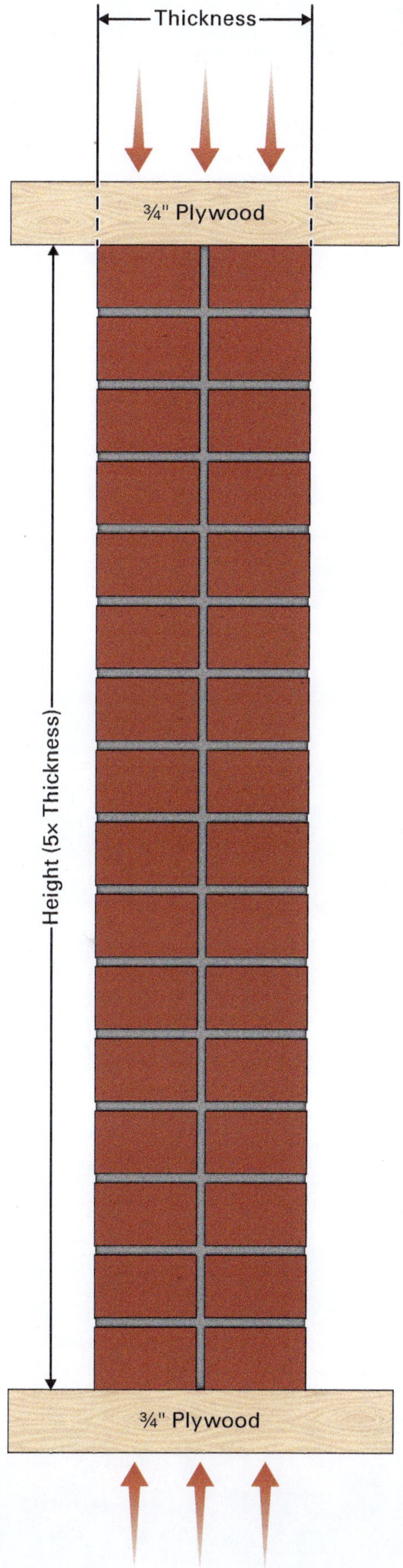

Figure 4 Compressive prism test.

Step 6 Finish filling the void area with grout, and vibrate or rod again at least 25 times. Rod at least 1½" into the first layer.

Step 7 Level the top of the prism with a trowel.

After the prism cures for 24 hours, it can be bundled with another piece of wood on top and sent to the laboratory.

2.4.0 Preparing and Testing Mortar and Grout Prisms

Mortar prisms, or test cubes, are normally 2 in^2. They are tested in the laboratory to check the strength characteristics of the mortar to be used on the job. A cube compressive-strength test is shown in *Figure 5*.

Figure 5 Cube compressive-strength test.
Source: Shutterstock

These cube tests are made only on laboratory-mixed mortar if the mortar design has been specified by proportion according to *ASTM C270.* If the mortar design has been specified by strength according to *ASTM C780,* both lab-mixed and job-mixed mortar will be tested referencing *ASTM C1019.*

Some projects may require you to periodically make mortar prisms to check the field mortar for uniform batching procedures. However, compressive strength is not specified for field-tested mortar. If you must make test cubes of mortar at the jobsite, a mortar cube mold or test tray should be used. The mortar should be taken from the latest batch mixed, and the cubes should be labeled.

Grout prisms are also tested. The job specifications will set the number of grout test prisms and the stages of the job at which they are to be made. Typically, grout is sampled whenever there is a change in the mix, materials, or mixing method. About ½ ft^3 of grout is taken from work in progress to make three prisms. Make the grout test prism rectangles twice as high as they are wide. The usual size is 3½" × 3½" × 7" high.

These are the steps to follow in preparing grout test prisms, according to *ASTM C1019*:

Step 1 Select a flat location where the prism can harden undisturbed for two days.

Step 2 Place a flat piece of nonabsorbent wood, ⅝" thick and 3½ in^2, on the flat surface.

Step 3 Tape some sheets of paper towel to one face on each of the four blocks.

Step 4 Place the block flush to the edges of the wood to make a cavity, with the papered faces on the inside of the cavity, as shown in *Figure 6*.

Step 5 Pour grout halfway up into the cavity.

Step 6 Rod the cavity with a wood or metal rod 25 times to eliminate bubbles.

Step 7 Pour grout to fill the cavity completely.

Step 8 Rod the cavity 25 times to eliminate bubbles and ensure that the rod also penetrates the lower level.

Step 9 Trowel the top of the prism so that it is flat, and cover it with wet burlap.

Step 10 After two days, separate the block to remove the prism. Pack it in sand or sawdust and send it to the laboratory for further aging and testing.

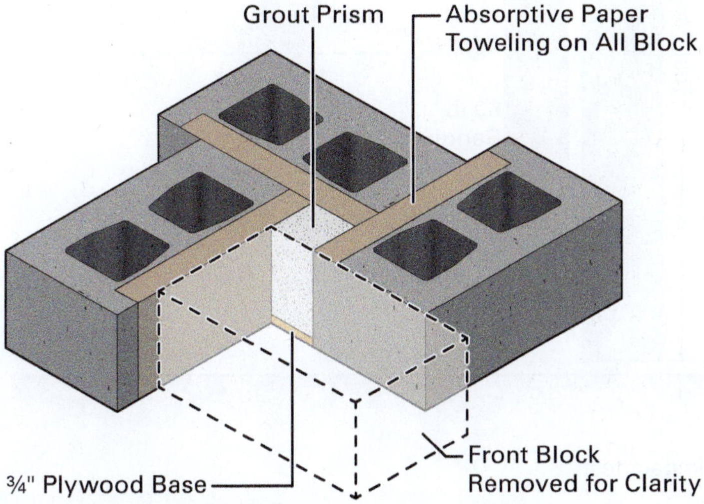

Figure 6 Grout prism mold of block.

2.5.0 Testing Masonry Prisms

Masonry prisms are tested to determine the assembled masonry's flexural, tensile, and compressive strength. The mortar bedding, the thickness and tooling of joints, the grouting, the moisture content of the masonry, and the bonding all affect the strength of the constructed wall.

The stress tests discover the prism's breakpoint. A change in the mortar formula and application will probably change the breakpoint more than any other factor. Tests have shown that increasing mortar strength by 160% increases the strength of the entire structure by 25%.

2.5.1 Impact Test

The impact test measures bond strength by testing a masonry prism's ability to withstand impulsive stresses that occur at different points and at random times. The test requires dropping a 60 lb sandbag directly on the face of the test prism, as shown in *Figure 7*. The sandbag reduces the risk of damaging the masonry units.

The test prisms are 48" high and are supported by a base 6" less than the panel's height; as such, a prism 48" high would be supported along 42" of its base. The drop distance is increased until failure occurs. However, the bag is never dropped or swung from above 20'. The reported value is the drop distance at which the prism breaks.

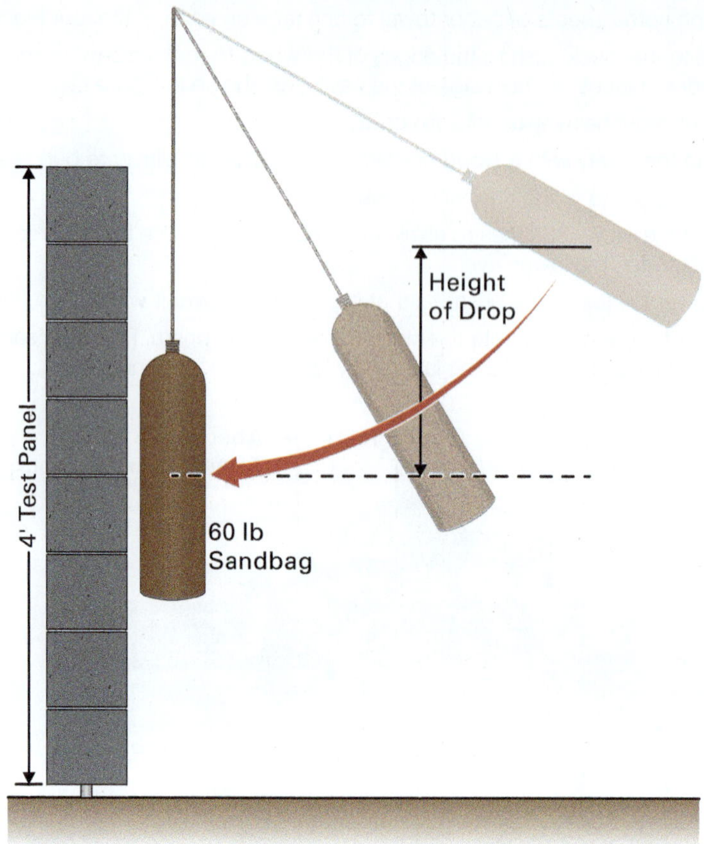

Figure 7 Impact test.

2.5.2 Flexural Test

The flexural strength bond test measures the force a masonry beam or wall can resist when the force is applied against the face of the masonry. This test provides important information for structures subject to high winds or other types of lateral loads. Under these conditions, walls bend or flex to some degree, which can crack the mortar joints if the strength of bond is not high enough.

There are three methods for testing flexural strength: using a one-third-point concentrated load, using a uniform load, or using a uniform transverse load. With the one-third-point load or the uniform load method, the prism is loaded in the test machine as if it were simply a supported masonry beam.

As shown in *Figure 8*, the specimen should be placed with the finish face down and the backup side on the top, or load, side. The load is applied at the two points equidistant from the ends and each other.

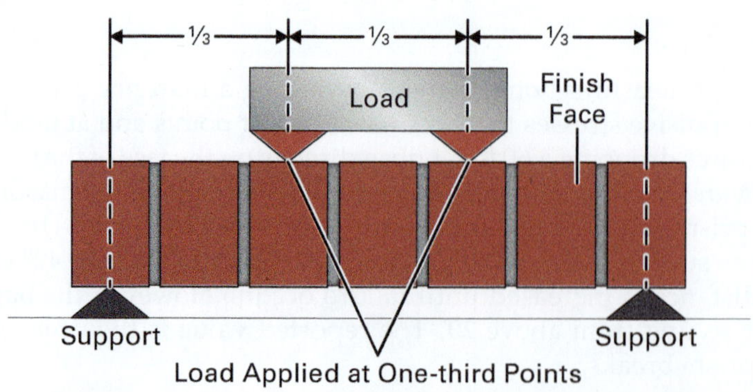

Figure 8 One-third-point flexural strength test.

With the transverse load method, the prism is placed upright. A plastic or rubber bag filled with air is placed between the wall face and a backboard to apply various amounts of pressure against the prism. The amount of pressure on the wall and the degree of movement that results are recorded for use in a detailed design analysis.

2.5.3 Tensile and Other Tests

The tensile strength bond test measures a prism's resistance to twisting apart. The top bearing block turns the prism in one direction, while the lower bearing block turns the prism in the opposite direction.

The bond wrench test tests prisms of full-size masonry units rather than cubes. The test machine wrenches the bonded units apart, measuring the bond's flexural tensile strength.

A diagonal shear test measures the amount of stress a prism will resist along its diagonals. A test machine applies compressive force along the diagonal so that the shear stress is applied to the mortar joints, as shown in *Figure 9*. The prism fails when the applied loads exceed the strength of bond of the mortar and masonry.

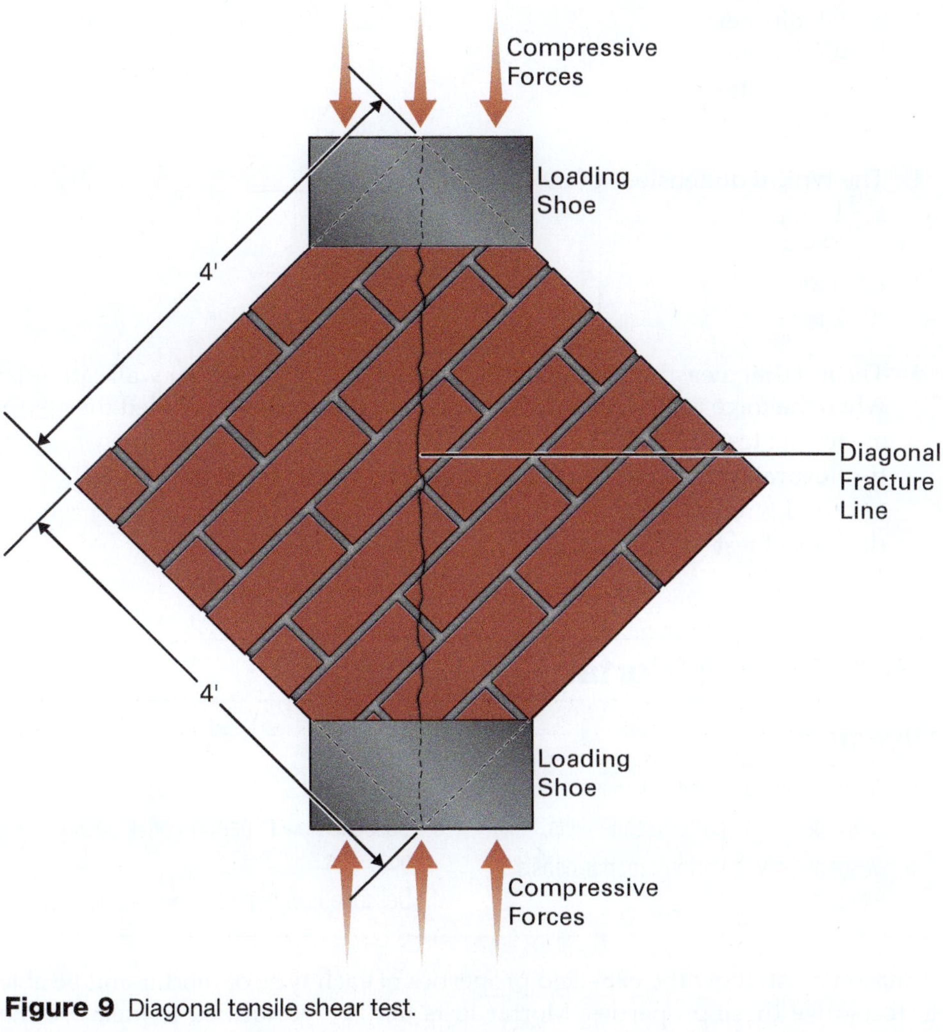

Figure 9 Diagonal tensile shear test.

2.0.0 Section Review

1. Which masonry units should be selected for the test panel?
   a. A random selection of units
   b. The best-quality units
   c. Units from the first delivery
   d. Units from the last delivery

2. For a project that will involve the construction of 20,000 ft^2 of wall area, the typical minimum number of test prisms to be made will be _____.
   a. 12
   b. 9
   c. 6
   d. 3

3. When building a grouted masonry prism, after rodding or vibrating the grout, allow the units to absorb water from the grout by letting the prism stand for approximately _____.
   a. 60 minutes
   b. 45 minutes
   c. 30 minutes
   d. 15 minutes

4. The typical dimensions of mortar test cubes are _____.
   a. 1" × 4"
   b. 2" × 3"
   c. 3 in^2
   d. 2 in^2

5. The test that measures the amount of force a masonry beam or wall can resist when the force is applied against the face of the masonry is called the _____.
   a. tensile test
   b. flexural test
   c. bond wrench test
   d. impact test

3.0.0 Testing Mortar

Performance Task

2. Perform a slump test.

Objective

Describe how mortar is tested to ensure quality control on a project.

a. Describe how to perform sand tests.
b. Describe how to perform mortar consistency tests.
c. Describe how to perform brick absorption tests.
d. Describe how to perform laboratory tests.

A mason must know the expected properties of each type of mortar and be able to recognize these properties. Mortar tests may be performed to measure the following qualities of mortar:

- *Consistency or wetness*—Consistency or wetness is determined by the amounts of both water and sand added to the mixture. The consistency of mortar should vary according to the ASTM mortar type. Mortar with a good consistency (*Figure 10*) should be workable but should not flow freely.

- *Water retention*—Water retention affects all other qualities. Mortars with high water retention stiffen more slowly. Mortars with low water retention may dry so quickly that they do not allow the mason to adjust the masonry units and may cause weak joints that are not weather tight. Water retention is especially important on hot, dry, and windy construction sites where mortar may quickly lose its workability because of water evaporation.

- *Plasticity*—Plasticity is related to water retention and consistency. The mortar must be plastic enough to stay on a trowel, must spread easily but hold a uniform thickness, and must stay soft enough to adjust the masonry units before it sets up.
- *Strength of bond*—Strength of bond gives the mortar its ability to tie units together. Strength of bond enables masonry to withstand compressive, tensile, and shear stresses.
- *Compressive strength*—Compressive strength in mortar increases the masonry unit's resistance to downward force or load. The ASTM classifies mortars according to their compressive strength.
- *Durability*—Durability is the ability to withstand weathering and freeze-thaw cycles without deterioration.
- *Flexural strength*—Flexural strength is the ability to withstand bending forces. Masonry structures may expand or contract due to weather conditions, due to alternate wetting and drying, or as a result of water loss. Horizontal forces from winds or earthquakes also tend to bend or flex the masonry. These stresses must be resisted by the mortar's flexural strength.

Figure 10 Mortar with good consistency.
Source: McGee Brothers

Mortar will have all desired properties only if the proper materials are correctly proportioned and mixed. Some mortar properties can only be determined through laboratory testing, while others can be tested using quick field tests. You should know how to perform the field tests used to check the quality of materials or mortar mixes and how to prepare samples used in laboratory testing.

3.1.0 Performing Sand Tests

Because the ingredients of mortar directly affect its quality, field tests are used before the mortar is mixed. Three field tests can be used to judge the quality of sand:

- *Standard screen sieve test*—This test ensures that the sand is of a grade suitable for mortar.
- *Siltation test*—This test determines the purity of the sand.
- *Colorimetric test*—This test indicates the amount of salt or organic materials in the sand.

3.1.1 Sieve Test

The standard screen sieve test checks the gradation of the sand. Properly graded sand contains a variety of particle sizes and results in a very uniform and workable mortar.

Small particles will fill the gaps between the larger particles to help maintain the consistency of the mortar. Poorly graded sand has particles all the same size or does not have a variety of sizes. This results in a mixture with a large proportion of void spaces between the particles, resulting in an inconsistent mortar.

The sieve test determines the percentage of sand in different size classifications. It is performed by passing the sand through a series of sieves or screens with progressively smaller openings. A size 4 sieve, for example, allows particles ¼" and smaller to pass through, while a size 200 sieve allows the passage of only very fine material.

ASTM C144, Standard Specification for Aggregate for Masonry Mortar, establishes gradation limits for mortar sand; these should be followed to ensure well-graded sand. *Table 1* shows the recommendations for the percent of sand by grain size.

TABLE 1 Recommended Sand Gradation Limits

	Percent Passing	
Sieve Size	**Natural Sand**	**Manufactured Sand**
4	100%	100%
8	95%–100%	95%–100%
16	60%–100%	60%–100%
30	35%–70%	35%–70%
50	15%–35%	15%–35%
100	2%–15%	10%–25%
200	–	0%–10%

The proportions of the different sizes of sand can be checked using the specifications set by *ASTM C144*. The process calls for sifting a measured amount of sand through various sizes of sieves and measuring the sand that passes through each. The procedure is as follows:

Step 1 Fill a box with sand and weigh the box and the sand. Record this information as the initial or before weight, or *B*.

Step 2 Beginning with the largest sieve size, sift the sand through the screen into another container.

Step 3 Place all sand that passes through the sieve back into the first box. Weigh the box and the sand that passed through the sieve. Record this information as the after weight, or *A*, and record the size of the sieve.

Step 4 Take the next smaller sieve and repeat the procedure and recordings, using each sieve size called for in the job specifications. Different jobs will have different requirements for the fineness of the sand.

Step 5 Determine the percentage of sand passing through each sieve size by dividing the after weight, *A*, by the before weight, *B*, for each sieve size. The formula is:

$$(\text{weight } A_n \div \text{weight } B_n) \times 100 = \text{percentage passing}$$

For example, if the box and sand weigh 25 lb before sifting and 20 lb after sifting, the percentage of sand passing through this sieve is 80%:

$$(20 \div 25) \times 100 = 80\%$$

The sieve sizes recommended for grading mortar sand are given in *Table 1*. It specifies that the gradation of the material from any one source shall be reasonably uniform and shall not be permitted to vary over the extreme range. It also recommends that sand for any one job be supplied from the same source. Sand from different sources may result in different mortar qualities.

3.1.2 Siltation Test

The siltation test determines the presence of **silt**, clay, or other earth particles in the sand. The sand must be clean, or the strength of bond, color, and adhesive strength of the mortar will be affected. The following simple process can be performed to measure the presence of loam, soil, silt, clay, mica, or humus:

Silt: A sedimentary material consisting of very fine particles intermediate in size between sand and clay.

Step 1 Fill a clear, quart-size glass container with 3" of sand.

Step 2 Partially fill the container with potable water (2" or 3" above the sand) and shake it vigorously.

Step 3 Let the mixture stand for at least one hour. The sand will settle to the bottom. Any organic material, such as silt or loam, in the sample will settle just above the sand, as shown in *Figure 11*.

Step 4 Measure the layer of silt or loam. The sand is not clean enough for mortar if the layer is more than $\frac{1}{8}$" thick.

Figure 11 Siltation test container.

3.1.3 Colorimetric Test

While the siltation test determines the presence of earth contaminants in the sand, the colorimetric test shows the amount of salts and organic material in the sand. Organic material can reduce the mortar's strength of bond. The following steps can be used for determining the presence of organic material in aggregates:

Step 1 Make a 3% sodium hydroxide solution by dissolving 1 oz of sodium hydroxide in 1 qt of distilled water. Sodium hydroxide can burn skin or clothes, so take appropriate care to avoid injury.

Step 2 Use a container with graduated-ounce markings. Fill the container with the sand to the $4\frac{1}{2}$ oz mark.

Step 3 Add the sodium hydroxide solution to the 7 oz mark. Seal the container tightly and shake it vigorously.

Step 4 Let the container stand for at least 12 hours. The color of the liquid will show the amount of organic material in the sand.

WARNING!

Sodium hydroxide is caustic and can burn your skin. Always wear the required personal protective equipment (PPE) when working with sodium hydroxide or proprietary cleaners that include this ingredient.

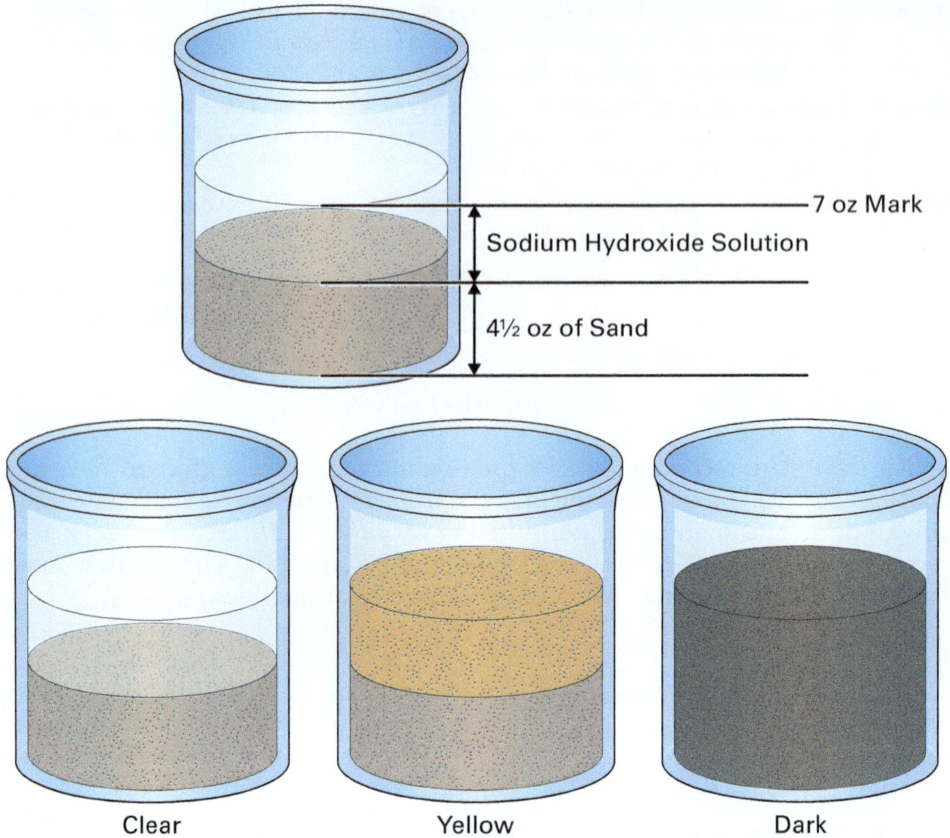

Figure 12 Colorimetric test.

Figure 12 depicts the colorimetric test and results. If the liquid is clear, the sand is free from impurities and may be used for mortar. If the liquid is slightly yellowed, the sand has some organic content and should be thoroughly washed before use. If the liquid is darker, the sand should not be used, because the dark color indicates a high degree of foreign matter in the sand.

3.2.0 Performing Mortar Consistency Tests

Slump: A measure (0" to 12") of the consistency or fluidity of a cementitious product such as concrete, masonry, or grout. Lower slump (1" to 4") indicates a stiff mix; higher slump (8" to 12") indicates a fluid mix.

Mortar can be tested for plasticity and color consistency immediately after mixing. To test mortar plasticity, simply turn the trowel upside down. The mortar should not fall off immediately. Mortar that easily slides off the trowel with a rolling or throwing motion but clings to an upside-down trowel is considered to have the correct plasticity.

To test color consistency or uniformity, flatten some mortar under a trowel. If streaks of color appear, additional mixing is needed. To test color consistency after the mortar has set, make, label, and place test cubes outside at the jobsite.

Make these cubes from the same type of mortar and with the same materials used on the job. Allow the cubes to cure at the jobsite for at least five days before checking the effects of weather on the final mortar color.

A **slump** test may be used to ensure that each mortar batch has the same consistency. Specifications are normally not established for the slump of mortar, but a slump test will ensure that the mortar consistency is constant throughout the job.

Specifications are usually set for grout slump, so grout must be tested and records maintained. *Figure 13* shows the relative stiffness of concrete, mortar, and grout. The stiffer concrete and mortar will keep the shape of the cone better than the more fluid grout.

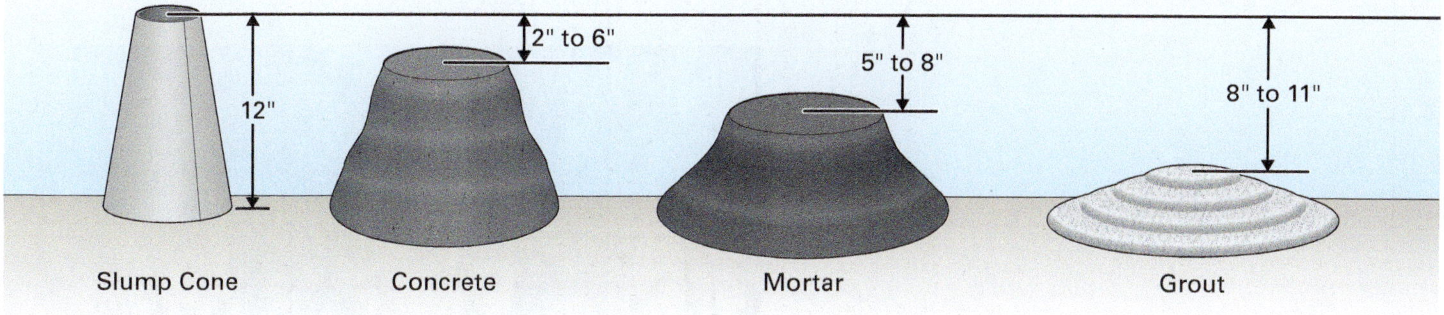

Figure 13 Slump test comparison.

The slump test procedure is as follows:

Step 1 Gather the slump test supplies, the most important of which is the moistened cone mold (*Figure 14*). Place the cone mold on a firm, level, moistened, nonabsorbent surface and hold it in place by standing on the foot pieces.

Figure 14 Slump test supplies.
Source: Shutterstock

NOTE

The entire slump test from the start of the filling to the completion of the cone lift should be accomplished in two and a half minutes.

Step 2 Fill the cone with fresh mortar or grout to one-third full by volume, and then vibrate or rod it evenly over the surface area 25 times using a standard steel ⅝" smooth tamping rod with a rounded end (*Figure 15*).

Figure 15 Cone mold one-third full.

Step 3 Add the next layer of fresh mortar or grout until the cone is two-thirds full by volume (*Figure 16*). Vibrate or rod this layer evenly 25 times, with the rod just penetrating the first layer.

Figure 16 Cone mold two-thirds full.

Step 4 Heap the cone to overflowing, and vibrate or rod this third layer evenly 25 times, with the rod just penetrating the second layer (*Figure 17*). If the mortar or grout sags below the top of the cone, add more mortar or grout to keep the mortar or grout above the top of the cone.

Figure 17 Cone mold full and overflowing.

Step 5 After rodding or vibration, use the rod to scrape excess mortar or grout off the top of the cone (*Figure 18*). Clean any spilled mortar or grout from around the base of the cone.

Figure 18 Leveling top of cone mold.

Step 6 Using the hand grips, lift the cone vertically. This must be done slowly and carefully within three to seven seconds, while avoiding rotational movement or bumping the molded mortar or grout. Then, place the cone upside down next to, but not touching, the slumped mortar or grout. Lay the rod across the cone over the mortar or grout (*Figure 19*).

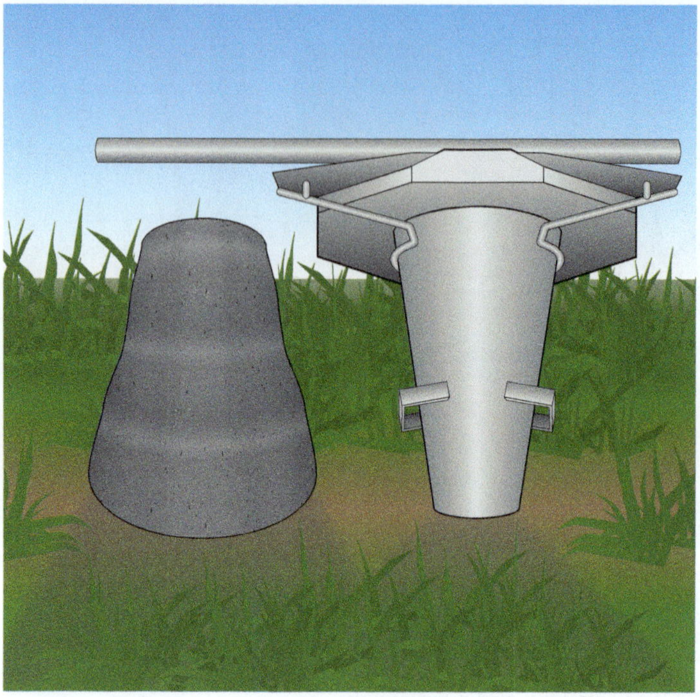

Figure 19 Inverted cone with rod over slumped mortar.

Step 7 To determine the slump, measure the distance from the bottom of the tamping rod to the sagged top of the mortar or grout at the original center of the cone, to the nearest $\frac{1}{2}$" (*Figure 20*). Record the slump and discard the test sample.

Figure 20 Measuring the slump.
Source: Shutterstock

A 5" to 8" fall, or slump, is usually specified for mortar. The more fluid in the mortar, the looser the consistency and the greater the slump. Grout for use in reinforced masonry should have a slump of 8" to 11".

The slump for both grout and mortar should be consistent throughout the span of the job. A minimum of $^1/_2$ ft^3 of grout or mortar should be taken for slump or strength tests whenever sampling is specified.

NOTE

The use of rebar as a tamping rod may cause inaccurate test results, because the deformations in the rebar will introduce air into the mortar or grout.

3.3.0 Performing Brick Absorption Tests

CMUs are cured and dried at the manufacturing plant. When delivered to the jobsite, their moisture content should be within specified limits. Concrete units should never be moistened before or during laying, because they shrink as they dry. The resultant stress will likely crack the wall.

Brick, on the other hand, is not controlled for moisture during manufacturing or curing. The pores (or small openings) in fired clay products tend to draw or suck water into the unit. This action in a brick is referred to as its *suction*, or *initial absorption rate*.

The suction of brick has little bearing on its water resistance but does have an important effect on the bond between brick and mortar. If the absorption rate is too high, too much water is diverted from the mortar between the units. Excessive water absorption reduces hydration, changes the strength of the mortar, and results in a weak bond that is not watertight.

To prevent these conditions, brick can be tested for absorption. The method for determining the initial rate of absorption of brick is described in *ASTM C67*. It consists of immersing a dry unit in water to a depth of $^1/_8$" for one minute, removing the unit from the water, weighing the unit, and then comparing its final weight with its dry weight.

Brick produced commercially has suctions ranging from 3 g per minute to more than 112 g per minute. There is no consistent relationship between total absorption and suction. Some brick with high total absorption has low suction, and vice versa. The suction of the brick when laid can be controlled at the jobsite by wetting. Suction is not controlled by product specifications.

ASTM C67 specifications call for wetting the brick at the time of installation if the absorption rate is greater than 30 g of water per minute over 30 in^2, or 0.035 oz per in^2, per minute, at the time of installation. A simple rule-of-thumb absorption test is as follows:

Step 1 Mark a circle 1" in diameter on one of the brick to be used in the job. A quarter will serve as a good pattern. Make certain that the surface to be mortared is used for the test.

Step 2 Inside the circle, drop 20 drops of water from a common medicine dropper.

Step 3 Using a watch with a second hand, note the time it takes for the masonry to absorb the water.

If the brick's absorption time is longer than 90 seconds, the units need no further preparation before laying. If the absorption time is less than 90 seconds, the units are very porous and need to be thoroughly wetted before laying. The water can be applied by spray, drip, or soaker hose. In wetting high-suction brick, sprinkling is not enough. A hose stream should be sprayed on the brick pile until water runs from all sides.

The brick's surface should be allowed to dry before it is laid in the wall. A surface film of water will cause the brick to float on the mortar bed and prevent proper bonding. Checking a broken unit, as in *Figure 21*, will show whether moisture is evenly distributed.

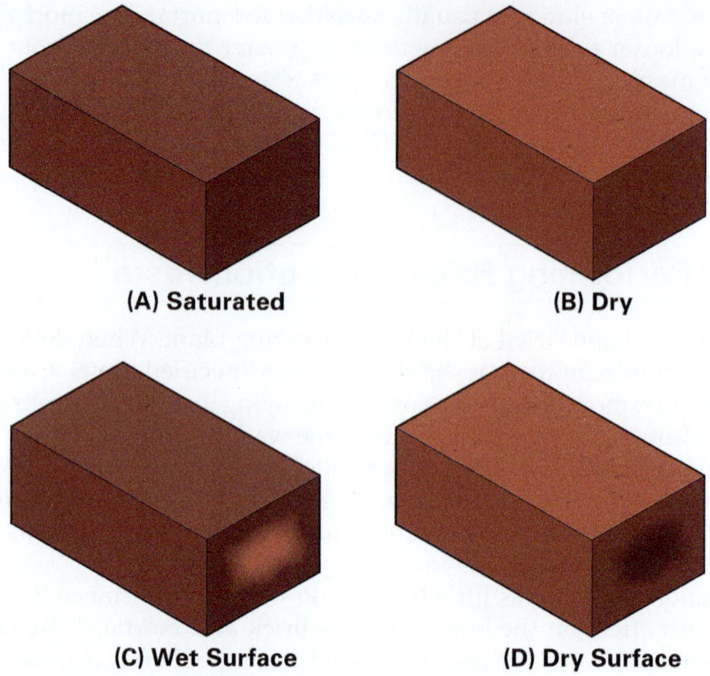

Figure 21 Moisture condition of brick.

Masonry units with too much moisture take more time to set because they cannot absorb much water from the mortar. Units with no moisture prevent the mortar from curing properly. Hard-finished brick should be covered on the job so that they do not get too wet. Soft-finished or high-suction brick, on the other hand, must usually be wetted before laying to obtain the proper bond.

3.4.0 Performing Laboratory Tests

Laboratory tests are important when mortar is specified by strength rather than by proportions. Mortar is tested when hardened and when plastic. For strength testing, mortar is put into a 2" cube mold and cured for 24 hours. It is then unmolded and stored for 27 days to get its 28-day strength. Occasionally, grout is set in a 2" cube mold for testing in addition to the larger grout prisms. Grout cubes are moist cured for 48 hours, then tested.

It is recommended that three mortar cubes be tested for every 100,000 masonry units used. Tests of plastic mortar should also be made on an ongoing basis as the job progresses.

3.4.1 Flow and Flow-After-Suction Tests

The flow-after-suction test measures the ability of plastic mortar to retain water and cohesiveness after the masonry units have absorbed water from it. The test is done in the laboratory using a flow table, flow plate, and mortar made from the same design and materials as those at the jobsite.

The mortar is put into a mold, rodded, and unmolded on a flow table. The spread of the mortar is measured before it is disturbed. Then, a weighted flow plate is dropped on the mortar from $\frac{1}{2}$" above. The dropping is done 25 times in 15 seconds, and the resultant spread of the mortar is measured. This is the measure of flow.

The mortar is suctioned to simulate the effect of highly absorbent brick. The weighted flow plate is dropped on the mortar another 25 times in 15 seconds from $\frac{1}{2}$" above. The resultant spread of mortar is measured again. This is the measure of flow after suction.

For laboratory testing, the allowable initial flow ranges from 105% to 115% of the undisturbed mortar. Testing in the field gives results ranging from 130% to 150% of the undisturbed mortar. After suctioning and the second disturbance, the flow after suction must exceed 75% of the original undisturbed diameter.

3.4.2 Cone Penetration Test

The cone penetration test is another method of testing the consistency of plastic mortar. It measures the depth of penetration of a cone-shaped plunger into a mortar sample.

For the test, a metal cup measuring 3" in diameter by 3½" in depth is filled with the mortar to be tested. The mortar is evenly tamped or spaded so that no air pockets are trapped in the sample. The prepared sample is placed under a cone penetrometer, which drops a metal cone into the mortar and measures the depth the cone penetrates the mortar. The measurement after 30 seconds is the measure of mortar consistency and is reported in millimeters.

The cone penetration test is also used to measure how long mortar will remain consistent. To measure consistency retention, the cone penetration test is done on several samples at intervals of 15 minutes, taken directly from the mixer. The laboratory first sets limits on desirable consistency and then measures the time, usually in minutes, during which the mortar keeps that same consistency.

3.0.0 Section Review

1. Which test determines the purity of sand?
   a. Colorimetric
   b. Standard screen sieve
   c. Siltation
   d. Tensile

2. Which test measures mortar consistency?
   a. Colorimetric
   b. Tensile
   c. Siltation
   d. Slump

3. The initial rate of absorption of clay brick has an important effect on _____.
   a. mortar bond
   b. water resistance
   c. crack resistance
   d. hardening time

4. The method of testing the consistency of plastic mortar that measures the depth of penetration of a plunger into a mortar sample is called the _____.
   a. volumetric air test
   b. aggregate-to-cement ratio test
   c. flow-after-suction test
   d. cone penetration test

4.0.0 Field Inspections and Observations

Performance Tasks

There are no Performance Tasks in this section.

Objective

Describe how field inspections and observations are used to ensure quality control on a project.

a. Describe why and how inspections of standards and codes are performed.
b. Describe why and how inspections of materials are performed.
c. Describe the types of observations that are undertaken during construction.
d. Describe why and how construction tolerances are monitored.

The job contract lists in detail, directly or by incorporation of standards, the specifications for submittals, materials, mixes, testing, preparation, installation, cleaning, and inspections. Field inspections may also include reviewing the level of skill exhibited by craft professionals. For masonry especially, the skill and thoroughness of the worker can mean the difference between acceptance or rejection of the finished product. Inspections may include the following:

- Storing and protecting materials
- Mortar mixing, tempering, placing, tooling, and pointing
- Wetting brick with high initial rates of absorption, unit blending, and unit placing
- Keeping CMUs dry
- Installing connectors, joint reinforcement, control joints and/or expansion joints, flashing, and weep holes
- Placing reinforcement and grouting
- Installing temporary bracing and shoring
- Curing and protecting masonry during construction
- Cleaning masonry

Know what inspectors look for, and inspect your own work first. Knowing and previewing what the inspector will check is an excellent way to ensure that your work meets the standards.

4.1.0 Performing Standard and Code Inspections

Field observations and inspections have become more common and more important with the increase in construction litigation. Field inspection intends to ensure that the finished work complies with the contract specifications and that the craftwork meets the required standards.

Good craftwork affects masonry performance over time and is essential to high-quality construction. Masonry construction requires skilled craftworkers working cooperatively with the architect and engineer to execute the design. The goal of quality craftwork is common to all concerned parties.

The contractor is ultimately responsible for project quality. The architect, engineer, or any independent inspection agencies or testing laboratories are there to verify compliance with standards, designs, and performance criteria; in addition to the contractor, they may also perform field observations and inspections.

Supervisors should see every part of the job at least twice a day and continually inspect ongoing work to ensure that the procedures are safe, the work is sound, and the progress is satisfactory.

Federal and state inspectors visit public jobs to check craftwork and progress. County and/or city inspectors visit jobsites to check conformance to building codes. Inspectors check for conformance to general construction, electrical, plumbing, fire safety, occupational safety, and public health codes. Architects or project engineers check that the work conforms to the specifications.

Many inspectors come unannounced and on an irregular basis. Compliance inspectors usually come when notified of the completion of a stage of work. Routine inspections will take less time if the inspectors or supervisors tracking the work remember the following points:

- Check the specifications for notes of inspections or stages of work that require certificates from officials or insurance companies.
- Check the specifications for the submittals list; note the inspection reports required.
- Contact the agency that granted the building permit for a list of required inspections.
- Ask the mechanical contractors, before they cover their work, if all required inspections are complete. Intermittent subcontractors may forget to tell the supervisor that an inspector is needed before the work can be covered.
- Learn which inspectors insist on seeing the work and which ones allow covering the work without inspection if they are late.
- Call the local fire department to determine what inspections and permits they require.
- Ask the architect or engineer for a list of inspections they have performed and those they have scheduled, such as with mortgage holders, the Federal Housing Administration (FHA) or US Department of Veterans Affairs (VA), tenants, or independent inspecting services.
- Call the health department for regulations about temporary toilets, water supply, sewage, and swimming pools, and ask if any inspections are needed.
- If necessary, ask the permitting agency about permits and inspections for working on public property, obstructing traffic, or connecting to utilities.
- Post a list so that involved supervisors and workers will know the inspection type and schedule.

4.2.0 Performing Material Inspections

An inspector must be familiar with the project specifications and verify compliance with standards. Material inspection is the first part of the process. Manufacturers must supply test certificates showing that materials meet or exceed referenced standards. The inspector must check the certificates and then inspect the materials to make sure they match the certificates and orders.

The inspector checks masonry units for color, texture, size, and match to samples. They are rejected if they are soiled, chipped, or broken in transit. If they do not have certificates, the inspector picks random samples and sends them to a testing laboratory. They will also check the moisture condition of clay units and the dryness of CMUs at the time of laying.

The inspector checks mortar and grout ingredients for contamination and rejects contaminated materials or bagged materials that show signs of water absorption. Packaged materials should be sealed with the manufacturer's labels legible and intact.

The inspector checks material certificates and bills of lading against the specifications and purchase orders to ensure that the material was ordered according to specifications and delivered as ordered. Accessories are also checked for design compliance and identified for location within the structure.

In addition, the inspector checks that the materials are stored properly. Masonry materials and units must be protected against weather and contamination, and CMUs and accessories must also be protected against moisture.

Acceptable mixing and batching procedures should have been established during preconstruction. The inspector checks that these procedures are being followed and monitors tempering time to ensure that the mortar is fresh.

Inspectors will also check mortar and grout for contamination, materials for appropriate storage, and accessories for design and location. Use the mortar mixing and batching procedures set in the contract, and monitor tempering time. Inspect and check all of these as you work.

4.3.0 Monitoring Structural Elements During Construction

The project engineer is responsible for structural inspections of load-bearing masonry. Field observation of masonry construction is not intended to certify structural characteristics, but to check the progress of the work.

NOTE

Mortar joints should be carefully inspected to ensure that there are no openings. Any defective joints must be tuck pointed.

Before masonry work begins, foundation beams and other structural elements must be checked for line and grade. Areas must be cleaned of dirt, grease, or other materials that could impair the mortar bond. Dimensions and layouts must be checked against drawings, and adjustments must be made to correct discrepancies. Steel reinforcing dowels must be checked for proper location.

Because they are so critical, the inspector checks for proper embedment and coverage of anchors, ties, and joint reinforcements. The inspector monitors vertical coursing and joint widths, the placing of reinforcement and ties, and grouting pours.

The inspector checks that cavities are clear of mortar droppings and that flashing, weep holes, control joints, expansion joints, lintels, sills, caps, copings, and frames are set properly. They also check that rebar is clean and its coating is intact. Windowsills, doorframes, landings, and framing are also checked to ensure that they are free of mortar.

Inspectors may observe methods or procedures that appear to conflict with the specifications. They will bring these items to the attention of the supervisor or contractor, who will make any adjustments. Masons observing specification conflicts should bring these items to the attention of their supervisor immediately.

4.4.0 Monitoring Construction Tolerances

Inspection of the work in progress is a continuing task. Because so many building components are prefabricated today, tolerances are extremely important. Different materials have different tolerances due to the nature of their physical properties and manufacturing methods. Masonry has room for adjustment but must meet a specific set of tolerances. *Figure 22* shows the tolerances set for mortar joints in block wall construction.

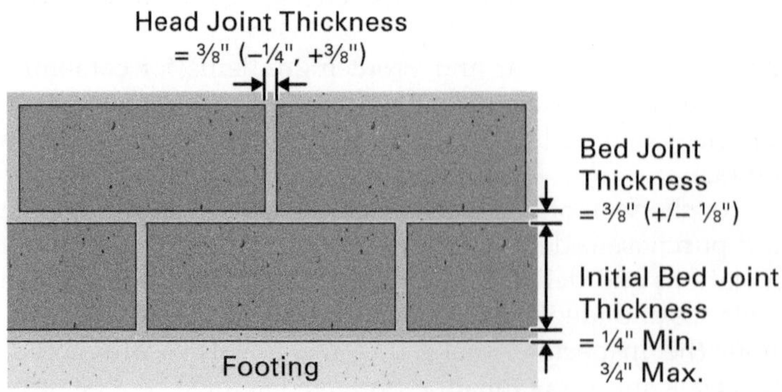

Figure 22 Allowable tolerances for block mortar joints.

Expansion or control joints filled with elastomeric sealants also have tolerances. A realistic minimum thickness is $^3\!/_8$", to match the width of the mortar joints.

Concealed concrete and steel structural framing members have greater tolerances than exposed structural members such as masonry panels, veneer, or curtain walls. The AISC sets the tolerance for steel frame, the ACI sets the tolerance for concrete frame, and the BIA sets the tolerance for brick veneer.

The adjustment necessary to keep a plumb line across the brick face must be taken up in the veneer anchors. If adjustments are not made carefully, unanticipated stresses will be put on the anchors, and the veneer may eventually detach as the anchor fails. Improper alignment of anchors with the frame and brick veneer will cause stress on the anchors.

The best solution is to have a variety of anchor lengths on hand to accommodate construction tolerances. Anchors must have a minimum $^5\!/_8$" mortar cover on the outside wall face and a minimum $1^1\!/_2$" embedment in masonry unit veneers. If the framing member is out of plumb, choose an anchor size that will safely span the gap.

OSHA Inspectors

Occupational Safety and Health Administration (OSHA) inspectors regularly visit construction sites unannounced. They can levy fines or stop work if conditions are dangerous. Keep the jobsite safe, and avoid costly fines or work stoppages.

4.0.0 Section Review

1. It is considered a best practice for supervisors to observe every part of the job at least _____.
   a. once a week
   b. three times a week
   c. twice a day
   d. once a day

2. An inspector will reject bagged mortar and grout materials that show signs of _____.
   a. exposure to sunlight
   b. water absorption
   c. settling
   d. having been stacked too high

3. The inspector is responsible for reviewing each of the following during masonry construction *except* the _____.
   a. proper embedment and coverage of anchors, ties, and joint reinforcements
   b. proper set of weep holes, control joints, expansion joints, lintels, sills, caps, copings, and frames
   c. cleanliness and coating of rebar
   d. structural integrity of load-bearing components

4. The minimum embedment that anchors must have in a masonry unit veneer is _____.
   a. $2^1\!/_4$"
   b. 2"
   c. $1^3\!/_4$"
   d. $1^1\!/_2$"

Module 28207 Review Questions

1. For most construction projects, quality standards are established _____.
   a. by the general contractor
   b. by the office secretary
   c. in the contract documents and specifications
   d. by the subcontractor

2. Test results, reports, and other documents to be provided for the architect and engineer during the course of a project are called _____.
   a. submittals
   b. periodic updates
   c. project progress documentation
   d. submissions

3. Documents that name exact properties of materials and methods of installation without using brand names are known as _____.
   a. proprietary specifications
   b. descriptive specifications
   c. performance specifications
   d. reference standard specifications

4. If a type of brick to be used is not specified in a contract, the default is Type _____.
   a. FBA
   b. FBB
   c. FBS
   d. FBX

5. Standards for masonry mortar cover which four types?
   a. R, RS, N, O
   b. M, S, N, O
   c. M, S, RN, RO
   d. RM, S, N, RO

6. Before construction begins, laboratory requirements may call for the building of up to how many test prisms?
   a. 3
   b. 5
   c. 8
   d. 10

7. To provide test results that more closely reflect a wall's performance, a prism's height should be what amount of its thickness?
   a. Half
   b. Twice
   c. Four times
   d. Five times

8. Laboratory stress tests are conducted to find a masonry prism's _____.
   a. breakpoint
   b. strength limit
   c. failure threshold
   d. index of disintegration

9. What is the name of the test shown in the following figure?

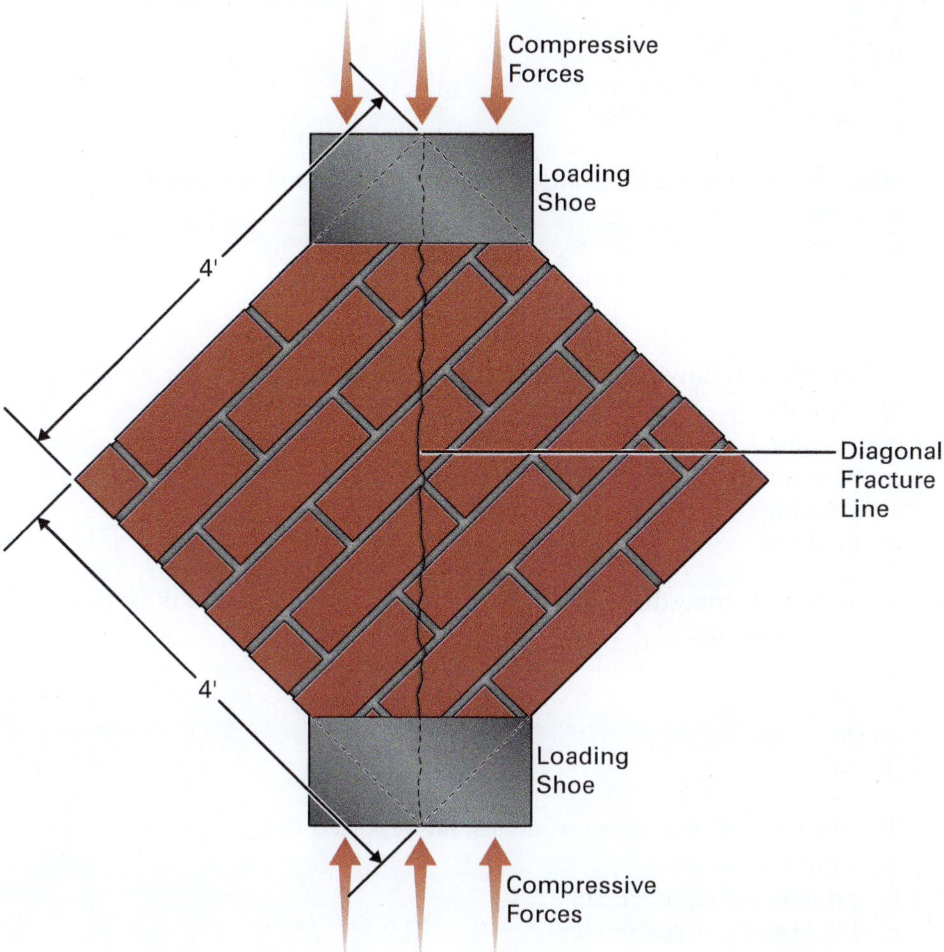

   a. Diagonal tensile shear test
   b. One-third-point flexural strength test
   c. Prism mold strength test
   d. Cube compressive strength test

10. Mortar's ability to tie units together results from its _____.
    a. compressive strength
    b. plasticity
    c. strength of bond
    d. durability

11. The percentage of sand in different size classifications is determined by a _____.
    a. sieve test
    b. sedimentation test
    c. settling test
    d. screen test

12. Sand is clean enough to use in mortar if the thickness of the siltation test layer is no more than _____.
    a. $\frac{1}{4}$"
    b. $\frac{3}{16}$"
    c. $\frac{1}{8}$"
    d. $\frac{1}{16}$"

13. To check the effects of weather and curing on the color consistency of mortar, make test cubes and expose them to weather for *at least* how many days?
    a. 2
    b. 5
    c. 7
    d. 28

14. When testing mortar or grout, slump is measured to the nearest _____.
    a. inch
    b. ¾"
    c. ½"
    d. ¼"

15. Which of the following is another term that describes a brick's initial rate of absorption?
    a. Suction
    b. Sponging
    c. Wicking
    d. Leakage

16. To reach full strength, a cube of mortar to be strength tested must be cured for how many days?
    a. 2
    b. 14
    c. 28
    d. 30

17. The results of a cone penetration test are reported in _____.
    a. eighths of an inch
    b. pounds per square inch
    c. kilograms per centimeter
    d. millimeters

Answers to odd-numbered questions are found in the Review Question Answer Key in the back of the book.

Answers to Section Review Questions

Answer	Section	Objective
Section One		
1. d.	1.1.0	1a
2. b.	1.2.0	1b
Section Two		
1. a.	2.1.0	2a
2. a.	2.2.0	2b
3. c.	2.3.0	2c
4. d.	2.4.0	2d
5. b.	2.5.2	2e
Section Three		
1. c.	3.1.0	3a
2. d.	3.2.0	3b
3. a.	3.3.0	3c
4. d.	3.4.2	3d
Section Four		
1. c.	4.1.0	4a
2. b.	4.2.0	4b
3. d.	4.3.0	4c
4. d.	4.4.0	4d

Module Review Answer Key

MODULE 01 (28208)

Answer	Section
1. a.	1.1.2
3. a.	1.1.9
5. d.	1.3.0
7. a.	2.2.0

MODULE 02 (28203)

Answer	Section
1. b.	1.0.0
3. c.	2.0.0
5. b.	2.0.0
7. c.	2.3.0
9. c.	3.0.0
11. a.	3.2.0

MODULE 03 (28204)

Answer	Section
1. c.	1.0.0
3. a.	1.0.0
5. d.	1.1.0
7. c.	1.2.0
9. d.	1.2.0
11. a.	1.3.0
13. b.	2.1.0
15. b.	3.0.0
17. b.	3.2.0
19. b.	4.2.0

MODULE 04 (28205)

Answer	Section
1. c.	1.1.0
3. a.	1.3.0
5. c.	1.3.3
7. b.	1.6.2
9. d.	2.2.0
11. a.	2.3.0
13. b.	3.2.0
15. b.	3.4.0

MODULE 05 (28206)

Answer	Section
1. b.	1.1.0
3. d.	1.2.4
5. a.	2.0.0
7. c.	2.2.0
9. c.	3.0.0
11. b.	3.2.1

MODULE 06 (28207)

Answer	Section
1. c.	1.0.0
3. b.	1.1.0
5. b.	1.1.3
7. d.	2.2.0
9. a.	2.5.3
11. a.	3.1.1
13. b.	3.2.0
15. a.	3.3.0
17. d.	3.4.2

APPENDIX 28207A

ASTM Standards for Masonry Construction

Brick

ASTM C27, Standard Classification of Fireclay and High-Alumina Refractory Brick
ASTM C32, Standard Specification for Sewer and Manhole Brick (Made from Clay or Shale)
ASTM C34, Standard Specification for Structural Clay Load-Bearing Wall Tile
ASTM C43, Standard Terminology of Structural Clay Products (note: withdrawn 2009)
ASTM C56, Standard Specification for Structural Clay Nonloadbearing Tile
ASTM C62, Standard Specification for Building Brick (Solid Masonry Units Made from Clay or Shale)
ASTM C106, Specification for Refractories and Incinerators (note: withdrawn 1972)
ASTM C126, Standard Specification for Ceramic Glazed Structural Clay Facing Tile, Facing Brick, and Solid Masonry Units
ASTM C155, Standard Classification of Insulating Firebrick
ASTM C212, Standard Specification for Structural Clay Facing Tile
ASTM C216, Standard Specification for Facing Brick (Solid Masonry Units Made from Clay or Shale)
ASTM C279, Standard Specification for Chemical-Resistant Masonry Units
ASTM C315, Standard Specification for Clay Flue Liners and Chimney Pots
ASTM C410, Standard Specification for Industrial Floor Brick
ASTM C416, Standard Classification of Silica Refractory Brick
ASTM C530, Standard Specification for Structural Clay Nonloadbearing Screen Tile
ASTM C652, Standard Specification for Hollow Brick (Hollow Masonry Units Made from Clay or Shale)
ASTM C902, Standard Specification for Pedestrian and Light Traffic Paving Brick
ASTM C1261, Standard Specification for Firebox Brick for Residential Fireplaces
ASTM C1272, Standard Specification for Heavy Vehicular Paving Brick

Concrete Masonry Units

ASTM C55, Standard Specification for Concrete Building Brick
ASTM C73, Standard Specification for Calcium Silicate Brick (Sand-Lime Brick)
ASTM C90, Standard Specification for Loadbearing Concrete Masonry Units
ASTM C129, Standard Specification for Nonloadbearing Concrete Masonry Units
ASTM C139, Standard Specification for Concrete Masonry Units for Construction of Catch Basins and Manholes
ASTM C744, Standard Specification for Prefaced Concrete and Calcium Silicate Masonry Units
ASTM C936, Standard Specification for Solid Concrete Interlocking Paving Units
ASTM C1319, Standard Specification for Concrete Grid Paving Units

Natural Stone

ASTM C119, Standard Terminology Relating to Dimension Stone
ASTM C503, Standard Specification for Marble Dimension Stone
ASTM C568, Standard Specification for Limestone Dimension Stone
ASTM C615, Standard Specification for Granite Dimension Stone
ASTM C616, Standard Specification for Quartz-Based Dimension Stone
ASTM C629, Standard Specification for Slate Dimension Stone

Mortar and Grout

ASTM C5, Standard Specification for Quicklime for Structural Purposes
ASTM C33, Standard Specification for Concrete Aggregates
ASTM C91, Standard Specification for Masonry Cement
ASTM C144, Standard Specification for Aggregate for Masonry Mortar
ASTM C150, Standard Specification for Portland Cement

ASTM C199, Standard Test Method for Pier Test for Refractory Mortars
ASTM C207, Standard Specification for Hydrated Lime for Masonry Purposes
ASTM C270, Standard Specification for Mortar for Unit Masonry
ASTM C330, Standard Specification for Lightweight Aggregates for Structural Concrete
ASTM C331, Standard Specification for Lightweight Aggregates for Concrete Masonry Units
ASTM C404, Standard Specification for Aggregates for Masonry Grout
ASTM C476, Standard Specification for Grout for Masonry
ASTM C595, Standard Specification for Blended Hydraulic Cements
ASTM C658, Standard Specification for Chemical-Resistant Resin Grouts for Brick or Tile
ASTM C887, Standard Specification for Packaged, Dry, Combined Materials for Surface Bonding Mortar
ASTM C1142, Standard Specification for Extended Life Mortar for Unit Masonry
ASTM C1329, Standard Specification for Mortar Cement

Reinforcement and Accessories

ASTM A36, Standard Specification for Carbon Structural Steel
ASTM A82, Standard Specification for Steel Wire, Plain, for Concrete Reinforcement
ASTM A153, Standard Specification for Zinc Coating (Hot-Dip) on Iron and Steel Hardware
ASTM A167, Standard Specification for Stainless and Heat-Resisting Chromium-Nickel Steel Plate, Sheet, and Strip
ASTM A185, Standard Specification for Steel Welded Wire Reinforcement, Plain, for Concrete
ASTM A496, Standard Specification for Steel Wire, Deformed, for Concrete Reinforcement
ASTM A615, Standard Specification for Deformed and Plain Carbon-Steel Bars for Concrete Reinforcement
ASTM A641, Standard Specification for Zinc-Coated (Galvanized) Carbon Steel Wire
ASTM A951, Standard Specification for Steel Wire for Masonry Joint Reinforcement
ASTM B227, Standard Specification for Hard-Drawn Copper-Clad Steel Wire
ASTM B766, Standard Specification for Electrodeposited Coatings of Cadmium
ASTM C915, Standard Specification for Precast Reinforced Concrete Crib Wall Members
ASTM C1089, Standard Specification for Spun Cast Prestressed Concrete Poles
ASTM C1242, Standard Guide for Selection, Design, and Installation of Dimension Stone Attachment Systems

Sampling and Testing

ASTM C67, Standard Test Methods for Sampling and Testing Brick and Structural Clay Tile
ASTM C97, Standard Test Methods for Absorption and Bulk Specific Gravity of Dimension Stone
ASTM C109, Standard Test Method for Compressive Strength of Hydraulic Cement Mortars (Using 2-in. or [50-mm] Cube Specimens)
ASTM C138, Standard Test Method for Density (Unit Weight), Yield, and Air Content (Gravimetric) of Concrete
ASTM C140, Standard Test Methods for Sampling and Testing Concrete Masonry Units and Related Units
ASTM C170, Standard Test Method for Compressive Strength of Dimension Stone
ASTM C173, Standard Test Method for Air Content of Freshly Mixed Concrete by the Volumetric Method
ASTM C231, Standard Test Method for Air Content of Freshly Mixed Concrete by the Pressure Method
ASTM C241, Standard Test Method for Abrasion Resistance of Stone Subjected to Foot Traffic
ASTM C267, Standard Test Methods for Chemical Resistance of Mortars, Grouts, and Monolithic Surfacings and Polymer Concretes

ASTM C426, Standard Test Method for Linear Drying Shrinkage of Concrete Masonry Units
ASTM C780, Standard Test Method for Preconstruction and Construction Evaluation of Mortars for Plain and Reinforced Unit Masonry
ASTM C880, Standard Test Method for Flexural Strength of Dimension Stone
ASTM C952, Standard Test Method for Bond Strength of Mortar to Masonry Units
ASTM C1006, Standard Test Method for Splitting Tensile Strength of Masonry Units
ASTM C1019, Standard Test Method for Sampling and Testing Grout
ASTM C1072, Standard Test Methods for Measurement of Masonry Flexural Bond Strength
ASTM C1093, Standard Practice for Accreditation of Testing Agencies for Masonry
ASTM C1148, Standard Test Method for Measuring the Drying Shrinkage of Masonry Mortar
ASTM C1194, Standard Test Method for Compressive Strength of Architectural Cast Stone
ASTM C1195, Standard Test Method for Absorption of Architectural Cast Stone
ASTM C1196, Standard Test Methods for In Situ Compressive Stress within Solid Unit Masonry Estimated Using Flatjack Measurements
ASTM C1197, Standard Test Method for In Situ Measurement of Masonry Deformability Using the Flatjack Method
ASTM C1262, Standard Test Method for Evaluating the Freeze-Thaw Durability of Dry-Cast Segmental Retaining Wall Units and Related Concrete Units
ASTM C1314, Standard Test Method for Compressive Strength of Masonry Prisms
ASTM C1324, Standard Test Method for Examination and Analysis of Hardened Masonry Mortar
ASTM D75, Standard Practice for Sampling Aggregates
ASTM E72, Standard Test Methods of Conducting Strength Tests of Panels for Building Construction
ASTM E447, Test Methods for Compressive Strength of Laboratory Constructed Masonry Prisms (note: withdrawn 1988)
ASTM E488, Standard Test Methods for Strength of Anchors in Concrete Elements
ASTM E514, Standard Test Method for Water Penetration and Leakage through Masonry
ASTM E518, Standard Test Methods for Flexural Bond Strength of Masonry
ASTM E519, Standard Test Method for Diagonal Tension (Shear) in Masonry Assemblages
ASTM E754, Standard Test Method for Pullout Resistance of Ties and Anchors Embedded in Masonry Mortar Joints

Assemblages

ASTM C901, Standard Specification for Prefabricated Masonry Panels
ASTM C946, Standard Practice for Construction of Dry-Stacked, Surface-Bonded Walls
ASTM C1283, Standard Practice for Installing Clay Flue Lining
ASTM E835, Standard Guide for Modular Coordination of Clay and Concrete Masonry Units (note: withdrawn 2011)
ASTM E1602, Standard Guide for Construction of Solid Fuel Burning Masonry Heaters

APPENDIX

Common Abbreviations

Common Abbreviations Used on Elevation Drawings (1 of 3)

Aluminum	AL
Asbestos	ASB
Asphalt	ASPH
Basement	BSMT
Beveled	BEV
Brick	BRK
Building	BLDG
Cast iron	CI
Ceiling	CLG
Cement	CEM
Center	CTR
Centerline	C or CL
Clear	CLR
Column	COL
Concrete	CONC
Concrete block	CONC B
Copper	COP.
Corner	COR
Detail	DET
Diameter	DIA
Dimension	DIM.
Ditto	DO.
Divided	DIV
Door	DR
Double-hung window	DHW
Down	DN or D
Downspout	DS
Drawing	DWG
Drip cap	DC
Each	EA
East	E
Elevation	EL
Entrance	ENT
Excavate	EXC
Exterior	EXT
Finish	FIN.
Flashing	FL
Floor	FL
Foot or feet	' or FT
Foundation	FND
Full size	FS
Galvanized	GALV
Galvanized iron	GI
Gauge	GA
Glass	GL
Glass block	GL BL
Grade	GR

Common Abbreviations Used on Elevation Drawings (2 of 3)

Grade line	GL
Height	HGT, H, or HT
High point	H PT
Horizontal	HOR
Hose bibb	HB
Inch or inches	" or IN.
Insulating (insulated)	INS
Length	LGTH, LG, or L
Length overall	LOA
Level	LEV
Light	LT
Line	L
Lining	LN
Long	LG
Louver	LV
Low point	LP
Masonry opening	MO
Metal	MET. or M
Molding	MLDG
Mullion	MULL.
North	N
Not to Scale	NTS
Number	NO. or #
Opening	OPNG
Outlet	OUT.
Outside diameter	OD
Overhead	OVHD
Panel	PNL
Perpendicular	PERP
Plate glass	PL GL
Plate height	PL HT
Radius	R
Revision	REV
Riser	R
Roof	RF
Roof drain	RD
Roofing	RFG
Rough	RGH
Saddle	SDL or S
Scale	SC
Schedule	SCH
Section	SECT.
Sheathing	SHTHG
Sheet	SH
Shiplap	SHLP
Siding	SDG
South	S
Specifications	SPEC
Square	SQ
Square inch	SQ IN.
Stainless steel	SST

Common Abbreviations Used on Elevation Drawings (3 of 3)

Steel	STL
Stone	STN
Terra-cotta	TC
Thick or thickness	THK or T
Typical	TYP
Vertical	VERT
Waterproofing	WP
West	W
Width	W or WTH
Window	WDW
Wire glass	W GL
Wood	WD
Wrought iron	WI

Common Abbreviations Used on Plan View Drawings (1 of 4)

Access panel	AP
Acoustic	ACST
Acoustical tile	AT.
Aggregate	AGGR
Air conditioning	AIR COND
Aluminum	AL
Anchor bolt	AB
Angle	AN.
Apartment	APT
Approximate	APPROX
Architectural	ARCH.
Area	A.
Area drain	AD
Asbestos	ASB
Asbestos board	AB
Asphalt	ASPH
Asphalt tile	AT.
Basement	BSMT
Bathroom	B
Bathtub	BT
Beam	BM
Bearing plate	BRG PL
Bedroom	BR
Blocking	BLKG
Blueprint	BP
Boiler	BLR
Bookshelves	BK SH
Brass	BRS
Brick	BRK
Bronze	BRZ
Broom closet	BC
Building	BLDG
Building line	BL
Cabinet	CAB.
Casing	CSG
Cast iron	CI

Common Abbreviations Used on Plan View Drawings (2 of 4)

Cast stone	CS
Catch basin	CB
Caulking	CLKG
Cellar	CEL
Cement	CEM
Cement asbestos board	CEM AB
Cement floor	CEM FL
Cement mortar	CEM MORT
Center	CTR
Center line	C or CL
Center matched	CM
Center to center	C to C
Ceramic	CER
Channel	CHAN
Cinder block	CIN BL
Circuit breaker	CIR BKR
Cleanout	CO
Cleanout door	COD.
Clear glass	CL GL
Closet	C, CL, or CLO
Cold air	CA
Cold water	CW
Collar beam	COL B
Concrete	CONC
Concrete floor	CONC FL
Concrete masonry unit	CMU
Conduit	CND
Construction	CONST
Contract	CONT
Copper	COP.
Counter	CTR
Cubic feet	CU FT
Cutout	CO
Detail	DET
Diagram	DIAG
Dimension	DIM.
Dining room	DR
Dishwasher	DW
Ditto	DO.
Double acting	DA
Double-strength glass	DSG
Down	DN
Downspout	DS
Drain	D or DR
Drawing	DWG
Dressed and matched	D & M
Dryer	D
Electric panel	EP
End to end	E to E
Excavate	EXC
Expansion joint	EXP JT

Common Abbreviations Used on Plan View Drawings (3 of 4)

Exterior	EXT
Library	LIB
Light	LT
Limestone	LS
Linen closet	L CL
Lining	LN
Linoleum	LINO
Living room	LR
Louver	LV
Main	MN
Marble	MR
Masonry opening	MO
Material	MATL
Maximum	MAX
Medicine cabinet	MC
Minimum	MIN
Miscellaneous	MISC
Mixture	MIX.
Modular	MOD
Molding	MLDG
Mortar	MOR
Nosing	NOS
Not to scale	NTS
Obscure glass	OBSC GL
On center	OC
Opening	OPNG
Outlet	OUT.
Overall	OA
Overhead	OVHD
Pantry	PAN.
Partition	PTN
Plaster	PL or PLAS
Plastered opening	PO
Plate	PL
Plate glass	PL GL
Platform	PLAT.
Plumbing	PLBG
Porch	P
Precast	PRCST
Prefabricated	PREFAB
Pull switch	PS
Quarry tile floor	QTF
Radiator	RAD
Random	RDM
Range	R
Recessed	REC
Refrigerator	REF
Register	REG
Storage	STG
Switch	SW or S
Telephone	TEL

Common Abbreviations Used on Plan View Drawings (4 of 4)

Terra-cotta	TC
Terrazzo	TER
Thermostat	THERMO
Threshold	TH
Toilet	T
Tongue-and-groove	T&G
Tread	TR or T
Typical	TYP
Unexcavated	UNEXC
Unfinished	UNF
Utility room	URM
Vent	V
Vent stock	VS
Vinyl tile	V TILE
Warm air	WA
Washing machine	WM
Water	W
Water closet	WC
Water heater	WH
Waterproof	WP
Weatherstripping	WS
Weep hole	WH
White pine	WP
Wide flange	WF
Wood	WD
Wood frame	WF
Yellow pine	YP

APPENDIX

Architectural Terms Commonly Found on Plans

Window Terms

Apron—A plain or molded piece of finish below the stool of a window that is installed to cover the rough edge of the wall finish.

Drip cap—A projection of masonry or wood on the outside top of a window to protect the window from rain.

Head jamb—The horizontal top post used in the framing of a window or doorway.

Light—A pane of glass.

Lintel—The horizontal structural member supporting a wall over a window or other opening.

Meeting rail—The horizontal center rail of a sash in a double-hung window.

Mullion—A large, vertical division of a window opening.

Muntin—A strip of wood or metal that separates and supports the panes of glass in a window sash.

Sash—The part of a window in which panes of glass are set; it is generally movable, as in double-hung windows. The two side pieces are called stiles and the upper and lower pieces are called rails.

Side jambs—Vertical side posts used in the framing of a window or doorway.

Sill—The horizontal member at the bottom of a window.

Stool—A flat, narrow shelf forming the top member of the interior trim at the bottom of a window.

Stop bead—The strip on a window frame against which the sash slides.

Pitched Roof Terms

Flashing—Sheet metal, copper, lead, or tin that is used to cover open joints to make them waterproof.

Gable—The end of a ridged roof as distinguished from the front or rear side.

Louver—An opening for ventilation that is covered by sloping slats to exclude rain.

Ridge—The top edge of a roof where the two slopes meet.

Ridgeboard—A board that is placed on the edge at the ridge of a roof to support the upper ends of rafters.

Saddle —A tent-shaped portion of a roof between a chimney and the main part of the roof; built to support flashing and to direct water away from the chimney.

Valley—The intersection of two inclined sides of a roof.

Cornice Terms

Cornice—The part of a roof that projects beyond a wall.

Cornice return—The short portion of a cornice that is carried around the corner of a structure.

Crown molding—The molding at the top of the cornice and just under the roof; can also be used as interior trim.

Fascia—The outside flat member of a cornice.

Frieze—A trim member used just below the cornice.

Soffit—The underside of a cornice.

Stair Terms

Headroom—The distance between flights of steps or between the steps and the ceiling above.

Landing—The horizontal platform in a stairway.

Nosing—The overhanging edge of a stair tread.

Rise—The vertical distance from the top of a tread to the top of the next highest tread.

Riser—The vertical portion of a step.

Run—The horizontal distance that is covered by a flight of steps.

Stringer—The supporting member at the sides of a stairway.

Tread—The horizontal part of a step on which the foot is placed.

Structural Terms

Anchor bolt—A bolt with the threaded portion projecting from a structure, generally used to hold the frame of a building secure against wind load. Anchor bolts may also be referred to foundation bolts or sill bolts.

Batt—A type of insulation designed to be installed between framing members.

Battens—Narrow strips of wood or metal used to cover vertical joints between boards and panels.

Beam— One of the principal horizontal members of a building.

Bridging—The process of bracing floor joists by fixing lateral members between them.

Camber—The concave or convex curvature of a surface.

Expansion joint—The separation between adjoining parts to allow for small relative movements, such as those caused by temperature changes.

Footing—The foundation for a column or the enlargement at the bottom of a wall to distribute the weight of the superstructure over a greater area to prevent settling.

Furring—Strips of wood or metal applied to a wall or other surface to make it level, form an air space, or provide a fastening surface for a finish covering.

Girder—The main supporting beam (either timber or steel) that is used to support a superstructure.

Header—A wood beam that is set at a right angle to a joist to provide a seat or support over an opening.

Joist—The horizontal member used to support the subfloor or ceiling finish material. Joists are laid edgewise to form a floor support; they rest on the wall or on girders.

Monolithic concrete—A continuous mass of concrete that is cast as a single unit.

Rake—Trim members that run parallel to the roof slope and form the finish between the roof and the wall at the gable end.

Reinforced concrete—Concrete containing metal rods, wires, or other slender members. It is designed in such a manner that the concrete and metal act together to resist forces.

Sill plate—A horizontal member that is supported by a foundation wall or piers, and which in turn bears the upright members of a frame.

Slab—A cast-in-place concrete floor.

Stem wall—That portion of a foundation that rests on the footing.

Studs—The vertical, slender wood or metal members that are used to support the elements in walls and partitions.

Vapor barrier—A material that is used to retard the flow of vapor or moisture into the walls or floors and thus prevent condensation within them.

Veneer—The covering layer of material for a wall or facing materials applied to the external surface of steel, reinforced concrete, or frame walls.

ADDITIONAL RESOURCES

28208—Design Documents

Construction Terms Simply Defined: A Dictionary of Construction and Design You Can Actually Understand. Dominique and Greg Nelson. 2020. Independent publisher.

International Building Code 2021. Washington, DC: International Code Council.

International Residential Code 2021. Washington, DC: International Code Council.

MasterFormat 2020 Edition. Alexandria, VA: The Construction Specifications Institute (CSI) and Construction Specifications Canada (CSC).

Plan Reading & Material Takeoff. Wayne J. Del Pico. 2015. RSMeans.

The Construction Specifications Institute. This organization seeks to facilitate communication among everyone involved in the building process. **www.csinet.org**.

International Code Council. This membership organization is dedicated to building safety and fire prevention through developent of building codes. **www.iccsafe.org**.

28207—Construction Inspection and Quality Control

ACI 530/ASCE 5/TMS 402, Building Code Requirements for Masonry Structures, latest edition. Reston, VA: American Society of Civil Engineers.

ASTM A36, Standard Specification for Carbon Structural Steel, latest edition. West Conshohocken, PA: ASTM International.

ASTM A82, Standard Specification for Steel Wire, Plain, for Concrete Reinforcement, latest edition. West Conshohocken, PA: ASTM International.

ASTM A153, Standard Specification for Zinc Coating (Hot-Dip) on Iron and Steel Hardware, latest edition. West Conshohocken, PA: ASTM International.

ASTM A167, Standard Specification for Stainless and Heat-Resisting Chromium-Nickel Steel Plate, Sheet, and Strip, latest edition. West Conshohocken, PA: ASTM International.

ASTM A615, Standard Specification for Deformed and Plain Carbon-Steel Bars for Concrete Reinforcement, latest edition. West Conshohocken, PA: ASTM International.

ASTM A951, Standard Specification for Steel Wire for Masonry Joint Reinforcement, latest edition. West Conshohocken, PA: ASTM International.

ASTM C5, Standard Specification for Quicklime for Structural Purposes, latest edition. West Conshohocken, PA: ASTM International.

ASTM C55, Standard Specification for Concrete Building Brick, latest edition. West Conshohocken, PA: ASTM International.

ASTM C62, Standard Specification for Building Brick (Solid Masonry Units Made from Clay or Shale), latest edition. West Conshohocken, PA: ASTM International.

ASTM C67, Standard Test Methods for Sampling and Testing Brick and Structural Clay Tile, latest edition. West Conshohocken, PA: ASTM International.

ASTM C90, Standard Specification for Loadbearing Concrete Masonry Units, latest edition. West Conshohocken, PA: ASTM International.

ASTM C91, Standard Specification for Masonry Cement, latest edition. West Conshohocken, PA: ASTM International.

ASTM C126, Standard Specification for Ceramic Glazed Structural Clay Facing Tile, Facing Brick, and Solid Masonry Units, latest edition. West Conshohocken, PA: ASTM International.

ASTM C129, Standard Specification for Nonloadbearing Concrete Masonry Units, latest edition. West Conshohocken, PA: ASTM International.

ASTM C138, Standard Test Method for Density (Unit Weight), Yield, and Air Content (Gravimetric) of Concrete, latest edition. West Conshohocken, PA: ASTM International.

ASTM C140, Standard Test Methods for Sampling and Testing Concrete Masonry Units and Related Units, latest edition. West Conshohocken, PA: ASTM International.

ASTM C144, Standard Specification for Aggregate for Masonry Mortar, latest edition. West Conshohocken, PA: ASTM International.

ASTM C150, Standard Specification for Portland Cement, latest edition. West Conshohocken, PA: ASTM International.

ASTM C173, Standard Test Method for Air Content of Freshly Mixed Concrete by the Volumetric Method, latest edition. West Conshohocken, PA: ASTM International.

ASTM C207, Standard Specification for Hydrated Lime for Masonry Purposes, latest edition. West Conshohocken, PA: ASTM International.

ASTM C216, Standard Specification for Facing Brick (Solid Masonry Units Made from Clay or Shale), latest edition. West Conshohocken, PA: ASTM International.

ASTM C231, Standard Test Method for Air Content of Freshly Mixed Concrete by the Pressure Method, latest edition. West Conshohocken, PA: ASTM International.

ASTM C270, Standard Specification for Mortar for Unit Masonry, latest edition. West Conshohocken, PA: ASTM International.

ASTM C476, Standard Specification for Grout for Masonry, latest edition. West Conshohocken, PA: ASTM International.

ASTM C595, Standard Specification for Blended Hydraulic Cements, latest edition. West Conshohocken, PA: ASTM International.

ASTM C652, Standard Specification for Hollow Brick (Hollow Masonry Units Made from Clay or Shale), latest edition. West Conshohocken, PA: ASTM International.

ASTM C780, Standard Test Method for Preconstruction and Construction Evaluation of Mortars for Plain and Reinforced Unit Masonry, latest edition. West Conshohocken, PA: ASTM International.

ASTM C1019, Standard Test Method for Sampling and Testing Grout, latest edition. West Conshohocken, PA: ASTM International.

ASTM C1314, Standard Test Method for Compressive Strength of Masonry Prisms, latest edition. West Conshohocken, PA: ASTM International.

ASTM C1329, Standard Specification for Mortar Cement, latest edition. West Conshohocken, PA: ASTM International.

International Building Code, latest edition. Falls Church, VA: International Code Council.

NFPA 5000, Building Construction and Safety Code, latest edition. Quincy, MA: National Fire Protection Association.

TEK 3-8A, Concrete Masonry Construction. 2001. Herndon, VA: National Concrete Masonry Association. **www.ncma.org**

GLOSSARY

Beam pocket: An opening in a vertical masonry wall that allows an intersecting structural beam to bear on or pass through the wall.

Benchmark: A point established by the surveyor on or close to the building site. It is used as a reference for determining elevations during the construction of a building.

Blowout: The swelling or rupture of a cavity wall from too much pressure caused by pouring liquid grout into the cavity.

Blueprints: The traditional name and method of reproducing technical drawings.

Bond beam: A course of masonry units with steel rebar inserted and held in place by a solid fill of grout or mortar; used as a lintel or reinforcement beam to distribute stress.

Breaking the bond: Starting a course with a cut so as to center the header unit over the head joints in the course below.

Bridging: The mounding of grout or cement over an obstruction, creating a void under the obstruction.

Cap: Masonry units laid on top of a finished wall.

Chase: A continuous recess built into a wall to receive pipes, wires, or heating ducts.

Chases: Continuous recesses built into walls to receive pipes, wires, or heating ducts.

Civil plans: Drawings that show the location of the building on the site from an overhead view, including contours, trees, construction features, and dimensions.

Columns: A vertical reinforced masonry element designed to support a load, the width of which never exceeds three times its thickness and the height of which always exceeds four times its thickness.

Computer-aided design (CAD): The making of a set of construction drawings with the aid of a computer.

Construction plans: Detailed documents, both written and visual, that outline how to complete a construction project.

Coping: The materials or masonry units used to form a cap or finish on top of a wall, pier, chimney, or pilaster to protect the masonry below from water penetration. Coping is usually projected from both sides of the wall to provide a protective covering as well as an ornamental design.

Easement: A legal right-of-way provision on another person's property (e.g., the right of a neighbor to build a road or a public utility to install water and gas lines on the property). A property owner cannot build on an area where an easement has been identified.

Elevation view: A drawing providing a view from the front or side of a structure.

Finish floor elevation (FFE): The uppermost surface of a floor once construction is complete and prior to any finishes or coverings being applied. It is used as a reference plane for other vertical dimensions.

Flexural: Capable of turning, folding, or bending.

Floor plan: A drawing that provides an overhead view of the layout of each room.

Footing: An enlargement at the bottom of a wall that distributes the weight of the superstructure over a greater area to prevent settling.

Foundation plan: A drawing that shows the layout and elevation of the building foundation.

Front setback: The distance from the property line to the front of the building.

Humored: A slang term for gradually adjusting a section of masonry to correct alignment issues.

Hydrostatic pressure: The pressure at any point in a liquid at rest, equal to the depth of the liquid multiplied by its density.

Jambs: The side of an opening, or the vertical framing member on the side of the opening, usually for doorframes and window frames.

Key: A recess or groove in one placement of grout or concrete that is later filled with a new placement of grout or concrete so that the two lock together in a tongue-and-groove configuration.

Life safety plan: A floor plan of a building that identifies each fire and life safety system.

Membrane: A layer of thin, pliable material used to waterproof masonry.

Monuments: Physical structures that mark the location of every survey point.

Nominal size: Approximate or rough size (commercial size) by which products such as lumber and CMUs are commonly known and sold. Nominal sizes are normally slightly larger than the actual sizes (e.g., 2 × 4s).

Not to Scale (NTS): Describes drawings that show relative positions and sizes only, without scale.

Panel: A section of wall between control joints, wall ends, or a control joint and wall end.

Pencil rod: A type of metallic tie that is similar to the shape of a straight wooden pencil; used for control joints in concrete masonry construction.

Piers: A vertical reinforced masonry element typically shorter than a column or pilaster. Piers may be designed to carry loads or may be purely ornamental.

Plan view: A drawing that represents a view looking down on an object.

Plasticity: A material's ability to be easily molded into various shapes.

Priority wall: A full-height wall (usually a fire-rated wall) where drywall cannot be put in place if ductwork or other specialty trade work has already been installed. Drywall contractors must work on a priority wall before other trades can do their work.

Prism: A sample prepared under controlled conditions specifically for testing that uses the same materials, techniques, and craftsmanship as those used at the jobsite.

Property lines: The recorded legal boundaries of a piece of property.

Quality: The totality of features and characteristics of a product or service that bear on its ability to satisfy stated or implied needs.

Quality assurance: All planned and systematic actions necessary to provide adequate confidence that an item or a facility will perform satisfactorily in service.

Quality control: Procedures for testing, inspecting, checking, and verifying to ensure that the work meets the required standard of quality established by the contract documents.

Rebar: Reinforcing bar embedded in concrete, mortar, or grout to act together with the other components to resist loads.

Request for information (RFI): A document used during the construction process to clarify and resolve questions and information gaps in plans, drawings, specifications, and agreements.

Reveal: The side of an opening in a wall for a window or door. This is the part of the masonry jamb around a window or doorframe that can be seen from the frame to the face of the masonry wall.

Reveal: The visible portion of a masonry jamb between the face of the wall and the frame.

Roof plan: A drawing of the view of the roof from above the building.

Scale: The ratio between the size of a drawing of an object and the size of the actual object.

Section drawing: A cross-sectional view of a specific location, showing the inside of an object or building.

Segmental retaining wall (SRW): A wall made of segmental block stacked on top of each other without mortar bonding.

Sills: Horizontal members under doors or windows. Slip sills fit inside the doorframe or window frame; lug sills extend beyond the frame and into the masonry on the jamb sides of the frame.

Silt: A sedimentary material consisting of very fine particles intermediate in size between sand and clay.

Skew: The condition when two parts come together at an angle that is not 90 degrees, or perpendicular, to each other.

Slump: A measure (0" to 12") of the consistency or fluidity of a cementitious product such as concrete, masonry, or grout. Lower slump (1" to 4") indicates a stiff mix; higher slump (8" to 12") indicates a fluid mix.

Specifications: Precisely written presentation of the details of a plan.

Structural plans: A set of engineered drawings used to support the architectural design.

Symbols: Drawings that represent a material or component on a plan.

Tensile: Capable of being stretched or extended, ductile.

Thermal mass: The ability of a building wall to alternately absorb and release heat energy in response to temperature changes.

Title block: A part of a drawing sheet that includes some general information about the project.

Toothing: Construction of a temporary end of a wall with the end stretcher of every alternate course projecting. The projecting units are called *toothers*.

Topographical survey: An accurate and detailed drawing of a place or region that depicts all the natural and man-made physical features, showing their relative positions and elevations.

Wall type: Wall types are defined by their different construction materials and methods.

Waterproofing: The process of treating masonry with a material that will retard penetration of moisture.

Wick: A bundle of fibers that are loosely twisted or braided and woven together to form a cord that can carry water away from an area by capillary action, which occurs as long as the drip end is lower than the absorption end.

Wicks: Bundles of loosely twisted or braided fibers woven together to form cords that can carry water away from an area by capillary action, which occurs as long as the drip end is lower than the absorption end.